W9-CNK-410

The Path
of the Pole

Charles H. Hapgood

Write for our free catalog of exciting books and tapes.

The Path
of the Pole

Charles H. Hapgood

The Path of the Pole
by Charles H. Hapgood

ISBN 0-932813-71-2

©1958, 1970, 1999 Charles Hapgood
& the Hapgood Estate

All rights reserved

Printed in the United States of America

First printing September 1999

Published by
Adventures Unlimited Press
One Adventure Place
Kempton, Illinois 60946 USA
auphq@frontiernet.net

TO THE SPIRIT OF FREDERICK S. HAMMETT

Late Director, Cancer Research Center
of the Lankenau Hospital of Philadelphia.

A Pioneer in Science, a Champion of the Free Mind;

An Artist in Intellectual Inquiry;

A Fighter Against Those Blights of the
Scientific Community: Smugness, Intolerance
and Materialism.

FOREWORD
to the Second Edition

LIKE many people, I was introduced to *Earth's Shifting Crust* not by reading the original text nor through discussion in the technical journals, but through reading an abridged version in *The Saturday Evening Post*. This was an unusual experience, to read something of scientific interest in a family magazine, but what I read was even more unusual. I found myself reading a reasonably plausible explanation—the first ever printed—of the major deformations that have racked the earth's crust. The abbreviated version so intrigued me that I acquired a copy of the complete work at the earliest opportunity. The full text proved to be even more stimulating than its abbreviated predecessor. That first edition of the book was introduced to the public, through its foreword, by the eminent scientist Albert Einstein. I must confess that this fact impressed me to a considerable degree. At the time it never occurred to me that I might be asked to present the second edition; in fact, this still strikes me as somewhat incongruous.

Perhaps at this point I should briefly introduce myself to the reader. I am a mining geologist and a passable mineralogist, engaged in recent years in teaching these subjects. Geology, like all branches of science, has become separated into a maze of specializations. The adherents of one specialization are certainly more than dimly aware of what is going on in other fields, but can hardly consider themselves expert in any but their chosen field. I should not care to be accused of implying, through failure to admit the contrary, that I am a competent critic of Hapgood and Campbell's work. I most emphatically am not.

After carefully reading *Earth's Shifting Crust*, I began searching through the technical journals and other likely sources for the discerning criticism that I felt should be forthcoming from experts in the field. I should have known better than to expect it, I suppose, but hope springs eternal. A reaction came, of course, and largely it came from men who under ordinary circumstances are both rational and competent, but their reaction could hardly be described as rational; hysterical would be a better description. One observed, indignantly, that Hapgood was not a geologist. Admittedly this is a cardinal sin but hardly one punishable by scientific excommunication. Another cited, but failed to name, a scientist whose findings conflict with those of several world-renowned authorities selected by Hapgood as sources of technical data, and used this lack of agreement as an incontrovertible condemnation of the entire book.

I could continue with numerous examples, but this would be pointless. The fact is that almost without exception Americans commenting on the book couched their discussion in thick and unwarranted sarcasm, selecting trivia and factors not subject to verification as the bases for condemnation, seeking in this way to avoid the basic issues. Only the European reviewers were gracious enough to be fair, not that they accepted the theory without question, but they were prepared to offer it its day in court. Nowhere, in all that has been written about the book, have I found a single authority who has calmly and rationally offered a clear and documented criticism of the basic theory involved: that uncompensated masses on or in the earth may cause the earth's crust to slip over its core. Frankly, I wish someone would.

In the years since publication of the first edition of this work we have had, among other things, the benefit of the research of the International Geophysical Year. Incorporation of these and other data has had two extremely important effects upon Hapgood's theory: first to force a revision of the theory in relation to the mechanism of crustal displacement, and secondly to add tremendously to the weight of evidence supporting the thesis that crustal displacement has occurred. Regarding the first of these I believe that the author is to be congratulated for having the flexibility to adapt to new facts as they have become available. For the second, whereas there may have been a time when the occurrence of dislocations of the crust with respect to the earth's rotational poles could have been questioned, I personally feel that in the light of the data presented by Hapgood in this, the second edition of his book, such dislocations are no longer a matter of question.

Like many another engaged in teaching, I have grown weary of apologizing to my students for teaching time-worn theories whose logic, to use a kind word, is indefensible. The plain fact is that the logic of all previous theories of the earth's deformation is so obviously contrived, the holes are so gaping, that one is inclined to suspect that danger lurks there for the unwary. Now at last in Hapgood and Campbell's theory, actually a coalition of several older and poorly enunciated ideas, we find the first outwardly reasonable explanation of the observed facts in several major geological fields. Now I ask—no, I implore—my colleagues, those most competent to assume the task, to attack this theory with the weapons of well-documented proof. Or, failing this, let them build upon it to a better, clearer understanding of the forces that have deformed this planet we live upon. Let us not bury this idea prematurely through prejudice, as so many valuable ideas of the past have been buried, only to be sheepishly exhumed in later years. If it is an unworthy thing let it be properly destroyed; if not, let it receive the nourishment that it deserves.

F. N. Earll

DEPARTMENT OF GEOLOGY
Montana College of Mineral
Science and Technology

The Mather and Einstein Forewords
to the First Edition

THE most significant change in this book since Albert Einstein wrote his Foreword for the American edition and Professor Kirtley Mather wrote one for the British, Spanish, and Italian editions is directly related to the question on which they both expressed their strongest doubts: the ice-cap "mechanism" by which I proposed to account for displacements of the earth's outer shell. Their doubts have been vindicated by the progress of earth studies in the past decade. Advancing knowledge of conditions of the earth's crust now suggests that the forces responsible for shifts of the crust lie at some depth within the earth rather than on its surface.

Despite this change in the character of the proposed explanation of the movements, the evidence for the shifts themselves has been multiplied many fold in the past decade. The main themes of the book—the occurrence of the crust displacements even very recently in geological history, and their effects in forming the features of the earth's surface—therefore remain unchanged.

THE AUTHOR

Foreword to the First Edition

BRITISH, SPANISH, AND ITALIAN VERSIONS

by Kirtley F. Mather
Professor of Geology, Emeritus,
Harvard University; former
president, The American Association
for the Advancement of Science

THE idea that the history of the earth involves the shifting of its thin "crust" from time to time and place to place is certain to receive increased attention in the next few years. Knowledge is rapidly accumulating concerning the spatial relations of the crust and the underlying "mantle." Information regarding the physical properties of these parts of the stratiform planet is being secured by geophysicists. Many specific facts are now available concerning local changes of level and of geographic position of points on the earth's surface. The geologic records of the past are replete with items that suggest significant differences between the latitude and longitude of many places in earlier epochs and those of the present time.

The need is clearly apparent for a synthesis of all these many data that would integrate them in a broadly inclusive scheme and give them unified meaning in relation to a general principle. In geology, indeed in all scientific disciplines, analysis must lead to synthesis which in turn must be followed by further analytical studies in the repetitive cycles of advancing knowledge and understanding. This is evidently the aim of this thought-provoking book. Its greatest value will be found in the stimulus it should give to discussion, debate and controversial argument.

The concept of crustal shifting as an important and frequently repeated episode in earth history is not new. But the marshaling of data from many diverse fields of study and their interpretation in causal terms are sufficiently novel to make the authors' ideas worthy of careful study and appraisal. Indeed, certain aspects of their application of the general concept are radically new and will undoubtedly lead to healthy controversy. I cannot, for example, accept as valid certain interpretations made by the authors of some of the facts they cite, but these are minor matters and do not necessarily invalidate their major argument. My own confidence in the principle of isostasy leads me moreover to discount the computation of tangential forces resulting from "off-center" ice caps, but this is certainly a matter for further study. The results of geophysical research must accord with the facts of earth history if they are to be accepted as completely trustworthy.

All of which means that the authors of this novel interpretation of crustal movements have made a distinctive contribution to geological lore which should be of interest to all geologists. The numerous unsolved problems to which Mr. Hapgood directs attention should be the subjects of intensified debate among scientists in every part of the world. It should moreover be noted that this book is written in clear, nontechnical language. Mr. Hapgood has succeeded in bringing the thought within the reach of every educated layman. It is a readable survey of geological problems that too long have been the province of specialists alone.

Kirtley F. Mather
JULY 1, 1959

Foreword to the First Edition
by Albert Einstein

I FREQUENTLY receive communications from people who wish to consult me concerning their unpublished ideas. It goes without saying that these ideas are very seldom possessed of scientific validity. The very first communication, however, that I received from Mr. Hapgood electrified me. His idea is original, of great simplicity, and— if it continues to prove itself—of great importance to everything that is related to the history of the earth's surface.

A great many empirical data indicate that at each point on the earth's surface that has been carefully studied, many climatic changes have taken place, apparently quite suddenly. This, according to Hapgood, is explicable if the virtually rigid outer crust of the earth undergoes, from time to time, extensive displacement over the viscous, plastic, possibly fluid inner layers. Such displacements may take place as the consequence of comparatively slight forces exerted on the crust, derived from the earth's momentum of rotation, which in turn will tend to alter the axis of rotation of the earth's crust.

In a polar region there is continual deposition of ice, which is not symmetrically distributed about the pole. The earth's rotation acts on these unsymmetrically deposited masses, and produces centrifugal momentum that is transmitted to the rigid crust of the earth. The constantly increasing centrifugal momentum produced in this way will, when it has reached a certain point, produce a movement of the earth's crust over the rest of the earth's body, and this will displace the polar regions toward the equator.

Without a doubt the earth's crust is strong enough not to give way proportionately as the ice is deposited. The only doubtful assumption is that the earth's crust can be moved easily enough over the inner layers.

The author has not confined himself to a simple presentation of this idea. He has also set forth, cautiously and comprehensively, the extraordinarily rich material that supports his displacement theory. I think that this rather astonishing, even fascinating, idea deserves the serious attention of anyone who concerns himself with the theory of the earth's development.

To close with an observation that has occurred to me while writing these lines: If the earth's crust is really so easily displaced over its substratum as this theory requires, then the rigid masses near the earth's surface must be distributed in such a way that they give rise to no other considerable centrifugal momentum, which would tend to displace the crust by centrifugal effect. I think that this deduction might be capable of verification, at least approximately. This centrifugal momentum should in any case be smaller than that produced by the masses of deposited ice.

AUTHOR'S NOTE

UNTIL a decade ago the idea that the poles had often changed their positions on the earth's surface was regarded as extreme, improbable, and unsound. It was advocated strictly by cranks. Nobody who was anybody in the scientific world would have anything to do with it.

Fashions change. Today every other book dealing with the earth sciences devotes space to polar wandering and continental drift.

Polar wandering is based on the idea that the outer shell of the earth shifts about from time to time, moving some continents toward and others away from the poles, changing their climates. Continental drift is based on the idea that the continents move individually.

Many scientists have come to the point of accepting *both* these ideas. The evidence on hand now seems to them to require that the earth's surface has shifted as a whole and that continents have also changed their positions relative to one another.

Up to the present those who have accepted both ideas have not connected them. They think of them as independent processes acting simultaneously. A few writers have suggested that perhaps continental drift causes polar wandering.

This book advances the notion that polar wandering is primary and causes the displacement of continents.

Those geologists who have accepted polar wandering and continental drift, or only continental drift, put the last such change at a long time ago.

This book will present evidence that the last shift of the earth's crust (the lithosphere) took place in recent time, at the close of the last ice age, and that it was the cause of the improvement in climate.

Two kinds of evidence are responsible for these changing ideas. New knowledge of geomagnetism, or the polarization of rocks of the earth's crust by the earth's magnetic field, has led to the discovery that the poles have changed their places on the surface of the earth at least 200 times since geological history began. There is little doubt now but that when we have the complete list it will be twice as long, or even longer.

The other new body of knowledge has come from new methods of dating events in the past by the use of radioactive isotopes of a number of elements. An isotope of carbon (C^{14}, called radiocarbon) has enabled us to find reliable dates for geological events back to about 65,000 years ago. Isotopes of other elements are good for dating events two or three hundred thousand years in the past. Still others date rocks hundreds of millions of years old.

With these radioactive dating methods it has been possible to reconstruct the climatic history of the earth in great detail for the last hundred thousand years. That is what I shall try to do in this book.

Some of the results of the chronology of the glacial epoch worked out here are surprising. For example, I have found evidence of three different positions of the North Pole in recent time. During the last glaciation in North America the pole appears to have stood in Hudson Bay, approximately in Latitude 60° North and Longitude 83° West. It seems to have shifted to its present site in the middle of the Arctic Ocean in a gradual motion that began 18,000 or 17,000 years ago and was completed by about 12,000 years ago.

The radioactive dating methods further suggest that the pole came to Hudson Bay about 50,000 years ago, having been located before that time in the Greenland Sea, approximately in Latitude 73° North and Longitude 10° East. Thirty thousand years earlier the pole may have been in the Yukon District of Canada.

These ideas are new, and they will at first seem strange, but if the reader will plow through the necessary factual detail presented in this book, he may find sufficient proof.

ACKNOWLEDGMENTS

WHEN it comes time to write an acknowledgment of the assistance received from others in the preparation of a book, this job is sometimes accomplished in a perfunctory way; it is a job to be got over with but, at the same time, turned to advantage. I do not think that this is fair to the essentially social nature of science. The implication is usually obvious that the book is, in fact, the work of one or two perspiring and inspired persons, who, by themselves alone, have persevered against odds to complete an imperishable product. This distorts the process by which scientific and, indeed, all original work is done. Scientific research is essentially and profoundly social. Discoveries are not the product of single great minds illuminating the darkness where ordinary people dwell; rather, the eminent individuals of science have had many predecessors; they themselves have been merely the final organizers of materials prepared by others. The raw materials, the component elements that have made these great achievements possible, have been contributed by hundreds or thousands of people. Every step in the making of this book has been the result of contact with other minds. The work done by hundreds of writers over a number of centuries has been exploited, and the contributions of contemporary writers have been carefully examined. The product represents, I should like to think, a synthesis of thought; at the same time I hope its original elements will prove valid additions to the common stock of knowledge in the field.

Credit for the initiation of the research that led to this book belongs, in the first instance, to students in my classes at Springfield College, in Springfield, Massachusetts. A question asked me by Henry Warrington, a freshman, in 1949 stimulated me to challenge the accepted view that the earth's surface has always been subject only to very gradual change, and that the poles have always been situated precisely where they are today. As the inquiry grew, many students made valuable contributions to it in research papers. Among these I may name, in addition to Warrington, William Lammers, Frank Kenison, Robert van Camp, Walter Dobrolet, and William Archer.

Our inquiry first took organized form as an investigation of the ideas of Hugh Auchincloss Brown, and I am deeply indebted to him for his original sensational suggestion that ice caps may have frequently capsized the earth, for many suggestions for research that proved to be productive, for his generosity in sharing all his research data with us, and for his patience in answering innumerable letters.

In this early stage of our inquiry, when I was in every sense an amateur in many fields into which the inquiry led me, I received invaluable assistance from many specialists. These included several members of the faculty of Springfield College, especially Professor Errol Buker, without whose kindly sympathy our inquiry would have been choked in its infancy. Assistance with many serious problems was received from Dr. Harlow Shapley, of the Harvard Observatory, Dr. Dirk Brouwer, of the Yale Observatory, Dr. G. M. Clemence, of the Naval Observatory, and a number of distinguished specialists of the United States Coast and Geodetic Survey.

Our inquiry, in its third year, was involved in a difficulty that appeared to be insuperable, and from this dilemma it was rescued by an inspired suggestion made by my old friend the late James Hunter Campbell, who thereafter became my constant associate in the research project, and my collaborator. I must give credit to him for having taken hold of a project that was still an amateur inquiry, and transformed it into a solid scientific project.

When Mr. Campbell had developed his ideas far enough to assure us that the idea we had in mind was essentially sound, it became feasible to submit the results of our joint efforts to Albert Einstein, and we found him, from then on, a most sympathetic and helpful friend. Throughout an extended correspondence, and in personal conference, his observations either corroborated our findings or pointed out problems that we should attempt to solve. With regard to our inquiry, Einstein made an exception to his usual policy, which was to give his reactions to new ideas submitted to him, but not to offer his suggestions for their further development. In our case, with an uncanny sense, he put his finger directly upon problems that were, or were to be, most baffling to us. We had the feeling that he deeply understood what we were trying to do, and desired to help us. Our association with him represented an experience of the spirit as well as of the mind.

In the later stages of our inquiry, many distinguished specialists and friends helped us with particular problems. Suggestions were contributed by Professor Frank C. Hibben, of the University of New Mexico, Professor P. W. Bridgman, of Harvard, Dr. John M. Frankland, of the Bureau of Standards, Dr. George Sarton, Professors Walter Bucher and Marshall Kay of Columbia University, Dr. John Anthony Scott, Mrs. Mary G. Grand, Mr. Walter Breen, Mr. Stanley Rowe, Dr. Leo Roberts, Mr. Ralph Barton Perry, Jr., Mrs. Mary Heaton Vorse, Mr. Heaton Vorse, Mr. Chauncey Hackett, Mrs. Helen Bishop, and Mrs. A. Hyatt Verrill. To Dr. Harold Anthony, of the American Museum of Natural History, our debt is enormous. It was he who afforded Mr. Campbell and me our first opportunity to discuss our theories with a group of specialists in the earth sciences, when he invited us to talk to the Discussion Group of the Museum. In addition, Doctor Anthony has helpfully criticized parts of the manuscript and has helped me to get criticism from other experts. Captain Charles Mayo, of Provincetown, Massachusetts, in many long discussions over the years, has contributed innumerable valuable suggestions.

One farsighted scientist without whose generous help this book in its present form would have been impossible is David B. Ericson, marine geologist of the Lamont Geological Observatory. He has contributed many vitally important bibliographical suggestions, has corrected numerous technical errors, and has provided needed moral support. I am equally indebted to Professor Barry Commoner, of Washington University, who not only read the manuscript to suggest improvements of content and style but also helped me in the preparation of special articles for publication in the technical journals. Mr. Norman A. Jacobs, editor of the Yale Scientific Magazine, published the first of these articles.

During the last year I have received enormous assistance from Mr. Ivan T. Sanderson, who, as a biologist, has read the manuscript with a critical eye for misuse of technical vocabulary and for weaknesses in presentation. I have received invaluable help from Professor J. C. Brice, of Washington University, who has criticized the whole manuscript from a geological standpoint. I am deeply indebted to my aunt, Mrs. Norman Hapgood, for the first complete translation from the Russian of the report of the Imperial Academy of Sciences on the stomach contents of the Beresovka Mammoth, to Mrs. Ilse Politzer for the translation from the German of Einstein's letter of May 3, 1953, and to Mrs. Maely Dufty for assistance with the translation of his foreword into English. To many personal friends, in addition to those mentioned, I owe thanks for encouragement and for suggestions that often turned out to have major importance. I am indebted to John Langley Howard for his assistance with the illustration of this book, to Mr. Coburn Gilman, my editor, for his innumerable constructive suggestions and his understanding spirit, to Mr. Stanley Abrons for his painstaking work in preparing the glossary, and to Mr. Walter Breen for preparing the index.

In the final typing of the manuscript Miss Eileen Sullivan has had to encounter and survive difficulties and frustrations that only she and I can have an idea of. I am very grateful for her help.

Grateful thanks are extended to all publishers and individuals who have consented to the use of selections or illustrations, and in particular to the following:

Columbia University Press, for quotations from George Gaylord Simpson, *Major Features of Evolution;* Thomas Y. Crowell Co., for a passage from Frank C. Hibben, *The Lost Americans;* Dover Publications, Inc., for quotations from Beno Gutenberg, *Internal Constitution of the Earth* (paperbound, $2.45); W. H. Freeman & Co., for quotations from Krumbein and Sloss, *Stratigraphy and Sedimentation;* Alfred A. Knopf, Inc., for quotations from Hans Cloos, *Conversations With the Earth;* N. V. Martinus Nijhoff's Boekhandel en Uitgeversmaatschappij, The Hague, for quotations from J. H. F. Umbgrove, *The Pulse of the Earth;* Prentice-Hall, Inc., for quotations from R. A. Daly, *The Strength and Structure of the Earth; Science,* for quotations from various issues of this magazine; William Sloane Associates, Inc., for quotations from Thomas R. Henry, *The White Continent* (Copyright 1950 by Thomas R. Henry); University of Chicago Press, for quotations from various issues of the *Journal of Geology.*

Charles H. Hapgood
KEENE TEACHERS COLLEGE
OCTOBER, 1957.

In the new, revised edition of EARTH'S SHIFTING CRUST under the title *The Path of the Pole* I am especially indebted to Mr. Oppé and Mr. Delair for the great contribution they have made in researching the difficult field of Pleistocene history in South America, and to Fred Earll for his foreword, to say nothing of his helpful advice in preparing this edition. I should also thank those hundreds of research workers who have, by their efforts in geophysical research and in the extension of our knowledge of the past through the new techniques of absolute dating, made possible my own work.

C. H.
August, 1970

CONTENTS

ILLUSTRATIONS

THE
PATH
OF THE
POLE

"We know that there is no absolute knowledge, that there are only theories; but we forget this. The better educated we are, the harder we believe in axioms. I asked Einstein in Berlin once how he, a trained, drilled, teaching scientist of the worst sort, a mathematician, physicist, astronomer, had been able to make his discoveries. 'How did you ever do it,' I exclaimed, and he, understanding and smiling, gave the answer:

" 'By challenging an axiom!' "

Lincoln Steffens, *Autobiography* (p. 816)

chapter **1**

GEOMAGNETISM, CONTINENTAL DRIFT AND POLAR WANDERING

1. DO THE POLES MOVE?

WHAT do we mean by polar wandering? This phrase may mean several different things. It may be thought to mean a shift in the position of the axis of the earth. Everyone has seen pictures of the solar system, with the earth, the other planets, and the sun shown in relationship to one another. The earth is always shown slightly tipped. Its axis does not run straight up and down at right angles to the plane of the sun's equator, but slants at an angle. Any change in this angle, in the position of this axis, would be very important for us. It might mean, for example, that the South Pole would point directly at the sun. We would then have one hot pole and one cold

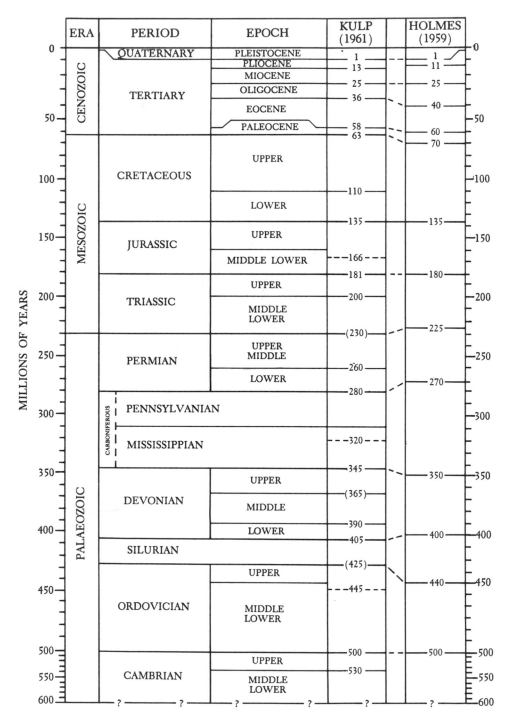

Fig. 1. *The geological table.*

pole. The hot pole would never have any night, and the cold pole would never have any day. The occurrence of this kind of polar shift has seldom been suggested, for the reason that no force capable of shifting the axis has ever been imagined, other then, possibly, a major planetary collision.

A second way of shifting the poles would be not to change the direction of the axis (let it point to the same stars as before) but merely to shift the earth around on its axis so that different places on its surface would be at the poles. This suggestion also has run into difficulties. The main obstacle to such a shift is the stablizing effect of the equatorial bulge of the earth. The earth is not a true sphere, of course, but a slightly flattened spheroid. The polar flattenings are balanced by a bulge at the equator. The diameters of the earth through the poles and through the equator differ by about 13 miles. The extra mass around the equator, rotating at a very rapid rate, acts like the rim of a gyroscope to keep the earth steady on its axis. The earth is not absolutely steady—it wobbles a little, its off-center path having an average radius of about 50 feet and completing a cycle in about 14 months. But this wobble is not important for us in the present discussion.*

During the nineteenth century some of the titans of geology, including James Clerk Maxwell (296)† and Sir George Darwin (son of Charles Darwin), considered this problem, and they decided that the stabilizing effect of the equatorial bulge was so great that no conceivable force originating within the earth could make it shift on its axis. They therefore dismissed the idea of any shift of the poles as impossible and, in fact, not worth discussing. Their influence was sufficient to make sure that nobody until now would seriously consider polar shifts.††

* The cause of the greater part of the wobble is attributed to differential attractions of the sun and moon on the equatorial bulge, and, in fact, its discovery was anticipated by Newton. A part of it is unexplained. Two recent writers (Mansinha and Smylie, 291e, f, g) have presented evidence that the wobble can be explained as the result of major earthquakes. However, it seems at least possible that the relationship is the reverse, that the wobble is the cause of some of the major earthquakes. (See Note 8.)

† Note: Figures in parentheses throughout the text refer to specific sources listed in the bibliography (page 372). The first figure indicates the correspondingly numbered work listed in the bibliography; the figure following the colon indicates the page reference in the cited work.

†† I once wanted to know the total stabilizing effect of the equatorial bulge and sought the aid of mathematicians. For the calculations see Note 8, p. 361 .

A third way to conceive of polar wandering is through the sliding of the earth's whole outer shell over semiliquid layers below. This idea was suggested long ago, like most good ideas, but the fellow who first advanced the notion, Damian Kreichgauer (256), did not have enough influence to impose the idea on his contemporaries. This theory requires only two things: first, that the material under the earth's rigid crust (more properly called the "lithosphere") be sufficiently liquid, and second, that there should be available a force sufficient to set the outer shell in motion and keep it moving for considerable distances.

Still another way to change the positions of the poles relative to the continents is to move the continents. This is now a very popular way of explaining things, and we shall discuss it below. However, first we must ask: What is the evidence that the poles really have changed their positions, in one way or another, relative to the various parts of the earth's surface, especially the continents? People have been claiming to have evidence of such changes for a hundred years, but only recently have we obtained evidence that really seems indisputable. The new evidence has come from studies of the past history of the earth's magnetic field.

2. THE GEOMAGNETIC EVIDENCE

The magnetic field of the earth is what influences the compass to point north. The compass needle does this because it is composed of iron, and iron is a magnetic substance; that is, a substance which will itself become magnetic when exposed to a magnetic field, and will therefore align itself with the lines of force of the earth's field. What is true of iron in the compass needle is also true of the iron in rocks composed of minerals containing iron.* The tiny iron particles in these rocks also take on a magnetization; they become miniature compasses lined up with the earth's magnetic field.

The compass needle, if it is free to move in all directions (vertically as well as horizontally), will start pointing gradually downward as the ship or plane bearing the compass approaches the north or south magnetic pole. At the north magnetic pole its north-pointing end will point straight down, and, of course, its south-seeking end will

---

* Four such minerals are magnetite, hematite, maghemite, and ilmenite (192:1114).

Fig. 2. *Pole positions determined from European rocks of Tertiary and Quaternary Periods.*

be pointing straight up. At the South Pole this would be reversed. At the magnetic equator the compass needle would lie level.

When iron-bearing rocks have become magnetized they are actually more informative than the mariner's compass. They indicate the direction of north as the ship's compass does. But by the angle of their dip they also indicate how far away the pole is. This means they indicate the *latitude*. The horizontal angle is called *variation*; the dip is called *inclination*. The variation gives the longitude of the sample relative to the present magnetic pole, and the inclination gives the latitude. The magnetic rocks adopt the direction of the

Fig. 3. *Pole positions determined from European rocks of Mesozoic Era.*

earth's field at the time that they are formed, and in many cases they preserve their directions of magnetization indefinitely.

Of course, there are many problems. Local movements of the crust may invalidate the evidence by moving the slab of rock containing the sample hundreds of miles along a major fracture, like the San Andreas Fault in California. The sample, through deep burial, might have been subjected to heat and pressure sufficient to destroy its magnetization or alter the direction of its field. There are other factors that may operate to invalidate a sample. Most of these, however, being local, may be eliminated by the simple means of basing estimates of the position of a pole at a particular time on many samples

Fig. 4. *Pole positions determined from European rocks of Late Paleozoic Era.*

taken from different places far apart, in different countries or continents.

Another problem arises from the fact that the magnetic field of the earth does not stay put. It is in constant motion, having, for one thing, a steady westward drift. Geophysicists studying this, however, have concluded that over a period of a few thousand years the earth's field returns to its original position, and that the average position of the magnetic pole over the whole period will coincide with the earth's axis of rotation.

It is simply a question, then, of taking samples from a rock thick enough to represent the sedimentation of several thousand years. If

Fig. 5. *Pole positions determined from European rocks of Precambrian and Cam-brian Periods.*

the samples come from lava flows, they have to be taken from successive lava flows indicated as having occured fairly close together in time.

When many samples are assembled, and the results are averaged to eliminate the errors due to local factors or to the "secular" variation of the earth's magnetic field, and when it has become reasonably certain that no factor has intervened to change the original direction of the magnetization, we begin to have a fairly reliable indication of

Fig. 6. *Pole positions determined from Asian rocks of the Ordovician to the Quaternary Period.*

the position of the pole at the time when the rocks were laid down. If we know approximately when they were laid down (dating them by the included fossils or by one of the methods of "absolute" dating now available) we can assign a date to the particular pole position.

It is obvious, from what I have said above, that there are booby traps, so to speak, for the unwary worker in this field. Nevertheless a vast amount of research has been done, and the state of the science has advanced until we can say that the present findings of pole

Fig. 7. *Pole positions determined from Indian rocks of Mesozoic Era and Tertiary Period.*

positions are reasonably reliable. Margins of error are usually allowed. They are indicated by "uncertainty ovals" drawn around each pole position. The reader will observe these ovals in Figure 8. Table 1 lists a great many positions for the North Pole, found from samples of magnetic rocks from all the continents (111a:4–46). Each of these positions has been found by averaging the directions from many samples in the same geological formation, or in geological formations of the same age from other localities. The uncertainty ovals are statistically determined and have 95% accuracy. It is the opinion of specialists that the findings are reliable. What the findings indicate is that the poles have changed their positions relative to the earth's surface many, many times throughout the history of the earth.

TABLE 1

Number of Positions of the North Pole

Epoch or Period	Dura- tion (Mil- lions of years)	Eur.	Asia	In- dia	N. Amer.	S. Amer.	Afr.	Aust.	Ant- arc.
Pleistocene	1	6	15	0	1	1	?	?	?
Tertiary	62	0	8	1	4	0	0	1	1
Cretaceous	73	0	3	10	1	0	1	1	0
Jurassic	46	6	2	1	0	1	7	1	3
Triassic	49	8	7	0	20	0	2	1	0
Permian	50	15	0	0	9	0	2	2	1
Carboniferous	65	15	1	0	5	0	4	2	0
Devonian	50	7	0	0	0	0	0	2	2
Silurian	20	0	2	0	2	0	1	3	0
Ordovician	75	0	2	0	1	0	1	3	0
Cambrian	100	1	0	0	3	0	0	2	0
Precambrian	?	3	0	0	22	0	0	3	0
		61	40	12	68	2	18	21	7

Grand Total of Poles...229

We see that 229 pole positions are indicated on this table. It is assumed that some of these may be duplications; that is, the same pole indicated from different continents, which is shown as a different pole because of continental drift. How many of these poles may be duplicates? In order to reduce the likelihood of duplicates I have listed, in Table 2, the poles found for each geological period from rock samples from one continent only. Since, obviously, the number of poles in any one geological period would have to be at least equal to the highest figure from any one of the continents, I have selected the continent showing the most pole positions for each particular period.

TABLE 2

Pole Positions Shown by Samples from One Continent

Epoch or Period	Duration (Years in Millions)	Number of Pole Positions	Continent
Pleistocene	1	15	Asia
Tertiary	62	8	Asia
Cretaceous	73	10	India
Jurassic	46	7	Africa
Triassic	49	20	North America
Permian	50	15	Europe
Carboniferous	65	15	Europe
Devonian	50	7	Europe
Silurian	20	3	Australia
Ordovician	75	3	Australia
Cambrian	100	3	North America
Precambrian	?	20	North America
		126	

It seems obvious that the number of pole positions so far determined, though large, is not nearly enough for any but tentative conclusions about continental drift. The following table illustrates this by giving the average time intervals between the poles so far determined for the different geological periods.

TABLE 3

Time Intervals Between Pole Positions

Epoch or Period	Duration (Years in Millions)	Number of Pole Positions	Average Interval
Pleistocene	1	15	66,666 yrs.
Tertiary	62	8	ca. 8,000,000 yrs.
Cretaceous	73	10	ca. 7,300,000 yrs.
Jurassic	46	7	ca. 6,500,000 yrs.
Triassic	49	20	ca. 2,500,000 yrs.
Permian	50	15	ca. 3,300,000 yrs.
Carboniferous	65	15	ca. 4,300,000 yrs.
Devonian	50	7	ca. 7,000,000 yrs.
Silurian	20	3	ca. 7,000,000 yrs.
Ordovician	75	3	ca. 25,000,000 yrs.
Cambrian	100	3	ca. 33,000,000 yrs.

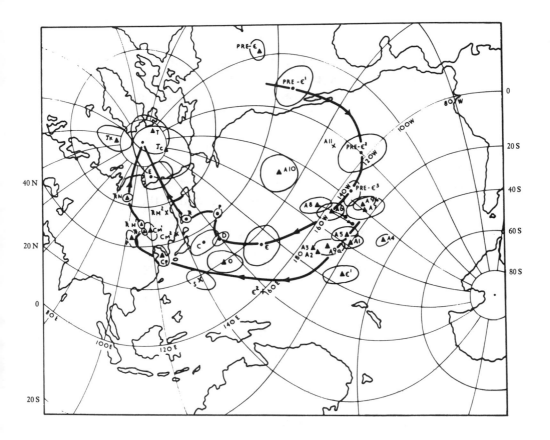

Fig. 8. Polar wandering curves based on European and North American rocks. (Mean poles based on all determinations are underlined. European poles are denoted by dots and American poles by triangles and crosses.)

It is obvious that such inequalities in average intervals between polar shifts can have no relation to reality. If there is any regularity (even approximate) in the intervals between polar shifts, the minimum period indicated in this table is more likely to reflect the truth than are the very long periods. It is perfectly obvious that for very long remote geological periods we have virtually no evidence. For the Pleistocene it seems we have an average interval of about 70,000 years between polar shifts. I shall show, in the following chapters, that this may not be very far from the mark. The very long intervals indicated for the older geological periods I think reflect merely our lack of information.

In this book I will give special attention to recent geological time; that is, to the Pleistocene Epoch. It is commonly thought that no polar shifts could have occurred in this recent period. It is the general impression that the poles have been in their present locations for at least several million years. Magnetic evidence in contradiction of this view, however, now exists. This evidence has resulted from studies of volcanic deposits in Japan by Nagata, Akimoto and others (318b) and from Soviet studies (249a). The Japanese samples were taken from the Omura-Yama group of volcanoes in the North Izu and Hakone volcanic region on the east coast of Honshu. Table 4, below, gives positions of the North Pole during the Pleistocene

TABLE 4
Pleistocene Positions of
Magnetic North Pole from Japanese Samples

Poles in Reverse Order	Positions	Radius of 95% Certainty Oval
1.	78°N, 80°W	14°
2.	80°N, 0 Long.	7°
3.	76°N, 14°W	6°
4.	75°N, 22°W	23°
5.	75°N, 7°E	20°
6.	50°N, 134°E	19°
7.	73°N, 46°W	15°
8.	75°N, 38°E	73°
9.	87°N, 90°W	10°
10.	81°N, 86°E	18°
11.	80°N, 72°W	7°
12.	84°N, 60°E	10°
13.	73°N, 112°W	16°
14.	57°N, 33°E	18°
15.	78°N, 144°E	16°
16.	75°N, 28°E	21°
17.	82°N, 24°W	7°
18.	72°N, 42°E	7°
19.	68°N, 55°E	5°
20.	54°N, 39°E	12°
21.	86°N, 97°E	17°
22.	82°N, 98°E	13°
23.	79°N, 65°E	9°
24.	54°N, 51°E	19°

Poles in Reverse Order	Positions	Radius of 95% Certainty Oval
25.	70°N, 26°E	28°
26.	73°N, 62°E	21°
27.	61°N, 108°E	6°
	Reversal?	
28.	68°S, 146°W	12°
29.	65°S, 155°W	7°
30.	47°S, 175°W	44°
31.	61°S, 169°W	13°
32.	43°S, 104°W	21°
33.	58°S, 3°W	31°
34.	25°S, 107°W	38°
35.	11°N, 113°W	15°
36.	62°S, 117°W	20°
37.	41°S, 157°E	24°
	Earlier Reversal?	
38.	85°N, 33°E	8°
39.	56°N, 86°E	7°
40.	67°N, 83°E	19°
41.	66°N, 93°E	7°
42.	79°N, 33°E	16°
43.	71°N, 36°E	49°

Epoch in reverse chronological order, beginning with the most recent based on Japanese rocks. Column 3·gives the radius of the oval of 95% certainty for each position. No precise datings of the different lava flows were possible. It is therefore not possible to average out the secular variations of the magnetic pole positions indicated, and they do not necessarily indicate the positions of the geographic poles. In some cases they may indicate the temporary position of the pole at the time of a lava flow, when the pole was actually in motion from one position to another.

It is interesting to note that Table 4 indicates that a complete reversal of the earth's magnetic field occurred twice during the period.

The authors, attempting to interpret the data of this table as a continuous curve, state: "On smoothing out the fluctuations whose periods are shorter than the period of each volcanic activity, the position of the north pole of the geomagnetic centered dipole can be estimated to have shifted from 72° N, 86° E to 81° N, 32° W during the whole Quaternary period" (318b:263). This interpretation of their data will be discussed below (Chapter VII, page 183).

The Soviet findings, as given in Table 5, are not nearly as complete, and their chronological order is uncertain. Nevertheless they indicate considerable displacements of the north magnetic pole.

TABLE 5

Pleistocene Positions of the North Magnetic Pole after Khramov et al. (249a)

Period	Pole Position	Linear Displacement* (approximate)
Early Quaternary	76°N, 166°W†	
Bakinskian	81°N, 166°E	300 miles
Quaternary	88°N, 135°W	480 "
Middle Quaternary	76°N, 146°W	720 "
Quaternary	80°N, 115°W	420 "
Upper Quaternary	75°N, 163°W	600 "
Upper Quaternary	77°N, 107°W	540 "

The geomagnetic data from older geological periods discussed above (Figs. 2-8, pp. 5-13) have inspired geologists to attempt to construct curves for the path of the pole for tens or hundreds of millions of years. It seems to me that the evidence is insufficient for the construction of curves in the older periods. It appears that even in the short period of the last million years such a curve can be constructed only at the cost of sacrificing much of the detail of the evidence. I do not think that those who assume this sort of progressive pole wandering over considerable periods have stopped to work out the logical consequences of their assumptions. What does the idea of a curve involve?

In the first place it involves a continuous motion, or, in the case of interruptions or standstills, the resumption of the motion in more or less the same direction. Changes of direction would have to be gradual in order to give us the long, smooth curves indicated by the authors of the diagrams. We must remember that if the lithosphere

---

* On the assumption that findings are in chronological order.
† This finding indicated reversed polarity.

is set in motion over the asthenosphere immense forces have been involved. A powerful momentum has been transferred to the lithosphere. An equal force will be required to stop it. But stop the lithosphere did, each and every time after a shift, long enough to form the sedimentary rocks that were to become magnetized in the direction of the earth's magnetic field at the time. It is obvious from the methods used in the fieldwork that the polar positions were stable for periods of thousands of years, just as they have been in our day ever since men began to make astronomical records in Egypt and in Babylon.

What is the probability that after the lithosphere had been brought to a halt, with each stand still lasting an unknown—perhaps considerable—length of time, that it would resume motion in the same or very nearly the same direction? And what sort of force within the earth could act intermittently in this way? The mechanism I propose to suggest a little later on in this chapter will be one that can account for the starting and stopping of the movements, but it is one that will not act continually in the same direction. According to it, the path of the pole would more likely be a sort of zigzag, with the direction of each new movement determined by the balance of forces within the earth at the particular time. The zigzag path would tend to insure the pole's remaining within the general area for a considerable time and taking long periods of time to migrate great distances. In Chapters IV, V, and VI and VII I shall present evidence to demonstrate this very sort of motion of the pole in the Pleistocene Epoch.

There are three possible interpretations of the geomagnetic evidence. We can assume polar wandering, continental drift, or a combination of the two. But in order to be able to judge the relevant evidence we must, before we proceed, take a look at what is known about the inside of the earth, where unknown forces have operated to bring these things about.

3. BRIEF SURVEY OF THE EARTH'S STRUCTURE

There is not much that we know about the inside of the earth, because we are unfortunately restricted to living on the outside of it. We must deduce what we know from mining and oil drilling activities, seismic waves from earthquakes or artificial explosions, and principles of physics. The subject is wrapped in uncertainty. Therefore, it is natural that there should be differences of opinion—opposing groups of scientists devoted to different notions. However, among the uncertainties there are some points on which most scientists are agreed. I will confine myself here to giving in a condensed form the consensus of geological opinion on the structure of the earth.

(a) There is an outer shell, composed of solid, crystalline rock about 30 to 40 miles thick, extending down to the melting points of the rocks. This layer is properly called the "lithosphere," although it is sometimes called the earth's "crust."

(b) The lithosphere is theoretically arranged in layers, which include loose sedimentary material on top, then rocks, called "sedimentary rocks," made of such sediments, then rocks originally made of such sediments, which have been melted and partly fused, called "metamorphic rocks." Generally below these sedimentary and metamorphic rocks are heavier granitic rocks (also partly sedimentary in origin) and finally still heavier basaltic rocks. These layers are not at all uniform in thickness; the layering is only a general tendency. Light rock exists at all depths in the lithosphere, even on the bottom of it. The lithosphere has been churned up all through its history by processes of mountain building, and the like.*

(c) At a depth of about five miles under the ocean bottoms and two or three times that under the continents, there is a discontinuity called the Moho,† that has been much in the news lately. Below this break the material differs in some ways from the material above, but just how we do not know. Some scientists believe there is a chemical difference. Others maintain that the difference is only a change of phase. It is, of course, true that heat and pressure can change the appearance, density and properties of a substance without changing the chemical composition. Everyone knows that graphite, if placed under sufficient pressure and heated, can be turned into diamond. There are some reasons for favoring the phase theory about the Moho. For one thing, it would be hard to reconcile a chemical

---

* According to Daly, mountain folding has involved the full depth of the lithosphere (97:399).

† Named after its Czechoslovak discoverer, Dr. Andrija Mohorovicic.

difference at the Moho discontinuity with the folding of the litho-sphere to its full depth in mountain building.

A few years ago some scientists conceived the idea of boring a hole down to the Moho. Congress approved a large appropriation for the operation, and there was hope for an answer to this question of chemical versus phase change. Unfortunately politics stepped in. The contract for the operation was taken away from a corporation that was used to oceanographic work and given to a Texas corporation less used to working with such problems. Difficulties developed and the project was abandoned. An interesting book, however, has been written about it (24a) and perhaps some day the project will be carried through (probably by the Russians).

This whole question of the Moho has produced some unfortunate confusion. Some scientists have fallen into the habit of regarding this Moho discontinuity as the bottom of the "crust." The crust used to be regarded as synonymous with the lithosphere; that is, the whole crystalline shell of the earth extending down to a depth of 30–40 miles. This definition has been abandoned. It has become the prac-tice to lump everything below the Moho under the general—and therefore meaningless—term "mantle." Of all the ambiguities in cur-rent scientific literature the confusion of the terms "crust" and "mantle" is the worst. It has confused the scientists themselves, as we shall see. One point must be made clear: The Moho does not mean the lower limit of the crystalline rocks. The rigidity and strength characteristic of those rocks continue to a much greater depth.

This confusion is related to another one, concerned with the strength of the lithosphere. Some scientists work from the premise that the lithospheric shell, the outermost shell of the earth, is weak; they even treat it mathematically as if it were a liquid! This is an error which comes from the failure to distinguish clearly between two different qualities of crystalline rocks, two different kinds of "strength." There is *tensile* strength, but there is also another kind, which we may call *crushing* strength. It is quite true that the rocks of the lithosphere do not have the tensile strength of steel. You could never make I beams of rock. It is plain that you can shatter rock with comparative ease. Crushing strength is an entirely different matter. The crushing strengths of rocks differ, naturally, with chemi-cal composition, but granite and basalt, for example, have enormous crushing strength, as anyone running a car into a granite cliff will discover. The lithosphere, according to Dr. Harold Jeffreys, the dean of British geophysicists, is rigid enough to transmit stresses across any distance (238:288). We shall see that this crushing strength of the lithosphere, including that part of it that lies under the oceans, will be very important for various aspects of polar wandering and continental drift.

It is important to note that the strength of rocks depends on their crystalline structure. The strength of crystals varies with their chemical composition. Ice crystals have little strength, but granite and basalt crystals are very strong. The rocks of the earth's crust are composed of billions of tiny, strong crystals that interlock with one another, pointing in all directions, not in neat arrangements. These strong, rigid, interlocking crystals prevent the molecules of the rocks from sliding past one another, as they do in liquids.

(d) Below the lithosphere there lies a thick layer of rock that is soft, amorphous, perhaps nearly liquid. This layer begins at the point where the earth's heat reaches the melting points of the rocks, and this is thought to be at a depth of 30 to 40 miles, as already mentioned, both beneath the continents and beneath the ocean basins. This soft layer is called the "asthenosphere" from the Greek word for "weakness."

The evidence for this soft layer is of various kinds. Daly cited geological evidence from mountain building and from the rebound of areas that were depressed under the ice load in Europe during the ice age (97); Gutenberg calculated that the soft layer extended from 60 to 120 miles down, but this was revised by Anderson to from 35 to 150 miles (1c:54). Vacquier has pointed out that magnetic anomalies, showing strike-slip displacements of the ocean floor (to be discussed later) suggest a soft layer underneath:

> The existence of displacements in the ocean floor of magnitude comparable to the distances that continents are presumed to have drifted, indicates that there must be a mechanism for lubricating, so to speak, these displacements, so that whole continents, and in the present case stretches of oceanic crust several hundred kilometres long remain virtually undistorted by the motion. (363a:143).

Chadwick presents various arguments for the existence of the soft layer (363a:215,225).

This soft material of the asthenosphere may be technically a liquid, but that doesn't mean that it is as liquid as water. It may be more like a stiff tar; the difference is a matter of viscosity. The viscosity of a liquid is its resistance to flow. Water has low viscosity. Melted rock, as we see it, for example, in lava flows, has higher viscosity. It is true also that pressure increases viscosity, because it presses molecules more closely together and makes it harder for them to flow past one another. The asthenosphere, then, may be fairly stiff, even right below the lower limit of the lithosphere, and probably grows stiffer with depth even though the heat, which is also assumed to increase with depth, tends to counteract the effects of pressure by reducing the viscosity. Studies of the velocity of earthquake waves suggest that the viscosity increases in the lower levels of the asthenosphere until at a depth of 180 to 250 miles the material may behave in some respects like a solid. Again, we shall find all these facts vital in evaluating theories of polar wandering and continental drift.

There are still other factors to be considered with respect to the asthenosphere, which, as we have just seen, in its lower parts may not be weak at all. So far as viscosity is concerned, chemical composition is just about as important as heat. Each chemical substance has its own melting point, its own degree of viscosity under the opposing influences of heat and pressure. This makes it impossible to be sure exactly where the lithosphere ends and the asthenosphere begins, or to know what the viscosity of the asthenosphere is. Such information is not at present available, though there are indications. We can make deductions from some lava flows as to the chemistry at the bottom of the lithosphere, but unfortunately lava flows differ in chemistry. Dr. Reginald Daly of Harvard reasoned that the asthenosphere right under the bottom of the lithosphere had to be virtually a liquid to account for the easy response of the lithosphere to the growth and melting of ice caps (97:19,389).

As if the situation were not already difficult enough, it appears that two other qualities of matter must be taken into account. One of these we call "plasticity," and the other, already mentioned, is change of phase. The term "plastic" is often used these days by writers on geophysics as if it were synonymous with viscosity. But the plastic behavior of rocks has nothing whatever to do with viscosity. It is, in fact, an *opposite* kind of behavior. The difference between them is all-important for the issues we are discussing. The difference between them is that if you apply pressure to a viscous material it will yield gradually at a speed proportional to the pressure applied. Therefore, with twice the pressure, the material will yield twice as fast. Plastic deformation works differently. Every chemical solid has a plastic limit. It will behave like a solid until the pressure reaches that limit. Then it yields suddenly and completely. It does not yield at a rate proportional to the applied force. It is almost like the breaking of the branch of a tree. When the branch breaks, its strength is suddenly and entirely gone.

This type of plastic behavior deep in the earth permits very sudden events to take place at a level where the viscosity is very high. According to Daly, who based his opinion on laboratory experiments carried out at Harvard by Bridgman (97:403), this is the true explanation of "deep focus" earthquakes. Completely unlike viscosity, the plastic strength of materials decreases steadily with increasing pressure. It therefore reaches a low point in the bottom levels of the asthenosphere, and this permits the fracture and slip of materials under great pressure, as were observed in the above-mentioned experiments in the laboratories at Harvard.

Some scientific writers, failing to take into account the plastic behavior of materials, have made the error of assuming that the deep-focus earthquakes mean that the crystalline structure of the lithosphere extends to depths of 300 to 430 miles, an impossibility unless the intensity of the earth's heat at these depths has been grossly exaggerated.

The earth's heat, unfortunately, is another mystery. Formerly it was thought that it originated with the planet, and was greatest in the earth's core. Now it is thought that most of the heat originates from the decay of radioactive materials in the lithosphere, and very little comes from greater depths. We have no direct evidence that can settle this matter. It is quite possible that the heat gradient we observe in deep mines, on which our estimates are based, does not extend all the way to the earth's center. If that is true it will pull the rug out from under many of our ideas.

(e) In addition to these various uncertainties, we must also consider change of phase. As already noted in the case of the Moho, a chemical substance under different conditions of heat and pressure may change state. It may expand or contract, or a change of phase may permit it to enter into chemical reactions with other minerals which could not have occurred in its original state.

It seems that such a layer has been discovered in the asthenosphere at a depth of about 100 miles. According to the Soviet geophysicist V. V. Beloussov (25a), chemical processes at this depth, made possible by change of phase, are changing heavier into lighter rock, thus causing gravitational instability as the lighter rock tries to rise to the surface. Beloussov has named this the "wave-guide layer." Observations by the American geophysicist Frank Press are in general agreement. Press finds (from satellite observations) that this layer is a very liquid one (349c). It seems that if the earth's outer shell does slide as a unit over the interior, this is the most likely level at which the movement can occur.

(f) Below the wave-guide layer the material continues to increase in viscosity, because of the increasing pressure, until it reaches the state that Professor Daly referred to as a "pressure solid." Below this point (at a depth of about 430 miles) earthquakes do not occur. It would seem that here viscosity has finally conquered plasticity. The exact depth at which the pressure solid begins is uncertain, but it is vital for the various issues involved in continental drift.

(g) Finally, at a depth estimated at 1800 miles, we arrive at the earth's core, containing, according to some experts, two layers: an outer one, solid, and an innermost one, liquid. The innermost core may or may not be an iron core; there may or may not be rapid currents in it; it may or may not have any connection with the earth's magnetic field. If we discuss these questions it is only to present differing views: There is no consensus.

4. CONTINENTAL DRIFT: THE THEORY OF ALFRED WEGENER

Alfred Wegener, who first drew attention to the idea of continental drift, was a good scientist, though not a geologist. He had found quantities of evidence of fossil flora and fauna that could not, in his opinion, be reconciled with the present positions of the poles. Inasmuch as the authorities, such as George Darwin and Maxwell, forbade any thought that the axis had moved, or that the earth had shifted on its axis, Wegener suggested that the continents had moved. This would have precisely the same effect as polar shift, for it would mean that at different times the same areas might be found in very different latitudes..

Wegener imagined that the continents, formed of light sedimentary and granitic rocks, were floating in the heavier basaltic material of the ocean bottoms. He thought he saw evidence, in the shapes of the continents, that they had once formed a single land mass, which had split so that pieces could go drifting off over the ocean bottoms. He thought of the ocean bottoms as composed of soft, amorphous or viscous rock. From a vast amount of fossil evidence of the plant and animal life of the past he imagined that he could reconstruct the paths of the continents over long periods of time. He proposed to explain the mystery of ice ages by this theory. He suggested that during the last ice age in the Northern Hemisphere, Europe and America had lain together near the North Pole, but that since then they had drifted away from the pole and each other.

Wegener's theory had great appeal—not because all the evidence supported it, not because the authorities were open-minded about it, not because its mechanics were very plausible, but because it was the *only* theory that, at the time, could make sense of the evidence of the fossil flora and fauna.

A number of weaknesses in this theory were gradually revealed. As knowledge of the ocean bottoms increased it was discovered that the rock under the oceans, which Wegener thought to be plastic enough for the continents to drift over it, was in fact very rigid, and indeed stronger than the rocks of the continents. This meant that the continents could not drift without displacing and pushing to one side a layer of rigid rock estimated to be twenty miles thick. It therefore seemed to be impossible for continents to drift. Jeffreys, basing his opinion on the evidence for a rigid and comparatively strong ocean floor, said, ". . . There is therefore not the slightest reason to believe that bodily displacements of continents through the lithosphere are possible" (239:346). The geophysicist F. A. Vening Meinesz, according to Umbgrove, conclusively proved the considerable strength of the crust under the Pacific (420:70).

Another weakness that appeared in the Wegener Theory was in its assumption that the ocean bottoms were smooth plains. The assumption seemed necessary to the theory, for otherwise the continents could hardly drift over them. However, as the result of the oceanographic work of recent years, it has been discovered that there are mountain ranges on the bottoms of all the oceans, and that some of these ranges are comparable to the greatest ranges on land. Furthermore, several hundred volcanic mountains have been discovered spread singly over the ocean bottoms, many of them of great age, others comparatively young.

The Wegener Theory involved the corollary that, as the continents had drifted very slowly over the smooth ocean floors, these floors would have accumulated sediment to great thicknesses. It was thought that the sediment laid down in the sea behind the moving continents would provide an unbroken record of life from the beginning of the geological record. However, the greatest surprise of recent oceanographic exploration has been the discovery that this supposed thick layer of sediment is nonexistent. The layer of sediment on the ocean bottom is uneven, in some places only a few feet or a few inches thick, and is rarely of great thickness.

Another startling contradiction to the original Wegener Theory is presented by recent data that have drastically changed our ideas regarding the date of the last ice age in North America. We have learned, through the new technique of radiocarbon dating, that this ice age ended only 10,000 years ago. In Wegener's time it was considered by geologists to have ended several times that long ago. Since Wegener supposed that Europe and North America had been situated close together and not far from the pole during the ice age, the new data have the effect of requiring an incredible rate of continental drift. Three thousand miles of drift in 10,000 years would amount to 1,500 feet a year. Furthermore, movement at something like this rate would have to be still going on, for the momentum of a continent in motion would be tremendous. And what would be the consequence of movement continuing at this rate? It would mean that oceanic charts would have to be revised every few years, while transatlantic cables would be breaking all the time.

To cap the case, Gutenberg has shown that the various forces that Wegener depended on to move the continents are either nonexistent or insufficient (194:209), while another geophysicist, Lambert, has stated that they amount to only one millionth of what would be required (64:162).

Despite the apparently overwhelming character of these objections, attempts to rehabilitate the Wegener Theory continued. Daly attempted to find a better source of energy for moving the continents (98); Hansen suggested that the centrifugal effects of ice caps might have moved the continents (199). A contemporary Soviet plant geographer, while recognizing the objections to the theory, nevertheless remarked of it that "it . . . constitutes the only plausible working hypothesis upon which the historical plant geographer can base his conclusions" (452). As recently as 1950 the British Association for the Advancement of Science divided about equally, by vote, for and against the Wegener Theory (351).

The difficulties with Wegener's concept of continental drift forced people to explore other possibilities of shifting the poles. The British astronomer Thomas Gold postulated that the earth's wobble on its axis could cause a plastic readjustment of its mantle sufficient to move the poles 90 degrees in a million years (176). The French geographer Jacques Blanchard suggested the possibility of extensive polar shifts due to more pronounced wobbling of the earth in the past (38). Professor Ting Ying H. Ma, of the National University at Taipai, Taiwan, suggested a combination of continental drift with displacement of the outer shells of the earth down to a depth of several hundred miles (285:290). Bain thought of displacements of the lithosphere to account for the facts of ancient plant geography and fossil soils, and suggested a mechanism to account for them (18). Pauly (342) revived a suggestion made by Eddington (124) that the lithosphere may have been displaced by the effects of tidal friction. Kelly and Dachille, in a provocative work on collision geology entitled *Target Earth*, offered the hypothesis of displacements of the lithosphere as the result of collisions with large planetoids (248). Recently Mansinha and Smylie have suggested that great earthquakes may have had the effect of shifting the poles (291 e, f, g).

5. CONTINENTAL DRIFT: THE NEW THEORY

None of these ideas was taken seriously until the development of the evidence of terrestrial magnetism. Then people began to take a new look at the theory of continental drift. Since changes in the positions of the poles relative to the continents now apparently had to be accepted, perhaps continental drift would provide a less sensational way out than displacements of the whole lithosphere. It is quite true that the geomagnetic evidence very early indicated clearly that at the very least both things had happened; nevertheless, such is the frailty of the human mind, scientific or not, that displacements of the lithosphere have been pushed far into the background, and all the attention has been paid to continental drift. Nothing has been done to develop the consequences or implications of polar shift, while on the other hand a large amount of research has gone into explaining how continental drift could have occurred.

The most vital problem, if the continental-drift theory was to be resuscitated, was to find an adequate source of power to move the continents. Professor J. Tuzo Wilson (445a, 445b), together with others, proposed such a source. They suggested that the continents could be moved by the operation of powerful currents under the lithosphere, powered by convection cells formed in what they call the "mantle" and we call the asthenosphere.

A convection current is set in motion in a liquid by a source of heat situated at some depth and concentrated more or less at some point. A Bunsen burner under a beaker of water will illustrate the principle. The heat is concentrated at some point under the beaker. It heats the water immediately above, which expands and rises to the surface. At the same time water begins to sink to take the place of the rising water. The rising water reaches the surface and flows horizontally to the point where the water is moving downward, and so a circular convection cell is created. Anyone can see a convection cell of this sort in operation if he throws a spoonful of cornmeal into a pan of boiling water.

Mr. Wilson and other geophysicists now argue that when a convection current in the mantle rises under the crust and spreads out in opposite directions it will tend to pull the crust apart. There will be a drag on the undersurface of the crust. Then when the current arrives at the place where it will descend again into the earth, it will tend to pull down and fold the crust over it. The effect is that a whole section of the crust has been moved laterally. The speeds of such currents have been estimated at from one to four centimeters (one and a half inches) a year. It is obvious that if such a movement should continue for a few hundred million years continents could be moved a long way.*

* At the rate of 4 centimeters per year, a continent could be moved 4 kilometers (2½ miles) in 100,000 years, and about 2,500 miles in 100,000,000 years. It is obvious that, with this rate of speed, the continental-drift theory cannot contribute to the explanation of the cause of the Pleistocene ice ages, which is the main theme of this book.

At this point the confusion of terminology already mentioned becomes serious. When some writers speak of currents rising under the crust they mean the lithosphere, but others mean the Moho. There is a tendency to forget all about the fact that the strong, rigid crystalline rocks of the lithosphere extend down some 40–45 miles under continents and oceans alike. There is a tendency to think of the viscous mantle (the asthenosphere) as coming right up to the Moho, only some five miles below the bottoms of the oceans. These writers even go so far as to include the ocean bottoms themselves, which of course are part of the rigid lithosphere. We shall examine a notable example of this confusion below.

There are some rather serious difficulties with the convection-current idea. It is not that much doubt exists that convection currents, or some kind of currents, do operate below the lithosphere. It is a question of the scale on which they can operate and of their ability to satisfy the complex requirements of the evidence.

The first problem is the motive power, the radioactive heat. In order to move whole continents, the convection currents would have to have amplitude. They must rise from the very bottom of the mantle, not far above its boundary with the core, and when they reach the bottom of the lithosphere, they must flow horizontally for thousands of miles. The source of power for this motion is supposed to be radioactive heat. But it has been learned—and the matter is not in dispute; it is part of the consensus—that most of the radioactive materials in the earth are confined to the outer forty miles of it— i.e., to the lithosphere! Most of the heat from the radioactivity is believed to be generated there. Logically the lithosphere ought to be nearly as hot as the bottom of the mantle. It would seem strange that the convection currents should be bringing up heat to the lithosphere when it is in the lithosphere itself that most of the heat is developed! Is not this something like bringing coals to Newcastle? Even if we allow that some localized heating in depth may produce convection currents, still this concentration of most of the radioactive heat in the lithosphere does most certainly remove much potential power (and therefore amplitude) from the convection currents.

Another factor militates against convection cells. We have seen, in our review of the earth's structure, that viscosity increases with pressure until, at a certain depth, the material of the earth is thought to assume a pressure-solid state. This situation develops long before the core is reached. It is not probable that convection currents can operate significantly in a pressure solid. It is therefore unlikely that convection cells can develop the amplitude necessary to move continents. A recent writer on this subject, Norman H. Sleep, has remarked that failure to take the effects of viscosity into considera- tion tends to vitiate much thinking about convection currents (384a:542). A laboratory experiment constructed by T. D. Foster has demonstrated that high viscosity in a liquid inhibits convec- tion circulation and reduces the amplitudes of the convection cells (165a:685).

Now I ask, what is the sense of using one term, "mantle," for a series of shells that includes part of the rigid lithosphere (below the Moho), the weak shell of the asthenosphere, the chemically active wave-guide layer, and the pressure-solid shell down to the boundary of the core? Let us be careful not to be confused by it ourselves.

Now we come to another difficulty: The rising convection current comes up under the lithosphere (not the Moho) and shears out, as they say, under it. Remember that the rigid shell of the lithosphere is 30–40 miles thick. Do you believe that this current of viscous rock moving gently under it *can tear it apart?* Do you believe that, having torn it apart, and having moved it a few hundred or a few thousand miles, this current can then pull the lithosphere down into the earth's depths, folding it? It all sounds extremely improbable to me.

6. THE HYPOTHESIS ACCORDING TO DIETZ

A contemporary scientist who has taken a leading part in the develop- ment of the present theory of continental drift is Robert S. Dietz. He has simplified our task of presenting his ideas by very clearly defining his assumptions and postulates. He also states that they should be regarded as tentative and suggestive rather than final in any sense of the word. To start with he gives his basic assumption:

STATEMENT No. 1
Large-scale thermal convection currents, fueled by the decay of radio- active minerals, do operate in the mantle. They do provide the primary diastrophic forces affecting the lithosphere (112a:289).

We can take this first statement to mean that he ascribes the formation of mountain chains and other features of the earth's surface in the last analysis to these convection currents. He then defines the earth's outer shell as follows:

STATEMENT No. 2

It is relevant to speak of the strength and rigidity of the earth's outer shell. The term "crust" has been effectively pre-empted from its classical meaning by seismological usage applying it to the layer above the Moho . . . For considerations of convective creep and tectonic yielding we must refer to a lithosphere and an asthenosphere. Deviations from isostasy prove that approximately the 70 outer kilometers of the earth (under the continents and oceans alike) is moderately strong and rigid even over time-spans of 100,000 years or more; this outer rind is the lithosphere. Beneath lies the asthenosphere separated from the lithosphere by the level of no strain or isopiestic level; it is a domain of rock plasticity and flowage where any stresses are quickly removed . . . If convection currents are operating "subcrustally," as is commonly written, they would be expected to shear below the lithosphere and not beneath the "crust" as this term is now used (112a:291).

It seems clear from the above passage that Dietz accepts the views I have presented in the "consensus" as to the strength and rigidity of the lithosphere and its thickness of about 30–40 miles. He even points out himself that the convection currents would shear beneath the lithosphere and not invade it. Then he makes a statement that appears to be in total contradiction to all this:

STATEMENT No. 3

. . . In summary, the model proposed here is that the mantle is ultramafic or peridotitic [composed of heavy rock]. The continents are buoyant rafts of sialic scum . . . the sea floor is essentially the exposed mantle of the earth covered only by an oceanic rind. Since the sea floor is fully invaded by the convection circulation the term oceanic crust is a misnomer (112a:292).

It seems that in statement (2) he allows for the existence of the lithosphere, but in statement (3) he disregards it. In statement (2) we have a rigid earth shell possessing some strength extending down 30–40 miles under both oceans and continents, while in statement (3) we have the convection current invading the sea floor itself. He makes his meaning very explicit:

STATEMENT No. 4

Owing to the small strength of the lithosphere and the gradual transition in rigidity between it and the asthenosphere, the lithosphere is not a boundary to convection circulation and neither is the Moho beneath the oceans because this is not a density boundary but simply a hydration of the mantle substance itself. Thus the oceanic "rind" is almost wholly coupled with the convective overturn of the mantle creeping at the rate of a fraction of a centimeter a year to as much as 1 or 2 centimeters per year. Since the sea floor is covered by only a thin veneer of sediments with some mixed in effusives, it is essentially an outcropping mantle. So the sea floor marks the tops of large convection cells and slowly spreads from zones of divergence to those of convergence. These cells have dimensions of several thousands of kilometers; some cells are quite active now while others are dead or dormant. They have changed position with geologic time causing new tectonic patterns . . . (112a: 292).

Dietz appears to have accomplished the feat of adopting two opposite positions simultaneously. His suggestion of a gradual transition between lithosphere and asthenosphere can be accepted only in a limited sense. The strength resulting from crystalline structure in the lithosphere will be essentially unmodified until, with increasing depth, the heat reaches the melting points of the rocks. Then the crystalline structure will disappear and the rocks will become viscous and weak, and we have the asthenosphere. The transition cannot be gradual in the sense that it is continuous from the top to the bottom of the lithosphere. It must be comparatively sudden, with variations from place to place because of variations in the heat and in the chemical composition of materials, which, of course, have different melting points. But these variations would average out within a range, let us guess, of about 6 miles. That, at least, seems reasonable to me.

7. SEA-FLOOR SPREADING

Fortunately for the resurrected continental-drift theory new discoveries have suggested a way out of the dilemma presented by Dietz.

A few years ago a great system of fissures or canyons was discoverd in the Atlantic Ocean, running down the crest of the Mid-Atlantic Ridge (see Fig. 9, pg. 32). As the progress of oceanographic exploration revealed the extension of this ridge to all the oceans, it also revealed that the system of fissures was worldwide. It has been

Fig. 9. Features of the Atlantic Ocean bottom.

charted for a length of 40,000 miles. The worldwide midoceanic ridge is a belt of mountain ranges several hundred miles wide, many of the ranges rising to heights under the sea of two miles or more. It has been found that the rocks of the ridge are of recent origin as compared with the rocks of the ocean floor on each side of the ridge. It seems that islands in the Atlantic are geologically older the farther they are from the ridge. The astonishing conclusion appears to be that magma, heavy basaltic molten rock, is continuously (or perhaps periodically) welling up through the 40,000-mile-long oceanic fissure, thus creating new ocean bottom and spreading the ocean floor. The sea bottom on each side of the ridge is being pushed away, and therefore the continents are being pushed apart. The Atlantic Ocean is therefore a "rift ocean," created by the pushing apart of the continents.

At the present time it is still believed that the sea-floor-spreading process is powered by convection currents. It is possible that convection currents are entirely unnecessary as a mechanism to cause sea-floor spreading. There may be quite a different way to account for the facts. But before we consider this alternative possibility, there are some other matters that demand attention. The oceanographer Bruce Heezen has discovered a number of serious difficulties with the ocean-floor-spreading hypothesis.

8. HEEZEN'S DIFFICULTIES

Heezen notes first (205b:235–289) that in the Atlantic the rock formations at the continental slopes on both sides of the ocean do not bend down but rather break off abruptly as if some terrific force had cracked the continental block and pulled the pieces apart. This of course is in accordance with the sea-floor-spreading hypothesis. However, it dismayed him to find the same thing on the Pacific coast of North America. He considers this contradictory to the hypothesis because according to it America has been moving westward into the Pacific, where there is no continental block and where no land mass was torn apart.

A second difficulty that occurs to him is that if the Atlantic Ocean is a recently formed rift ocean only a couple of hundred million years old, then Europe and Asia must have been moving eastward as America moved westward. Asia must therefore be encroaching on the Pacific. It seems to follow from this that the Pacific Ocean must be much older than the Atlantic. Accordingly there should be, Heezen suggests, thicker deposits of sediments on its bottom than on the Atlantic Ocean bottom. Instead he finds the sediments on the floor of the Pacific are similar in type and quantity to those on the floor of the Atlantic.

On the other hand, if the Pacific Ocean floor has been spreading from the Mid-Pacific Ridge, in the same way as the Atlantic floor, and the Pacific is also a rift ocean, as some enthusiasts now claim, then the difficulties are only increased.

Heezen points out that, according to the theory, the world-encircling rift canyon, as it broadens with the intrusion of new matter from below, is inevitably pushing the different segments of the earth's crust against one another. The Mid-Atlantic Ridge is pushing America in one direction while the Mid-Pacific Ridge is pushing it in precisely the opposite direction. It seems that this amounts to a battle between the two ridges. Which will win? And what will happen to the continent itself and to the sea floors caught in between? Obviously something has to give.

It might be argued that the folding of the Rockies and the Andes might represent a yielding of the continents to such pressure. But why are there not similar mountain ranges on all coasts? There are no coastal mountain ranges on either the western or eastern coasts of the Atlantic, nor on most of the coasts of the Indian Ocean (which has its own midoceanic ridge), nor on the coasts of Australia, nor on the northern coasts of North America and Asia. Coastal mountain ranges are, indeed, exceptional on the face of the earth.

Heezen observes that the 40,000-mile-long midoceanic ridge curls and twists all over the globe. If along its whole length convection currents are rising to crack the lithosphere, then these currents would have to be very long, sausage-shaped affairs, hard to visualize. And where are they sinking? The midoceanic ridge is highly active seismically and gives evidence of a high heat flow from the interior, facts which agree with the assumption that convection currents are rising under the ridge; but Heezen points out that corresponding areas where the currents should be sinking ought to show equal seismicity and a heat-flow deficit. But there is little evidence that the currents are sinking at the borders of continents or under them, as the theory requires. Oxburgh and Turcotte (340a:2645) have shown that the areas on the earth's surface where the geological facts are consistent with the assumption of sinking convection currents are not nearly so extensive as the midoceanic-ridge system. There is at present no sort of balance between them.

Heezen also points out evidence of two kinds that appears to contradict the assumption that the ocean floors are involved in convection cells. He mentions, first, the great systems of faults on the ocean bottoms (Figs. 9-10, pp. 32, 37), which suggest that great slabs of the ocean floors have been displaced laterally, relative to one another, for distances of hundreds of miles. This cracking of the ocean floors suggests that they are, in fact, rigid plates which could not be "carried along" by convection currents moving under the lithosphere. He views as further evidence of the strength of the ocean floors the existence of innumerable mountains (called "seamounts") scattered on the floors of all the oceans, which have not sunk into the mantle despite their frequently enormous mass and weight.

9. HIS SOLUTION

In view of this cataract of difficulties, what does Heezen propose? He suggests an entirely new mechanism to account for the lithosphere's being pulled apart so that new ocean bottom may be formed. *He suggests that the earth is growing,* that it has expanded enormously through geological time. This idea is naturally full of difficulties, and perhaps raises much more serious problems than it solves. We shall soon see, however, that it is not necessary; a far simpler solution exists.

10. FURTHER OBJECTIONS TO SEA-FLOOR SPREADING

Some oceanographers have found evidence that to them seems inconsistent with the assumptions of sea-floor spreading. Douglas J. Elvers (130a) finds objections to the "conveyor-belt mechanism" (the sea floors carried by convection currents) in much evidence derived from a study of magnetic anomalies in the floor of the Pacific. A magnetic anomaly is a deviation from the normal strength of the earth's magnetic field caused, apparently, by differences in the strengths, or conflicts in the directions, of magnetization of the rocks underlying, in this case, the ocean floor. The study covered an area of more than 400,000 square miles. Elvers suggests that the findings of the survey would require

> . . . a mechanism that could form the lineation patterns *in situ*, in the Pacific Ocean crust. This mechanism requires the existence of a crustal plate fractured by regional stresses. Igneous material would then be injected into the fracture patterns, forming a "mega-dike swarm . . ." (130a:3-4).

We shall see shortly that polar wandering can provide a solution for this problem as well as for Heezen's dilemmas. (See Chapter IX, Figs. 32, 33, and 34 and explanations, pp. 205, 207-08.)

Watkins and Richardson (437a) find that data from the Mid-Atlantic Ridge are not in agreement with the hypothesis of sea-floor spreading, and they complain rather bitterly that geophysicists at present, because of their enthusiasm for the theory, are forcing the evidence to fit it. I quote their abstract in full, despite the fact that it contains technicalities that do not concern us:

> The desirability of accurate delineation of areas of active crustal spreading is the motive for presentation of arguments against unrestricted application of the crustal spreading hypothesis to analyses of all linear magnetic anomalies associated with mid-oceanic rises. It is suggested that no clear delineation will result if spreading rate changes are invoked to force a fit of local observation to hypothetical geomagnetic polarity changes; if major crustal tectonic histories are proposed by the absence rather than the presence of the classic magnetic anomaly patterns; if minor characteristics of magnetic anomalies and available local geological data are ignored; and if, for the period prior to 4 m.y. ago a geomagnetic polarity time scale which is either hypothetical or insufficiently well defined is utilized. A summary is made of relevant rock properties and igneous thermal and structural histories before examining a published magnetic traverse and other data from the Mid-Atlantic ridge at 22.5°North. It is concluded that a series of flat finite prisms of alternating polarity, as required by the simple crustal spreading model, is inconsistent with the observations (437a:257).

11. A POSSIBLE SOLUTION: POLAR SHIFT

We have seen that the ocean-spreading hypothesis, based on a "conveyor-belt mechanism" powered by convection currents, is full of difficulties. There is, however, a way to remove most, if not all,

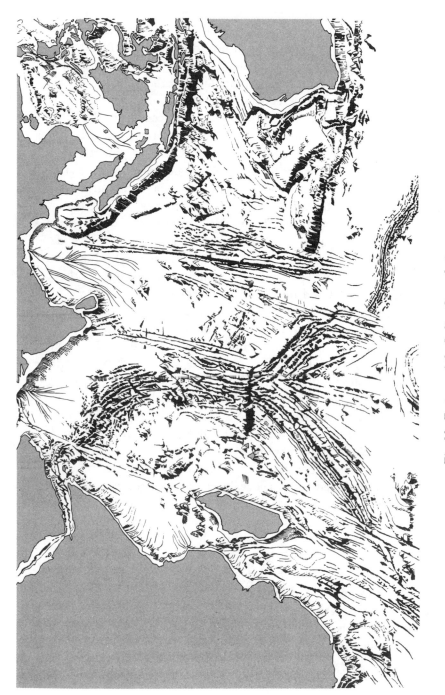

Fig. 10. Features of the Indian Ocean bottom.

of the difficulties. Let us suppose that the real engine to provide the force for sea-floor spreading is polar wandering, in the form of horizontal displacements of the entire lithosphere at the short intervals suggested by the geomagnetic evidence. We have already seen that many of those who favor the continental-drift theory have also accepted polar wandering as equally indicated by the evidence.

Let us visualize briefly the effects of the displacement of the lithosphere. Here we will simply point to the fact that the oblateness of the earth involves the consequence that in any displacement of the lithosphere parts of the earth's surface will be under compression, and others under extension. This is because such a movement will displace some areas toward the poles, and others toward the equator. Areas moved poleward will be under compression, while those moved toward the equator will be stretched, with consequent widespread fissuring. The enormous force of the moving lithosphere would produce folding of strata in some areas and fissuring in others, as is fully explained and illustrated in Chapter IX.

Let us suppose that the world-encircling midoceanic ridge and fissure system is a product of the periodical displacements of the lithosphere, a semipermanent feature in terms of the last few hundred million years. Let us now visualize a series of displacements occurring in scattered directions rather than along a polar-wandering path. We can observe that, no matter in what direction the lithosphere is displaced, some part of the midoceanic ridge is bound to be moved toward or across the equatorial bulge of the earth, and therefore to be subjected to extension, or stretching. Here the lithosphere would be pulled apart, perhaps a matter of a few miles, and the fissure would be filled from below by magma, which, when cooled, would form new sea floor.

At the same time that this area was stretched it would follow from the assumption that another area, simultaneously moved toward a pole, would be compressed, and the rock strata would be slightly folded as a result. In each displacement of the lithosphere the amounts of extension and compression would be exactly equal, so that no over-all increase of the earth's surface would be necessary, but the continents could, through many displacements, gradually be moved considerable distances relative to one another.

Of course, the gradual extension of the ocean floors first in one area and then in another would bring pressures against the sides of the continents. As Beloussov has remarked (25a), and as I pointed out in the first edition of this book, there is good geological evidence for the deep subsidence of continents or parts of continents (see pp. 235-48). The folding of the continental strata themselves might well provide room for the expanding sea floor. Uplifts of parts of the sea floor could also be a part of the process. After all, it is now agreed that the present oceans are not geologically very old. The continents that occupy the present ocean basins may not in every case be continents that have been pushed away by sea-floor spreading.

One of the problems that have puzzled geologists is the existence of the steep continental slopes, which plunge down abruptly from the edges of the continental shelves. Perhaps they have in fact resulted from the pressures of the expanding sea floors against all sides of the continents. Such pressures would naturally tend to steepen the edges of the continents. Ericson, Ewing, Heezen, and Wollin seem to have observed evidence of this process:

Older sediments from the vicinity of the continental slope give evidence of marked steepening of the slope through faulting or monoclinal folding in late Cenozoic time . . . (141:205).

12. THE SPEED OF POLAR SHIFT

We have seen that the geomagnetic evidence has generally been interpreted as suggesting that polar shift has occurred whether or not continental drift has also happened. Some geologists definitely prefer the idea of polar shift. Deutsch, for example, considers polar wandering more likely than continental drift (111a:8). Chadwick is in agreement:

. . . The balance of the evidence at present appears to favour displacements of the whole crust over the substratum rather than polar wandering of the whole earth. Vening Meinesz has suggested that displacements of this type might be produced by large scale convection currents in the mantle. It is usually supposed that the orogenic significance of polar wandering is slight, but the possible effects of the equatorial bulge in movements of the whole crust appear to merit further investigation (363a: 230).

At this point Chadwick refers the reader to this author's earlier work, *Earth's Shifting Crust*, of which the present work is a revision.

Deutsch cites evidence that polar shifts in the past may have occurred at such a speed as largely to escape notice by the method of geomagnetism. He states:

. . . We might take 50 meters a year as a typical value for very fast (hypothetical) polar wandering. At this rate a 45° route into the North Pacific, or say 10,000 km for the return journey, would take 200,000 years. I think the largest interval that could conceivably have been missed in the post-Eocene record is ten million years. Then the probability of pinning down a fast-moving pole is 2%, and for earlier periods the chances decline even farther. This should instill sobering thoughts regarding the extent of our ignorance a decade since the first spectacular results from paleomagnetism were reported (111a:37).

This is very important for my discussion, because in Chapters IV, V, VI and VII I am arguing for three displacements of the lithosphere in the last 100,000 years, for which but little geomagnetic confirmation is available. According to my assumptions, in the last instance, at the end of the most recent ice age in North America, the lithosphere was shifted some 30° or about 2000 miles in a period not exceeding 10,000 years. This would mean an average speed of about 1000 feet per year, a very high speed, about five times the maximum speed allowed by Deutsch. With such a speed as this, quite obviously, the mathematical chances of discovering a displacement by the geomagnetic method would fall to a fraction of one percent.

Deutsch cites evidence in the Precambrian rocks of Scotland of a displacement of the lithosphere that seems to have occurred at a speed "several orders of magnitude" faster than the hypothetical speed he refers to above. Since an order of magnitude means multiplication by a factor of 10, he is obviously contemplating a very high speed indeed (111a:37). At first glance such a speed may seem improbable, but I shall present a great deal of evidence to support the assumption. In any case it is very significant that Deutsch should say that

. . . it is still possible to argue that polar wandering was sufficiently rapid to be missed entirely by paleomagnetism (111a:36).

13. THE MECHANISM FOR POLAR SHIFT

It is necessary to admit, in the first place, that at the present time there is no satisfactory explanation of the *modus operandi* of displacements of the lithosphere. The purpose of this book is simply to present the case for the assumption that such shifts have occurred and to show how the assumption explains numerous unsolved problems in geology and in the evolution of life. In addition, I present evidence dating the last three displacements and placing them in the latter part of the last geological epoch, the Pleistocene.

While the explanation of the cause of displacements may have to wait for future developments, the requirements for a successful explanation may be suggested now. Any successful explanation must account for both the initiation and the termination of such a movement. It must suggest a mechanism to provide for travel by the lithosphere at several times the rate of speed now estimated for the assumed subcrustal currents. And it will have to explain the periodicity that seems to have emerged from the study of the evidence presented in Chapters IV–VIII.

Although a specific cause for displacements is not yet in sight, there are indications of the general direction in which we may have to look for an ultimate explanation. Almost certainly any displacement of the outer shells of the earth must be due to gravitational imbalances within the lithosphere or immediately below it, imbalances giving rise to centrifugal or centripetal effects such as those originally postulated by Campbell in his elaboration of his ice-cap mechanism. (See pp. 330.) That such imbalances now exist is unquestioned, and some of them are of considerable magnitude. There are many surface features scattered over the earth that are out of gravitational balance, but even larger anomalies appear to exist under the surface, within the lithosphere or below it.

The principle of gravitational balance of the earth's surface is referred to as the principle of "isostasy." Theoretically continents, ocean basins, mountain ranges and other features are always seeking equilibrium, which they attain by rising or sinking until (like a piece of wood floating on the water) they are at an elevation that is correct for their density. They can do this because the lithosphere, as we have already pointed out, does not have great *tensile* strength; it can fracture easily if there is much pressure in either direction, up or down, and this lets any section of the earth's surface find its natural elevation, as if it were floating in the asthenosphere below. I say "theoretically." In practice we see that the theory does not always work. The earth's surface gets out of balance. There are certainly movements of some sort going on within the body of the earth by which great masses of material are changing position, sometimes uplifting or depressing the surface. In these movements the balance of density both at the surface and below it (within the lithosphere) may be upset.

Another factor of possibly prime importance in this question is the great distortions of the earth's shape due to the existence of a so-called "third axis." These bumps and protuberances, of vast extent, are generally disregarded in calculations of isostasy, but they are relevant to the question of the dynamic balance of the lithosphere. (For a further discussion of triaxiality, see Note 5, p. 357 below.)

If I may suggest, despite our present state of ignorance, a specific example of a displacement, let us suppose that a large surplus (or deficiency) of mass is brought into existence somewhere near the surface of the earth. This could result from the rise of material under the lithosphere, uplifting it, or by the reverse process, a sinking causing a depression of the lithosphere. According to the principles of the dynamics of rotating bodies, surplus mass near the earth's surface will cause a centrifugal effect, operating outward at right angles to the earth's axis, while deficiencies of mass will cause a centripetal effect, operating inward. The tangential components of these forces (as shown by Campbell, Fig. 35, p. 331) would operate equatorward in the case of surplus masses and poleward in the case of deficiencies.

Let us assume that we have a large surplus mass lying in and under the lithosphere on one side of the earth, and that the centrifugal effect of this mass is sufficient to start a displacement of the outer shells of the earth over the wave-guide layer. The important thing to note is that this force is progressive. When a displacement once starts, the force increases by geometrical progression with the increasing distance of the anomalous mass from the earth's axis of rotation. The tangential component of the force, however, begins to decline (as Campbell shows) before the equator is reached, and reverses after that point, so that the movement is braked and finally brought to a stop. This example satisfies the first two requirements for an acceptable mechanism.

The wave-guide layer found by Beloussov is of great advantage for this concept of displacement. It suggests an easy zone of shear for the movement, wherein all frictional effects will be minimized. Actually the displacement would take place, according to this thinking, at a level where the viscosity of the asthenosphere would be reduced to its lowest point by the fluid wave-guide layer, and so the lithosphere would in effect be borne along on a stream flowing in a liquid, much as the Gulf Stream flows over the deeper waters of the ocean. The movement might be the equivalent of a flow of liquid over liquid. Friction would be minimized, while viscosity would present no bar to a comparatively rapid displacement.

The last point is one of great importance, for the field studies I am presenting below indicate that the shifts of the lithosphere have at times attained extraordinary speeds as compared with the speeds of subcrustal currents now estimated by geophysicists. The combination of the geometrical progression of centrifugal effects with the zone of easy shear in the wave-guide layer opens up the possibility of extremely rapid movements of the earth's outer shell. In later chapters the reader will find much empirical field evidence in support of this.

It should be borne in mind that while the displacement of the earth's whole outer shell would be an event of gigantic magnitude, it might actually meet less resistance than the movement of a large convection current. The convection currents imagined by geophysicists involve the movement of trillions of tons of highly viscous rock against the viscous resistance of the earth's mantle. In a displacement of the lithosphere over the wave-guide layer, on the other hand, there would be minimum friction at the interface. It would involve only a very thin layer of the most liquid part of the asthenosphere. The movement would be one of gliding, acknowledged to be the most economical form of motion.

A possible new direction for investigation into the cause of displacements of the lithosphere was suggested by two articles in "Mines" magazine during 1968. The authors, Professors Ramon E. Bisque (35a) and George E. Rouse (359a), announced a new hypothesis linking the development of the features of the earth's surface with forces set in motion in the earth's core. For many reasons this is a most interesting development. Their hypothesis, if it finally proves out, may provide the explanation for the inequalities of the distribution of mass within the earth which I regard as the most probable cause of displacements.

14. IS THE POLE MOVING NOW?

Two rather curious pieces of evidence suggest that the lithosphere may be in motion at the present time. We have two observations of a movement of the North Pole with reference to the earth's surface. The first of these is cited by Deutsch (111a:37–38) on the authority of Munk and MacDonald. It suggests that the North Pole moved 10 feet in the direction of Greenland along the meridian of 45° West Longitude during the period from 1900 to 1960. This (according to Deutsch) would be at a rate of 6 centimeters (about two and a half inches) a year. The other finding, cited by Markowitz (292a), based on later data, suggests that the pole moved about 20 feet between 1900 and 1968 along the meridian of 65° West Long., and that it is now moving at the rate of about 10 centimeters (4 inches) a year. The difference between the two longitudes may not be particularly important, as the angular difference so near the pole is so small, but the difference in the two rates of motion may be very important. In the first place, it may be noted that a speed of 10 centimeters a year is two or three times the maximum speed usually estimated for subcrustal convection currents. This appears to imply that the displacement indicated as now occurring is not powered by convection currents. There is the suggestion of another mechanism at work.

A second point, possibly even more interesting, is that if both these observations were accurate when made, as we have every right to expect (in view of the eminence of the scientists involved), then we may have here evidence of a geometrical acceleration of the rate of motion. If the pole moved 10 feet between 1900 and 1960, but 20 feet between 1900 and 1968, then it moved 10 feet between 1960 and 1968, which would suggest an acceleration by a factor of about 8. The mechanism I have suggested above is based on a formula involving the geometrical progression of centrifugal effects, that is, the formula for calculating centrifugal force, which is a simple one (see p. 338).

chapter 2

THE FAILURE
TO EXPLAIN THE
ICE AGES

The evidence for displacements of the earth's outer shell is scattered over many parts of the earth and comes from several fields of science. It would not be justifiable to disregard this other evidence simply because the evidence from geomagnetism seems so strong. No other field furnishes so dramatic a confirmation of displacements as glacial geology. Here we review the facts that have led geologists, at various times during the last hundred years, to consider ideas of polar shift.

1. THE FAILURE OF THE OLDER THEORIES

A little more than a hundred years ago people were astonished at the suggestion that great ice sheets, as much as a mile thick, had once lain over the temperate lands of North America and Europe. Many ridiculed the idea, as happens with new ideas in every age, and sought to discredit the evidence produced in favor of it. Eventually the facts were established regarding an ice age in Europe and in North America. People later accepted the idea of not one but a series of ice ages. As time went on evidences were found of ice ages on all the continents, even in the tropics. It was found that ice sheets had once covered vast areas of tropical India and equatorial Africa.

From the beginning, geologists devoted much attention to the possible cause of such great changes in the climate. One theory after another was proposed, but, as the information available gradually increased, each theory was found to be in conflict with the facts, and as a consequence had to be discarded. In 1929, Coleman, one of the leading authorities on the ice ages, wrote:

Scores of methods of accounting for ice ages have been proposed, and probably no other geological problem has been so seriously discussed, not only by glaciologists, but by meteorologists and biologists; yet no theory is generally accepted. The opinions of those who have written on the subject are hopelessly in contradiction with one another, and good authorities are arrayed on opposite sides . . . (87:246).

Recent writers, such as Daly (98:257), Umbgrove (419:285), and Gutenberg (194:205), agree that the situation described by Coleman is essentially unchanged. In January, 1953, Professor J. K. Charlesworth, of Queen's University, Belfast, expressed the opinion that

The cause of all these changes, one of the greatest riddles in geological history, remains unsolved; despite the endeavors of generations of astronomers, biologists, geologists, meteorologists and physicists, it still eludes us (75:3).

A volume on climatic change, edited by Dr. Harlow Shapley (375), while suggesting minor refinements for various older theories, proposes no new ones and in no way modifies the general effect, which is that down to the present time the theorizing about the causes of ice ages has led nowhere.

2. THE MISPLACED ICE CAPS

One problem that writers on the ice ages have attempted to solve, sometimes in rather fantastic ways, but without success, is that of the wrong location of the great ice caps of the past. These ice caps have refused to have anything to do with the polar areas of the present day, except in a quite incidental fashion.

Originally it was thought that in glacial periods the ice caps would fan out from the poles, but then it appeared that none of them did so, except the ones that have existed in Antarctica. Coleman drew attention to the essential facts, as follows:

> In early times it was supposed that during the glacial period a vast ice cap radiated from the North Pole, extending varying distances southward over seas and continents. It was presently found, however, that some northern countries were never covered by ice, and that in reality there were several more or less distinct ice sheets starting from local centers, and expanding in all dirctions, north as well as east and west and south. It was found, too, that these ice sheets were distributed in what seemed a capricious manner. Siberia, now including some of the coldest parts of the world, was not covered, and the same was true of most of Alaska, and the Yukon Territory in Canada; while northern Europe, with its relatively mild climate, was buried under ice as far south as London and Berlin; and most of Canada and the United States were covered, the ice reaching as far south as Cincinnati in the Mississippi Valley (87:7-9).

With regard to an earlier age (the Permo-Carboniferous), Coleman emphasized that the locations of the ice caps were even further out of line:

> Unless the continents have shifted their positions since that time, the Permo-Carboniferous glaciation occurred chiefly in what is now the southern temperate zone, and did not reach the arctic regions at all (87:90).

He is much upset by the fact that this ice age apparently did not affect Europe:

> Unless European geologists have overlooked evidence of glaciation at the end of the Carboniferous or at the beginning of the Permian, the continent escaped the worst of the glaciation that had such overwhelming effects on other parts of the world. A reason for this exemption is not easily found (87:96).

One of the most extraordinary cases is that of the great ice sheet that covered most of India in this period. Geologists are able to tell from a careful study of the glacial evidences in what direction an ice sheet moved, and in this case the ice sheet moved northward from an ice center in southern India for a distance of 1,100 miles. Coleman comments on this as follows:

Now, an ice sheet on level ground, as it seems to have been in India, must necessarily extend in all directions, since it is not the slope of the surface it rests on that sets it in motion, but the thickness of the ice towards the central parts. . . .

The Indian ice sheet should push southward as well as northward. Did it really push as far to the south of Lat. 17° as to the north? It extended 1,100 miles to the Salt Range in the north. If it extended the same distance to the south it would reach the equator (87:110-11).

The great South African geologist A. L. du Toit pointed out that the ice caps of all geological periods in the Southern Hemisphere were eccentric as regards the South Pole, just as the Pleistocene ice caps were eccentric with regard to the North Pole (87:262). Is it not extraordinary that the Antarctic ice cap, which we can actually see because it now exists, is the only one of all these ice caps that is found in the polar zone?

Coleman, who did a great deal of field work in Africa and India, studying the evidences of the ice ages there, writes interestingly of his experiences in finding the signs of intense cold in areas where he had to toil in the blazing heat of the tropical sun:

On a hot evening in early winter two and a half degrees within the torrid zone amid tropical surroundings it was very hard to imagine the region as covered for thousands of years with thousands of feet of ice. The contrast of the present with the past was astounding, and it was easy to see why some of the early geologists fought so long against the idea of glaciation in India at the end of the Carboniferous (87:108).

Some hours of scrambling and hammering under the intense African sun, in lat. 27° 5', without a drop of water, while collecting striated stones and a slab of polished floor of slate, provided a most impressive contrast between the present and the past, for though August 27th is still early Spring, the heat is fully equal to that of a sunny August day in North America. The dry, wilting glare and perspiration made the thought of an ice sheet thousands of feet thick at that very spot most incredible, but most alluring (87:124).

When these facts were established, geologists sought to explain them by assuming that, at periods when these areas were glaciated, they were elevated much higher above sea level than they are now.

Theoretically, even an area near the equator, if elevated several miles above sea level, would be cold enough for an ice sheet. What made the theory plausible was the well-known fact that the elevations of all the lands of the globe have changed repeatedly and drastically during the course of geological history. Unfortunately for those who tried to explain the misplaced ice caps in this way, however, Coleman showed that they reached sea level, within the tropics, on three continents: Asia, Africa, and Australia (87:129, 134, 140, 168, 183). At the same time, W. J. Humphreys, in his examination of the meterological factors of glaciation, made the point that high elevation means less moisture in the air, as well as lowered temperature, and is therefore unfavorable for the accumulation of great ice caps (231:612–13).

3. WORLDWIDE PHASES OF COLD WEATHER

A widely accepted assumption with which contemporary geologists approach the question of ice ages is that the latter have occurred as the result of a lowering of the average temperature of the whole surface of the earth at the same time. This assumption has forced them to look for causes of glacial periods only in such factors as would tend to cool the whole surface of the earth at once. It has resulted in the assumption that glacial periods have always been simultaneous in the northern and southern hemispheres.

It is remarkable that this assumption has been maintained over a long period of time despite the fact that it is in sharp conflict with basic principles of physics in the field of meteorology. The basic conflict was brought to the attention of science at least seventy years ago; it has never been resolved. It consists essentially of the fact that glacial periods were periods of heavier rainfall in areas outside the regions of the ice sheets, so that this, together with the deep accumulations of ice in the great ice sheets, apparently must have involved a higher average rate of precipitation during ice ages. There is a great deal of geological evidence in support of this. Only recently, for example, Davies has discussed the so-called "pluvial" periods in Africa and has correlated them with the Pleistocene glacial periods (107).

Now, meteorologists point out that if precipitation is to be increased, there has to be a greater supply of moisture in the air. The only possible way of increasing the amount of moisture in the air is to raise the temperature of the air. It would seem, therefore, that to get an ice age one would have to raise, rather than lower, the average temperature. This essential fact of physics was pointed out as long ago as 1892 by Sir Robert Ball, who quoted an earlier remark by Tyndall:

. . . Professor Tyndall has remarked that the heat that would be required to evaporate enough water to form a glacier would be sufficient to fuse and transform into glowing molten liquid a stream of cast iron five times as heavy as the glacier itself (20:108).

William Lee Stokes has again called attention to this unsolved problem in an article entitled "Another Look at the Ice Age" in a statement that strongly suggests crust displacement:

Lowering temperatures and increased precipitation are considered to have existed side by side on a world-wide scale and over a long period in apparent defiance of sound climatological theory. Among the many quotations that could be cited reflecting the need for a more comprehensive explanation of this difficulty the following seems typical.

"In the Arequipa region [of Peru], as in many others in both hemispheres where Pleistocene conditions have been studied, this period appears to have been characterized by increased precipitation as well as lowered temperatures. If, however, precipitation was then greater over certain areas of the earth's surface than it is at present, a corollary seems to be implied that over other large areas evaporation was greater than normal to supply increased precipitation, and hence in these latter areas the climate was warmer than normal. This seems at first to be an astonishing conclusion. . . . We might propose the hypothesis that climatic conditions were far from steady in any one area, but were subject to large shifts, and that intervals of ameliorated conditions in some regions coincided with increased severity in others. The Pleistocene, then, may have been a period of sharper contrasts of climate and of shifting climates rather than a period of greater cold" (395:815-16).

From a number of points of view, the foregoing passage is extremely remarkable. Stokes recognizes the fact that the basic assumption of contemporary geologists regarding the glacial periods is in conflict with the laws of physics. Then, in the passage he quotes, he draws attention to the implications, which if the theory of continental drift is rejected seem to point directly to crust displacement, for in what other way can we explain how one part of the earth's surface was colder and another, at the same time, warmer than at present?

One of the arguments that are advanced in support of the assumption of worldwide periods of colder weather (which remains the generally accepted assumption of glaciologists) has its basis in geological evidence purporting to prove that ice ages occurred simultaneously in both hemispheres. A decade ago, however, Kroeber pointed to the essential weakness of this geological evidence when he showed the difficulty of correlating stratified deposits of different areas:

. . . There is plenty of geologic evidence, in many parts of the earth, of changes of climates, especially between wet and dry areas; and some of these happened in the Pleistocene. But the correlation of such changes as they occurred in widely separated regions, and especially as between permanently ice-free and glaciated areas, is an intricate, tricky, and highly technical matter, on which the anthropological student must take the word of geologists and climatologists, and these are by no means in agreement. They may be reasonably sure of one series of climatic successions in one region, and of another in a second or third region; but there may be little direct evidence on the correspondence of the several series of regional stages, the identification of which then remains speculative (257:650).

At the time that Kroeber remarked on the difficulty of correlating climatic changes in different parts of the world, we were not yet in possession of the data recently provided by the new techniques of radiocarbon and ionium dating. The effect of these new data has been to shorten very greatly our estimate of the duration of the last North American ice age. This estimate has been reduced, in the last few years, from about 150,000 years to about 50,000 years. Now, if we adopt the view that ancient glaciations, of which we know little, may reasonably be considered to have been the results of the same causes that brought about the North American ice age, then we must grant that they, too, may have been of short duration. But if this is true, how is it possible to establish the fact that they were contemporary in the two hemispheres? A geological period has a duration of millions of years. An ice age in Europe and one in Australia might both be, for example, of Eocene age, but the Eocene Epoch is estimated to have lasted about 15,000,000 years. We can discriminate roughly between strata dating from the early, middle, or late Eocene, but we have no way of pinpointing the date of any event in the Eocene. Even with the new techniques of radiodating now being applied to the older rocks, it is possible to determine dates only to within a margin of error of about a million years. How, then, is it possible to determine that an ice sheet in one hemisphere was really contemporary with an ice sheet or an ice age in the other?

The attempt to maintain the assumption of the simultaneousness of glaciations for the older geological periods is unreasonable. I shall show in what follows that it cannot be established even for recent geological time. It is my impression that the material evidence for the assumption was never impressive, and that the assumption was never derived empirically from the evidence but was borrowed a priori from the parent assumption; that is, the assumption of the lowering of global temperatures during ice ages, an assumption which is, as already pointed out, in conflict with the laws of physics.

If it is true that the fundamental assumption underlying most of the theories produced to explain ice ages is in error, we should expect that these theories, despite their many differences, would have a common quality of futility, and so it turns out. It is interesting to list the kinds of hypothetical causes that have been suggested to explain ice ages on the assumption of a worldwide lowering of temperature. They are as follows:

a. Variations in the quantity of particle emission and of the radiant heat given off by the sun.
b. Interception of part of the sun's radiation by clouds of interstellar gas or dust.
c. Variations in the heat of space; that is, the temperature of particles floating in space which, entering the earth's atmosphere, might affect its temperature.
d. Variations in the quantities of dust particles in the atmosphere, from volcanic eruptions or other causes, or variations in the proportion of carbon dioxide in the atmosphere.

There are serious objections to all these suggestions. So far as the variation of the sun's radiation is concerned, it is known that it varies slightly over short periods, but there is no evidence that it has ever varied enough, or for a long enough time, to cause an ice age. Evidence for the second and third suggestions is entirely lacking. The fourth suggestion is deprived of value because, on the one hand, no causes can be suggested for long-term changes in the number of eruptions or in the atmospheric proportion of carbon dioxide, and, on the other, there is insufficient evidence to show that the changes ever occurred.

I should make one reservation with regard to the fourth suggestion. There is at least one event that would provide an adequate cause for an increase in the atmosphere of both volcanic dust and carbon dioxide, and that is a displacement of the crust. The extremely far-reaching consequences of a displacement of the crust with respect to atmospheric conditions, and the importance of the atmospheric effects of a displacement for other questions, will be discussed in Chapter IX.

The theories listed above were attacked by Coleman, who complained that they were entirely intangible and unprovable. He said:

> Such vague and accidental causes for climatic change should be appealed to only as a last resort unless positive proof some time becomes available showing that an event of the kind actually took place (87:282).

Another group of theories attempts to explain ice ages as the results of changes in the relative positions of the earth and the sun. These are of two kinds: changes in the distance between the earth and the sun at particular times because of changes in the shape of the earth's orbit, and changes in the angle of inclination of the earth's axis, which occur regularly as the result of precession. The argument that precession was the cause of ice ages was advanced by Drayson in the last century (117). The argument based on these astronomical changes has been brought up to date in the recent work of Brouwer and Van Woerkom (375:147–58) and Emiliani (132). It now seems that these astronomical changes may produce cyclical changes in the distribution of the sun's heat, and perhaps in the amount of the sun's heat retained by the earth, but it is agreed, by Emiliani and others, that by itself the insolation curve or net temperature difference would not be sufficient to cause an ice age without the operation of other factors, and so Emiliani suggests that perhaps changes in elevation coinciding with the cool phases of the insolation curve may have caused the Pleistocene ice ages. One weakness of this suggestion is, of course, the necessity to suppose the accidental combination of two independent causes for ice ages.

There is another objection to be advanced against all theories supposing a general fall of world temperatures during the ice ages. We have seen that ice ages existed in the tropics and that great ice caps covered vast areas on and near the equator. This happened not once but several times. The question is, if the temperature of the whole earth fell enough to permit ice sheets a mile thick to develop on the equator, just where did the fauna and flora go for refuge? How did they survive? How did the reef corals, which require a minimum seawater temperature of 68° F. throughout the year, manage to survive? We know that the reef corals, for example, existed long before the period of the tropical ice sheets. Furthermore we know that the great forests of the Carboniferous Period, which gave us most of our coal, lived both earlier than and contemporarily with the glaciations of Africa and India, though in different places. Obviously this would have been impossible if the temperature of the whole earth had been simultaneously reduced, for the equatorial zone itself would have been uninhabitable, while all other areas were still colder. It is small wonder that W. B. Wright insisted, over a quarter of a century ago, that the Permo-Carboniferous ice sheets in Africa and India were proof of a shift of the poles (450).

4. THE NEW EVIDENCE OF RADIOCARBON DATING

The question of the causes of ice ages has been given increased importance by a recent revolution in our methods of dating geological events. In the course of the last twenty years all of our ideas regarding the chronology of the recent ice ages, their durations, and the speed of growth and disappearance of the great ice sheets have been transformed. This is altogether the most important new development in the sciences of the earth. The repercussions in many directions are most remarkable.

In order to get an idea of the extent of the change, let us see what the situation was only ten or fifteen years ago. As everybody is aware, geologists are used to thinking in terms of millions of years. To a geologist a period of 1,000,000 years has come to mean almost nothing at all. He is actually used to thinking that events that took place somewhere within the same 20,000,000-year period were roughly contemporaneous. As to the ice ages, the older ones were simply thrown into one of these long geological periods, but there was no way to determine their durations (except very roughly), their speeds of development, or precisely when they happened. It was convenient to assume that they had endured for hundreds of thousands or for millions of years, though no real evidence of this existed.

So far as the most recent division of geologic time, the Pleistocene, was concerned, geologists, with much more evidence to work from, saw that there had been at least four ice ages in a period of about 1,000,000 years. They consequently proposed the idea that the Pleistocene was not at all like previous periods. It was exceptional because it had so many ice ages. They may have been misled by failure to take sufficient account of the fact that glacial evidence is very easily destroyed, and that, as we go further back into geological history, the mathematical chances of finding evidences of glaciation, never very good, decrease by geometical progression.

Down to twenty years ago it was the considered judgment of geologists that the last ice age in North America, which they refer to as the Wisconsin glaciation, began about 150,000 years ago and ended about 30,000 years ago, as I have already said.

This opinion appeared to be based upon strong evidence. The estimates of the date of the end of the ice age were supported by the careful counting of clay varves (6) and by numerous seemingly reliable estimates of the age of Niagara Falls. As a consequence, experts were contemptuous of all those who, for one reason or another, attempted to argue that the ice age was more recent. One of these was Drayson, whose theory called for a very recent ice age (117). His followers produced much evidence, but it was ignored. When the Swedish scientist Gerard de Geer established by clay-varve counting that the ice sheet was withdrawing from Sweden as recently as 13,000 years ago, the implications were not really accepted, nor were his results popularly known. Books continued to appear, even thirty years afterward, with the original estimates of the age of the ice cap.

Then, following World War II, nuclear physics made possible the development of new techniques for dating geological events. One of these was radiocarbon dating.

The method of radiocarbon dating was developed by Willard F. Libby, nuclear physicist of the University of Chicago. It uses an isotope of carbon (Carbon 14) which has a "half-life" of about 5,570 years. A half-life is the period during which a radioactive substance loses half its mass by radiation. Among the very numerous artificial radioactive elements created in nuclear explosions some have half-lives of millionths of seconds; others, occurring in nature, have half-lives of millions of years. For geological dating it is necessary to have radioactive elements that diminish significantly during the periods that have to be studied, and that occur in nature.

Since radiocarbon exists in nature and has a relatively short half-life, the quantity of it in any substance containing organic carbon will decline perceptibly in periods of a few centuries. By estimating how much carbon was contained originally in the specimen and then measuring what still remains, the date of its geologic formation can be found to within a small margin of error.

When this method was first developed by Libby, it could date anything containing carbon of organic origin back to about 20,000 years ago. Since then the method has been improved, through the efforts of many scientists, and its range has been approximately tripled.

The first major result of the radiocarbon method was the revelation that the last North American ice sheet had indeed disappeared at a very recent date. Tests made in 1951 showed that it staged a re-advance in Wisconsin as recently as 11,000 years ago (272:105). When this date is compared with other dates showing the establishment of a climate like the present one in North America, it seems that most of the retreat and disappearance of the great continental ice cap, at least in the United States, can have taken little more than two or three thousand years. We shall examine these dates in detail in Chapter IV.

What was the significance of this new discovery, besides showing how wrong the geologists had been before? The fact is that so sudden a disappearance of a continental ice cap raises fundamental questions. It contradicts some basic assumptions of geological science. What has become of those gradually acting forces that were supposed to govern glaciation as well as all other geological processes? What factor can account for this astonishing rate of change? It seems self-evident that no astronomical change and no subcrustal change deep in the earth can occur at that rate.

When this discovery was made, I expected that the next revelation must be to the effect that the Wisconsin ice sheet had had its origin at a much more recent time than was suspected, and that the whole length of the glacial period was but a fraction of the former estimates. I had a while to wait, because radiocarbon dating in 1951 was not able to answer the question. By 1954, however, the technique had been improved so that it could determine dates as far back as

30,000 years ago. Many datings of the earlier phases of the Wisconsin glaciation were made, and Horberg, who assembled them, reached the conclusion that the ice cap, instead of being 150,000 years old, had appeared in Ohio only 25,000 years ago (222:278–86). This conclusion has been so great a shock that some writers have sought to evade the clear implications by questioning the reliability of the radiocarbon method. Horberg betrays evidence of the intensity of the shock to accepted beliefs when he says that the results of the evidence are so appalling from the standpoint of accepted theory that it may be necessary either to abandon the concept of gradual change in geology or to question the radiocarbon method.

In this book I am not going to question the general reliability of the radiocarbon method. I intend merely to question the theories with which the new evidence is in conflict. Doctor Horberg says that the necessity to compress all the later stages of the Wisconsin glaciation into the incredibly short period of 15,000 or 20,000 years involves an acceleration of geological processes—snowfall, rainfall, erosion, sedimentation, and melting—that seems to challenge the principle laid down by the founder of modern geology, Sir Charles Lyell, over a century ago. Lyell's principle, called "uniformitarianism," stated that geological processes have always gone on about as they are going on now.

The Wisconsin ice cap went through a number of oscillations, warm periods of ice recession alternating with cold periods of ice readvance. Horberg is at a loss to see what could cause them to occur at the velocity required by the radiocarbon dates. These seem to require an annual movement of the ice front of 2,005 feet, "two to nine times greater than the rate indicated by varves and annual moraines" (222:283).*

The fact that these new data call into question some basic ideas in geology is recognized by Horberg:

Probably only time and the progress of future studies can tell whether we cling too tenaciously to the uniformitarian principle in our unwillingness to accept fully the rapid glacier fluctuations evidenced by radiocarbon dating (222:285).

---

* Since he said this, earlier phases of the glaciation have been discovered (see Chapter IV), but this does not alter the fact that the new dates testify to an enormous acceleration of the glacial stages already known.

Recent geological literature shows that a rather desperate effort is being made to blur the significance of the new data. However, I would like to suggest some far-reaching implications. We have seen an ice sheet appear and disappear in—geologically speaking—a twinkling of an eye. There are three deductions to be made:

a. Any theory of ice ages must give a cause that can operate that fast.

b. If the last ice cap in North America appeared and disappeared in a short time, we cannot assume that the ancient ice caps lasted for longer periods.

c. If other geological processes are correlated with ice ages, then their tempo must also have been faster than we have supposed, and a cause must be found for their accelerated tempo.

5. CONCLUSION

It is clear that none of the great glaciations of the past can be explained by the theories hitherto advanced. The only ice age that is adequately explained is the present ice age in Antarctica. This is excellently explained. It exists, quite obviously, because Antarctica is at the pole, and for no other reason. No variation of the sun's heat, no galactic dust, no volcanism, no subcrustal currents, and no arrangements of land elevations or sea currents account for the fact. We may conclude that the best theory to account for an ice age is that the area concerned was at a pole. We thus account for the Indian and African ice sheets, though the areas once occupied by them are now in the tropics. We account for all ice sheets of continental size in the same way.

Stokes has provided an excellent list of specifications for a satisfactory ice-age theory, every one of which is met by the assumption of crust displacements as the fundamental cause (395:815–16):

a. An initiating event or condition.

b. A mechanism for cyclic repetitions or oscillations within the general period of glaciation.

c. A terminating condition or event.

d. It should not rely upon unprovable, unobservable, or unpredictable conditions when well-known or more simple ones will suffice.

e. It must solve the problem of increased precipitation with colder climate.

f. The facts call for a mechanism that either increases the precipitation or lowers the temperature very gradually over a period of thousands of years.

It is evident that a displacement of the crust could initiate an ice age by moving a certain region into a polar zone, while a later displacement could end the ice age by moving the same area away from the polar zone. The increased precipitation and the oscillations of the borders of the ice sheets can be explained by the atmospheric effects that would result from volcanism associated with the movement of the crust. These effects will be discussed in later chapters.

chapter **3**

THE FAILURE
TO EXPLAIN
CLIMATIC CHANGE

IN THE last chapter it was suggested that the ice ages can be explained by the assumption of frequent displacements of the earth's crust but that they cannot, at least for recent time (the Pleistocene Epoch) be explained by continental drift. The ice ages, however, represent only one side of the problem. If they are instances of extremely cold climates distributed in an unexplained manner on the earth's surface, there were also warm climates whose distribution is equally unexplained.

In connection with these warm climates in the present polar regions, there arises a contradiction of an especially glaring character. On the one hand there is evidence that the distribution of plants and animals in the past did not, as a rule, follow the present arrangements of the climatic zones. On the other hand, the trend of the new evidence is to show that climatic zones have always been about as clearly distinguished by temperature differences as they are today. This is in flat contradiction to the assumption, still widely held, that the earth, during most of geological history, did not possess clearly demarcated climatic zones. We are forced to conclude that, since many ancient plants and animals were not distributed according to the present climatic zones, the zones themselves have changed position on the earth's surface. This requires, as we have seen, that the surface shall have changed position relative to the axis of rotation. We shall now examine the evidence that supports this view.

1. AGES OF BLOOM IN ANTARCTICA

There have been many times during the history of the globe when the continent of Antarctica, now covered by a polar ice cap as much as two miles thick and covering an area of nearly 6,000,000 square miles, had warm climates.*

So far as we know at present, the very first evidence of an ice age in Antarctica comes from the Eocene Epoch (52:244). This was barely 60,000,000 years ago. Before that, for some billion and a half years, there is no suggestion of polar conditions, though very many earlier ice ages existed in other parts of the earth. Henry, in *The White Continent*, cites evidence of the passing of long temperate ages in Antarctica. He describes the Edsel Ford Mountains, discovered by Admiral Byrd in 1929. These mountains are of nonvolcanic, folded sedimentary rocks, the layers adding up to 15,000 feet in thickness. Henry suggests that they indicate long periods of temperate climate in Antarctica:

The greater part of the erosion probably took place when Antarctica was essentially free of ice, since the structure of the rocks indicates strongly that the original sediment from which they were formed was carried by water. Such an accumulation calls for an immensely long period of tepid peace in the life of the rampaging planet (206:113).

Most sedimentary rocks are laid down in the sea, formed of sediment brought down by rivers from nearby lands. The lands from

* In Chapter IV I shall present evidence to show that the most recent period of warm climate in Antarctica was contemporary with the last ice age in North America.

which the Antarctic sediments were brought seem to have disappeared without a trace, but of the sea that once existed where there is now land we have plenty of evidence. Brooks remarks:

. . . In the Cambrian we have evidence of a moderately warm sea stretching nearly or right across Antarctica, in the form of thick limestones very rich in reef-building Archaeocyathidae (52:245).

Millions of years later, when these marine formations had appeared above the sea, warm climates brought forth a luxuriant vegetation in Antarctica. Thus, Sir Ernest Shackleton is said to have found coal beds within 200 miles of the South Pole (71:80), and later, during the Byrd expedition of 1935, geologists made a rich discovery of fossils on the sides of lofty Mount Weaver, in Latitude 86° 58′ S., about the same distance from the pole, and two miles above sea level. These included leaf and stem impressions and fossilized wood. In 1952 Dr. Lyman H. Dougherty, of the Carnegie Institution of Washington, completing a study of these fossils, identified two species of a tree fern called Glossopteris, once common to the other southern continents (Africa, South America, Australia), and a giant tree fern of another species. In addition, he identified a fossil footprint as that of a mammallike reptile. Henry suggests that this may mean that Antarctica, during its period of intensive vegetation, was one of the most advanced lands of the world as to its life forms (207).

Soviet scientists have reported finding evidences of a tropical flora in Graham Land, another part of Antarctica, dating from the early Tertiary Period (perhaps from the Paleocene or Eocene) (364:13).

It is, then, little wonder that Priestly, in his account of his expedition to Antarctica, should have concluded:

. . . There can be no doubt from what this expedition and other expeditions have found that several times at least during past ages the Antarctic has possessed a climate much more genial than that of England at the present day . . . (349d:210)

Further evidence is provided by the discovery by British geologists of great fossil forests in Antarctica, of the same type that grew on the Pacific coast of the United States 20,000,000 years ago (206:9). This, of course, shows that after the earliest known Antarctic glaciation in the Eocene, the continent did not remain glacial but had later episodes of warm climate.

Umbgrove adds the observation that in the Jurassic Period the floras of Antarctica, England, North America, and India had many plants in common (420:263).*

There is one group of theories for explaining these facts to which we cannot appeal because of their inherent and obvious weaknesses. These are the theories that try to explain warm and cold periods in Antarctica by changes in land elevations, changes in the directions of ocean currents, changes in the intensity of solar radiation, and the like. It is obvious, for instance, that no hypothetical warm currents could make possible the existence of warm climates in the center of the great Antarctic continent if that continent were at the pole, and if by some miracle Antarctica did become warm, how could forests possibly have flourished there deprived of sunlight for half the year?

2. WARM AGES IN THE NORTH

The Arctic regions have been more accessible, and consequently they have been more thoroughly explored, than the Antarctic. It was from them that the first evidence came of warm-climate floras in a polar region. Most of the theories developed by those defending the theory of the permanence of the poles were specially designed to explain these facts.

One method of explaining the evidence was to suggest that the plants and animals of past geological areas, even though they belonged to similar genera or families as living plants and animals, and closely resembled them in structure, may have been adapted to very different climates. This argument often had effect, for no one could exclude the possibility that, in a long geological period, species might make successful adjustments to different climatic conditions. Where

* In *The New York Times* for December 6, 1969, Walter Sullivan reported comments by Drs. Laurence M. Gould and Grover Murray on a discovery of vertebrate fossils in the Alexandra Mountains only a few degrees from the South Pole. These fossils were those of reptiles resembling those of the Triassic Period on other continents. It is interesting that the evidence is interpreted as virtual proof of continental drift: that is, of the break-up of a supercontinent since the Triassic. However, such a conclusion is not justified. Only the dating of the rocks enclosing the fossils by some means of absolute dating could really establish the fact that the animals lived in the Triassic Period, rather than 50,000,000 years earlier or later. A number of shifts of the lithosphere, with successive alterations of geographical connections could have permitted the migration of the species without postulating continental drift.

single plants were involved such a possibility could not be dismissed. Where, however, whole groups of species, whole floras and faunas, were involved, there was increased improbability that they could all have been adjusted at any one time to a radically different environment from that in which their descendants live today. For this reason, and because the structure of plants has a definite relationship to conditions of sunlight, heat and moisture, biologists have abandoned this method of explaining the facts. Barghoorn, for example, says that fossil plants are reliable indicators of past climate (375:237–38).

It may be worthwhile to review, very briefly, some high points of the climatic history of the Arctic and sub-Arctic regions, beginning with one of the oldest periods, the Devonian, and coming down by degrees to periods nearer our own. (During this discussion the reader may find it helpful to refer to the table of geological periods, page 2.)

The Devonian evidence is particularly rich and includes both fauna and flora. Doctor Colbert, of the American Museum of Natural History, has pointed out that the first known amphibians have been found in this period in eastern Greenland, near the Arctic Circle, though they must have required a warm climate (375:256). Many species of reef corals, which at present require an all-year seawater temperature of not less than 68° F. (102:108), have been found in Ellesmere Island, far to the north of the Arctic Circle (389:2). Devonian tree ferns have been found from southern Russia to Bear Island, in the Arctic Ocean (177:360). According to Barghoorn, assemblages of Devonian plants have been found in the Falkland Islands, where a cold climate now prevails, in Spitzbergen, and in Ellesmere Island, as well as in Asia and America (375:240). In view of this, he remarks:

> The known distribution of Devonian plants, especially their diversification in high latitudes, suggests that glacial conditions did not exist at the poles (375:240).

In the following period, the Carboniferous, we have evidence summed up by Alfred Russel Wallace, co-author with Darwin of the theory of evolution:

> In the Carboniferous formation we again meet with plant remains and beds of true coal in the Arctic regions. Lepidodendrons and cala-

mites, together with large spreading ferns, are found at Spitzbergen, and at Bear Island in the extreme north of Eastern Siberia; while marine deposits of the same age contain an abundance of large stony corals (435:202).

In the Permian, following the Carboniferous, Colbert reports a find of fossil reptiles in what is now a bitterly cold region: "Large Permian reptiles . . . are found along the Dvina River of Russia, just below the Arctic Circle, at a North Latitude of 65°" (375:259). Colbert explains that these reptiles must have required a warm climate. In summing up the problem of plant life for the many long ages of the Paleozoic Era, from the Devonian through the Permian, Barghoorn says that it is "one of the great enigmas" of science (375:243).

Coming now to the Mesozoic Era (comprising the Triassic, Jurassic and Cretaceous Periods), Colbert reports that in the Triassic some amphibians (the labyrinthodonts) ranged all the way from 40° S. Lat. to 80° N. Lat. About this time the warm-water *Ichthyosaurus* lived at Spitzbergen (375:262–64). For the Jurassic, Wallace reports:

In the Jurassic Period, for example, we have proofs of a mild arctic climate, in the abundant plant remains of East Siberia and Amurland. . . . But even more remarkable are the marine remains found in many places in high northern latitudes, among which we may especially mention the numerous ammonites and the vertebrae of huge reptiles of the genera Ichthyosaurus and Teleosaurus found in Jurassic deposits of the Parry Islands in 77° N. Lat. (435:202).

For the Cretaceous Period, A. C. Seward reported in 1932 that "the commonest Cretaceous ferns [of Greenland] are closely allied to species . . . in the southern tropics" (373:363–71). Gutenberg remarks: "Thus, certain regions, such as Iceland or Antarctica, which are very cold now, for the late Paleozoic or the Mesozoic era show clear indications of what we would call subtropical climate today, but no trace of glaciation; at the same time other regions were at least temporarily glaciated" (194:195). This evidence, linked in this way with the problem of the ice ages we have already discussed, reveals the existence of a single problem. Ice ages in low latitudes, and warm ages near the poles, are, so to speak, the sides of a single coin. The correct explanation of one will probably involve the explanation of the other.

Following the Cretaceous, the Tertiary Period shows the same failure of the fauna and flora to observe our present climatic zones. Scott, for example, says: "The very rich floras from the Green River shales, from the Wilcox of the Gulf Coast and from the Eocene of Greenland show that the climate was warmer than in the Paleocene, and much warmer than today" (372:103).

In this Eocene Epoch we find evidence of warm climate in the north that is truly overwhelming. Captain Nares, one of the earlier explorers of the Arctic, described a twenty-five-foot seam of coal that he had thought was comparable in quality to the best Welsh coal, containing fossils similar to the Miocene fossils of Spitzbergen. He saw it near Watercourse Bay, in northern Greenland (319:II, 141–42). Closer examination revealed that it was, in reality, lignite. Nevertheless, the contained fossils clearly indicated a climate completely different from the present climate of northern Greenland:

> The Grinnell Land lignite indicates a thick peat moss, with probably a small lake, with water lilies on the surface of the water, and reeds on the edges, and birches and poplars, and taxodias, on the banks, with pines, firs, spruce, elms and hazel bushes on the neighboring hills . . . (319:II,335).

Brooks thinks that the formation of peat bogs requires a rainfall of at least forty inches a year and a mean temperature above 32° F. (52:173). This suggests a very sharp contrast with present Arctic conditions in Grinnell Land.

DeRance and Feilden, who did the paleontological work for Captain Nares, also mention a Miocene tree, the swamp cypress, that flourished from central Italy to 82° N. Lat., that is, to within five hundred miles of the pole (319:II, 335). They show that the Miocene floras of Grinnell Land, Greenland, and Spitzbergen all required temperate climatic conditions with plentiful moisture. They mention especially the water lilies of Spitzbergen, which would have required flowing water for the greater part of the year (319:II, 336).

In connection with the flora of Spitzbergen and the fauna mentioned earlier, it should be realized that the island is in polar darkness for half the year. It lies on the Arctic Circle, as far north of Labrador as Labrador is north of Bermuda.

Wallace describes the flora of the Miocene. He points out that in Asia and in North America this flora was composed of species that apparently required a climate similar to that of our southern states, yet it is also found in Greenland at 70° N. Lat., where it contained many of the same trees that were then growing in Europe. He adds:

> But even farther North, in Spitzbergen, 78° and 79° N. Lat. and one of the most barren and inhospitable regions on the globe, an almost equally rich fossil flora has been discovered, including several of the Greenland species, and others peculiar, but mostly of the same genera. There seem to be no evergreens here except coniferae, one of which is identical with the swamp-cypress (*Taxodium distichum*) now found living in the Southern United States. There are also eleven pines, two *Libocedrus*, two *Sequoias*, with oaks, poplars, birches, planes, limes, a hazel, an ash, and a walnut; also water lilies, pond weeds, and an Iris— altogether about a hundred species of flowering plants. Even in Grinnell Land, within 8¼ degrees of the pole, a similar flora existed . . . (446:182–84).

It has been necessary to dwell at length on the evidence of the warm polar climates, because this is important for the discussion that follows.

3. UNIVERSAL TEMPERATE CLIMATES— A FALLACY

The evidence I have presented above (and a great deal more, omitted for reasons of space) has long created a dilemma for geology. Only two practical solutions have offered themselves.* One is to shift the crust, and the other is to suggest that climatic zones like the present ones have not always existed. It is often suggested that the climates have been very mild, virtually from pole to pole, at certain times. The extent to which the latter theory is still supported is eloquent evidence of the theory of the permanence of the poles. When one inquires as to the evidence for the existence of such warm, moist climates, a peculiar situation is revealed. There is no evidence except the fossil evidence that the theory is supposed to explain. Could there be a better example of reasoning in a circle? Colbert cites evidence that the Devonian animals were spread all

---

* Continental drift is not acceptable because of the time factor. It cannot solve the problems of recent periods, that is, the Tertiary Period and the Pleistocene Epoch.

over the world, and then remarks that therefore ". . . it is reasonable to assume . . . that the Devonian Period was a time of widely
spread equable climates, a period of uniformity over much of the
earth's surface" (375:255). According to him, the same situation
held true through the Paleozoic and Mesozoic and even much later
periods (375:268). Other paleontologists reasoned in the same way.
Goldring, for example, remarked: "The Carboniferous plants had a
worldwide distribution, suggesting rather uniform climatic conditions" (177:362). She drew the same conclusions from the world-
wide distribution of Jurassic flora (177:363).

Is this theory of universal temperate climates inherently reasonable? The answer is that it is not. It involves, in the first place, ignor-
ing the astronomical relations of the earth and the sun. The theory
requires us to assume the existence of some factor powerful enough
to negate the variation of the sun's heat with latitude which, of
course, is due to the angle of inclination of the earth's axis of rotation. As Professor George W. Bain, of Amherst, has pointed out, the
result of this is that

. . . The thermal energy arriving at the earth's surface per day per
square centimeter averages 430 gram calories at the equator but declines
to 292 gram calories at the 40th parallel and to 87 gram calories at the
80th parallel . . . (18:16).

What force sufficiently powerful to counteract that fact of astronomy can be suggested, and, more important, supported by convincing evidence?

It was thought at first that universal temperate climates might be
accounted for by the theory of the cooling of the earth. Those who
favored this theory (253, 292) argued that since, in earlier ages, the
earth was hotter, the ocean water then evaporated much more rapidly, and it formed thick clouds that reflected the sun's radiant
energy back into space. The cloud blanket shut out the sun's
radiation but kept in the heat that radiated from the earth itself, and
this acted to distribute the heat evenly over the globe. The cloud
blanket must have been thick enough to make the earth a dark, dank,
and dismal place. Since, as Colbert shows, fossils are found outside
the present zones appropriate to them even in recent geological
periods, such conditions must have obtained during about 90 percent
of the earth's whole history, and most of the evolution of living
forms must have taken place in them.

For a number of reasons, including the difficulty of explaining how plants can have evolved in the polar regions without sunlight, this theory has been abandoned. We have also seen that the idea that the earth was ever hotter than now has recently been undermined. This has destroyed the dependability of the theory's basic assumption.

The fact that the theory never was reasonable is shown from Coleman's arguments against it, advanced more than a quarter of a century ago. He pointed out that not only are ice ages known from the earliest periods (from the Precambrian) but there is evidence that some of these very ancient ice ages were even more intensely cold than the recent ice age that came to an end 10,000 years ago (87:78). No less than six ice ages are known from the Precambrian (420:260). The evidence of one of these Precambrian or Lower Cambrian ice ages is interestingly described by Brewster:

In China, in the latitude of northern Florida, there is a hundred and seventy feet of obvious glacial till, scratched boulders and all, and over it lie sea-floor muds containing lower Cambrian trilobites, the whole now altered to hard rock (45:204).

It is obvious that such ice ages (and evidences of more of them are frequently coming to light) are in conflict with the theory of universal equable climates. Some of them are found right in the midst of periods thought to have been especially warm, such as the Carboniferous.

Coleman presents other geological evidence against the theory. The fact that most of the fossils found are those of warm-climate creatures is, he thinks, misleading. Plants and animals are more easily fossilized in warm, moist climates than they are in cold, arid ones. Fossilization, even under the most favorable conditions, is a rare accident. The fauna and flora of the temperate and arctic zones of the past were seldom preserved (87:252). Thus, while the finding of fossils of warm-climate organisms all over the earth is an argument against the permanence of the present arrangement of the climatic zones, it is not an argument for universal mild climates.

Another argument against such climates may be based upon the evidences of desert conditions in all geological periods. These imply worldwide variations in climate and humidity. Both Brooks (52:24–25, 172) and Umbgrove (420:265) stress the importance of this evidence. One of the most famous formations of Britain—the Old

Red Sandstone—is apparently nothing but a fossil desert. Coleman points to innumerable varved deposits in many geological periods as evidence of seasonal changes (87:253), which, of course, imply the existence of climatic zones.

Ample evidence of the existence of strongly demarcated climatic zones through the earth's whole history (at least since the beginning of the deposition of the sedimentary rocks) comes from other sources. Barghoorn cites the evidence of fragments of fossil woods from late Paleozoic deposits in the Southern Hemisphere that show pronounced ring growth, indicating seasons; he also points out that in the Permo-Carboniferous Period floras existed that were adapted to very cold climate (375:242). Colbert himself reports good evidence of seasons in the Cretaceous Period, in the form of fossils of deciduous trees (375:265).

Umbgrove cites the geologist Berry, who states that the fossilized woods from six geological periods, from the Devonian to the Eocene, show well-marked annual rings, indicating seasons like those of the present time. Furthermore, Berry goes on to say:

> Detailed comparisons of these Arctic floras with contemporary floras from lower latitudes . . . show unmistakable evidence for the existence of climatic zones . . . (420:266).

Brooks concludes, on the basis of Berry's evidence, that climatic zones existed in the Eocene (52:24). Ralph W. Chaney, after a study of the fossil floras of the Tertiary Period (from the Eocene to the Pliocene), concluded that climatic zones existed (72:475) during that whole period. The distinguished meteorologist W. J. Humphreys, whose fundamental work, The Physics of the Air, remains a classic, remarked in 1920 that there was no good evidence of the absence of climatic zones at any time from the beginning of the geological record. Finally Dr. C. C. Nikiforoff, an expert on soils (both contemporary and fossil soils), has stated that "in all geological times there were cold and warm, humid and dry climates, and their extremes presumably did not change much throughout geological history" (375:191). We will return, below, to the significance of fossil soils and present other evidence showing persistence of sharply demarcated climatic zones during the earth's whole history. But where, at this point, does the evidence leave us?

On the one hand, the evidence shows that the plants and animals of the past were distributed without regard to the present direction of the climatic zones. I have been unable to do more than suggest the immensity of the body of evidence supporting this conclusion. On the other hand, the attempt to deny the existence, in the past, of sharply demarcated climatic zones like those of the present has failed. It may even be said to have failed sensationally. There is no scrap of evidence for it except the evidence it is supposed to explain, while, on the other hand, it is in contradiction with both the fundamentals of astronomy and the preponderance of the geological facts.

So we are left with a clear-cut conclusion: Climatic zones have always existed, but they have followed different paths on the face of the earth. If changes in the position of the axis of rotation of the earth, and of the earth upon its axis, are equally impossible, and if the theory of continental drift provides no satisfactory solution for reasons already discussed (Chapter I), then we are forced to the conclusion that the surface of the earth must often have been shifted over the underlying layers.

4. THE EDDINGTON-PAULY SUGGESTION

Another suggestion for displacements of the earth's crust, to which I have already referred, is that of Karl A. Pauly, who has contributed new lines of evidence in support of such shifts. He has based his theory on Eddington's suggestion that the earth's crust may have been gradually shifted through time by the effects of tidal friction. The evidence for displacements presented by Pauly is most impressive.

Pauly finds, from a study of the elevations above sea level of the terminal moraines of mountain glaciers in all latitudes, that there is a correlation of elevation with latitude. While it is true that many factors influence the distance a mountain glacier may extend downward toward sea level, latitude is one of them, and by using a sufficient number of cases it is possible to average out the other factors and arrive at the average elevation of mountain glacier moraines above sea level for each few degrees of latitude from the equator toward the poles. This gives us a curve that makes it possible to compare the elevations of the terminal moraines of mountain glaciers that existed during the Pleistocene Epoch. Pauly finds that these moraines do not agree with the curve, indicating unmistakably a displacement of the earth's crust (342:89).

Pauly cites another impressive line of evidence in support of displacements of the lithosphere. He has compared the locations of the coal deposits of several geological periods (many of which are now in polar regions) with the locations of ice caps for the same periods. He lists 34 coal deposits regarded as of Jurassic-Liassic age and 17 of Triassic-Thaetic age, and finds that, if it is assumed that the centers of the ice caps of that time were located at the poles, then these coal deposits would have been located within or just outside the tropics, as would be correct. He says:

The very definite location of these coal deposits within the Trias-Jura tropical and subtropical zones cannot be mere coincidence. The distribution indicates the lithosphere has shifted (342:96).

Of the Permo-Carboniferous coal deposits, very widely distributed over the earth, he says that "95 out of 105 listed in *The Coal Resources of the World* lie within or just outside of the tropics as determined by the assumption that the North or South Pole lay under the center of one of the Permo-Carboniferous ice sheets" (342:97).

5. THE CONTRIBUTION OF GEORGE W. BAIN

Professor Bain has gone considerably beyond the categories of evidence that we have so far discussed. He has considered the specific chemical processes controlled by sunlight and varying according to latitude, and the remanent chemicals typical of soils developed in the different climatic zones. He has extended this sort of analysis also to marine sediments.

Bain's approach to the problem has many advantages. It circumvents, for one thing, the argument that plants of the past may have been adjusted to climates different from those in which their modern descendants live. He begins with a precise definition of each climatic zone in terms of the quantities of the sun's heat reaching the earth's surface. He points out that, as is known, the seasonal variation of this heat increases with distance from the equator (18:16). He then describes the global wind pattern resulting from this distribution of the sun's energy, defining clearly the conditions of the horse latitudes, in which most of the earth's deserts are found, and the meteorology of the polar fronts. He shows that there are distinct and different complete chemical cycles in each of these areas, and corresponding cycles in the sea. Many of the chemical compounds produced in each of these areas are included, naturally, in the rocks formed from the sediments, and they remain as permanent climatic records.

It is impossible, because of limitations of space, to do justice to Bain's comprehensive approach to this question. He establishes that great differences exist between the mineral components of the rocks in different climatic zones, resulting from the difference in the amount of the sun's radiant heat. With regard to the polar soils, he found that they are developed in circles on the earth's surface rather than in bands. Temperate and tropical soils are, of course, found in bands, since the zones are bands that encircle the earth.

It is clear that Bain has established a sound method for the study of the climates of the past. He has applied his method to the study of the climates of five periods, the Cambrian, Ordovician, Silurian, Devonian, and Permian (19a) (Figs. 11, 12, 13, 14, and 15, pp. 74-78), with significant results. He concludes, first, that climatic zones, representing the different distributions of solar heat, existed in those periods just as at present. This is proved by the specific remanent chemicals included in these rocks, which differ exactly as do the sediments of the different zones at the present time. This is, of course, fatal for the theory of universal equable climates.

His second conclusion is that the directions of the climatic zones have changed enormously in the course of time. He finds the equator running through the New Siberian Islands (in the Arctic Ocean) in the Permo-Carboniferous Period, and North and South America lying tandem along it (18:17). The evidence he uses seems to establish his essential point (and ours) that the climatic zones themselves have shifted their positions on the face of the earth.

Bain has drawn some interesting further conclusions. He states that the earth's crust must have been displaced over the interior layers and that "fixity of the axis of the earth relative to the elastice outer shell just is not valid. . . ." (18:46). He points to the fossil evidence of the cold zones (distributed in circular areas) and says, ". . . The recurrent change in position of these rings through geologic time can be accounted for now only on the basis of change

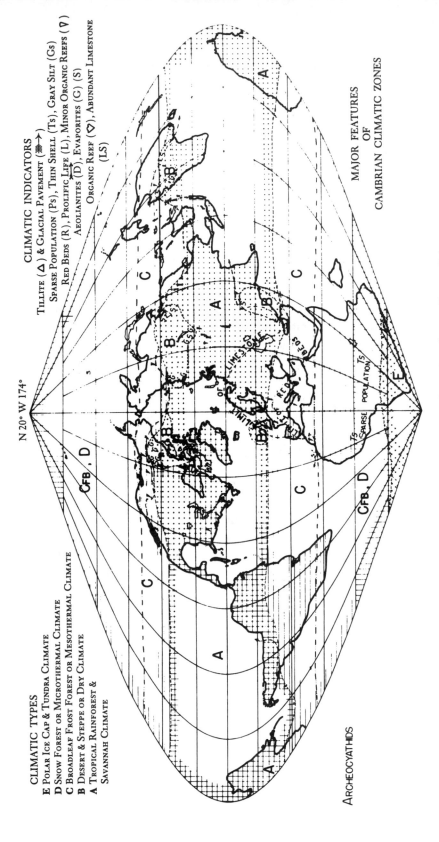

Fig. 11. *Climatic zones of the Cambrian Period, according to Bain.*

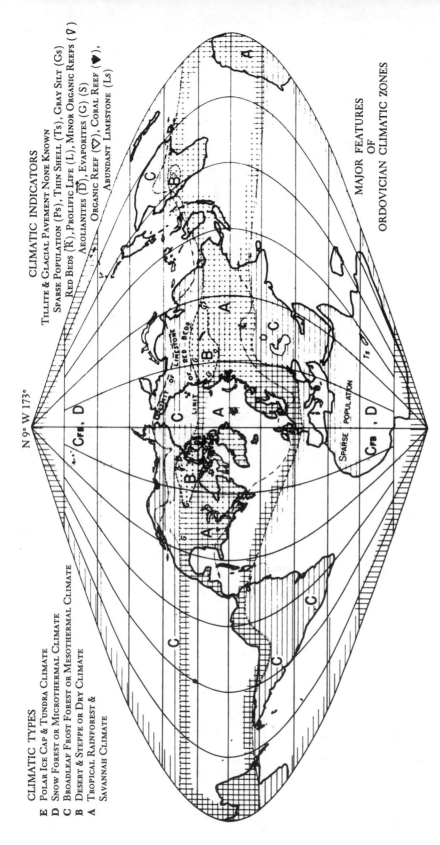

CLIMATIC TYPES

E POLAR ICE CAP & TUNDRA CLIMATE
D SNOW FOREST OR MICROTHERMAL CLIMATE
C BROADLEAF FROST FOREST OR MESOTHERMAL CLIMATE
B DESERT & STEPPE OR DRY CLIMATE
A TROPICAL RAINFOREST &
    SAVANNAH CLIMATE

CLIMATIC INDICATORS

TILLITE & GLACIAL PAVEMENT NONE KNOWN
SPARSE POPULATION (Ps), THIN SHELL (Ts), GRAY SILT (Gs)
RED BEDS (R̄), PROLIFIC LIFE (L), MINOR ORGANIC REEFS (∇̇)
AEOLIANITES (D̄), EVAPORITES (G), (S)
ORGANIC REEF (▽), CORAL REEF (♥),
    ABUNDANT LIMESTONE (Ls)

MAJOR FEATURES
OF
ORDOVICIAN CLIMATIC ZONES

Fig. 12. Climatic zones of the Ordovician Period, according to Bain.

N 9° W 173°

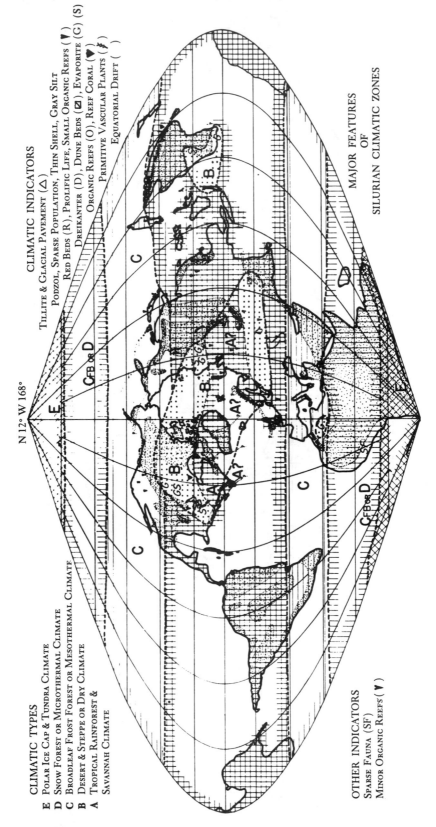

CLIMATIC INDICATORS

TILLITE & GLACIAL PAVEMENT (△)
PODZOL, SPARSE POPULATION, THIN SHELL, GRAY SILT
RED BEDS (R), PROLIFIC LIFE, SMALL ORGANIC REEFS (♥)
DREIKANTER (D), DUNE BEDS (☒) EVAPORITE (G) (S)
ORGANIC REEFS (O), REEF CORAL (♥)
PRIMITIVE VASCULAR PLANTS ( )
EQUATORIAL DRIFT ( )

MAJOR FEATURES
OF
SILURIAN CLIMATIC ZONES

CLIMATIC TYPES
E POLAR ICE CAP & TUNDRA CLIMATE
D SNOW FOREST OR MICROTHERMAL CLIMATE
C BROADLEAF FROST FOREST OR MESOTHERMAL CLIMATE
B DESERT & STEPPE OR DRY CLIMATE
A TROPICAL RAINFOREST &
SAVANNAH CLIMATE

OTHER INDICATORS
SPARSE FAUNA (SF)
MINOR ORGANIC REEFS (♥)

Fig. 13. Climatic zones of the Silurian Period, according to Bain.

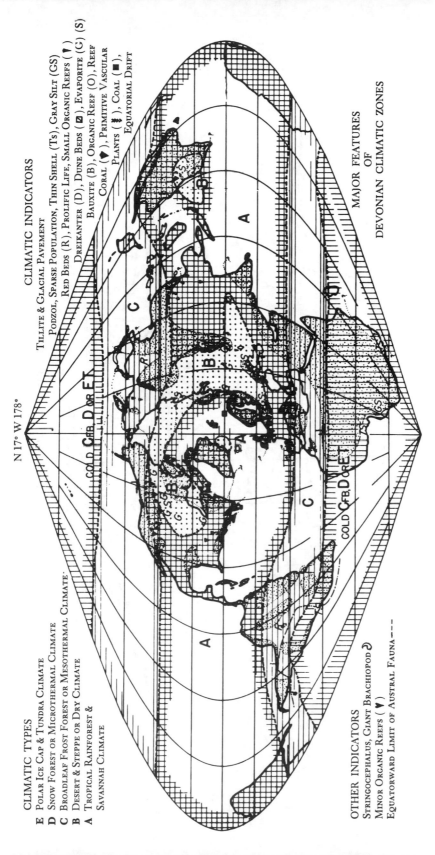

N 17° W 178°

CLIMATIC INDICATORS

Tillite & Glacial Pavement
Podzol, Sparse Population, Thin Shell (Ts), Gray Silt (GS)
Red Beds (R), Prolific Life, Small Organic Reefs (▼)
Dreikanter (D), Dune Beds (☒), Evaporite (G) (S)
Bauxite (B), Organic Reef (O), Reef
Coral (▼), Primitive Vascular
Plants (≢), Coal (■),
Equatorial Drift

MAJOR FEATURES
OF
DEVONIAN CLIMATIC ZONES

CLIMATIC TYPES

E Polar Ice Cap & Tundra Climate
D Snow Forest or Microthermal Climate
C Broadleaf Frost Forest or Mesothermal Climate
B Desert & Steppe or Dry Climate
A Tropical Rainforest &
   Savannah Climate

OTHER INDICATORS
Stringocephalus, Giant Brachiopod ᕑ
Minor Organic Reefs (▼)
Equatorward Limit of Austral Fauna – – –

Fig. 14. Climatic zones of the Devonian Period, according to Bain.

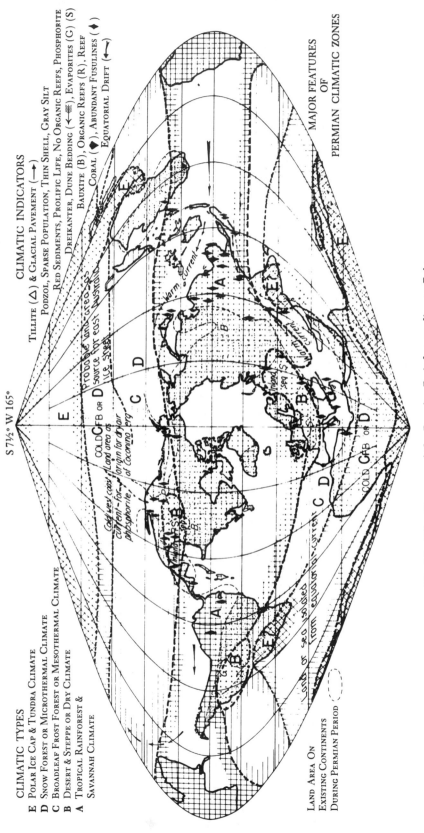

CLIMATIC TYPES

E POLAR ICE CAP & TUNDRA CLIMATE
D SNOW FOREST OR MICROTHERMAL CLIMATE
C BROADLEAF FROST FOREST OR MESOTHERMAL CLIMATE
B DESERT & STEPPE OR DRY CLIMATE
A TROPICAL RAINFOREST &
   SAVANNAH CLIMATE

CLIMATIC INDICATORS

TILLITE (△) & GLACIAL PAVEMENT (→)
PODZOL, SPARSE POPULATION, THIN SHELL, GRAY SILT
RED SEDIMENTS, PROLIFIC LIFE, NO ORGANIC REEFS, PHOSPHORITE
DREIKANTER, DUNE BEDDING (↙), EVAPORITES (G) (S)
BAUXITE (B), ORGANIC REEFS (R), REEF
CORAL (◆), ABUNDANT FUSULINES (◆)
EQUATORIAL DRIFT (↓)

MAJOR FEATURES
OF
PERMIAN CLIMATIC ZONES

LAND AREA ON
EXISTING CONTINENTS
DURING PERMIAN PERIOD (⌒)

S 7½° W 165°

Fig. 15. Climatic zones of the Permian Period, according to Bain.

in the position of the elastic shell of the earth relative to its axis of rotation" (18:46).

Even without the evidence of geomagnetism, or even if that evidence should someday be discredited, the evidence produced by Bain would be sufficient to establish the truth of displacements of the lithosphere. However, the mechanism he suggests does not seem satisfactory. He depends upon the effects of erosion. He points out that at the present time the balance of the sediment transfer by rivers is toward the equator. The mass thus added to the lithosphere on the equator has been given increased velocity by the fact of being moved equatorward, and this would tend to accelerate the rotation, but the gyroscopic effect of this, he thinks, would be to cause the rotating globe to precess in a direction at 90° to the direction of the rotation. The crust alone, however, not the entire globe, would be shifted (19a:128–129).

There seem to me to be three objections to this mechanism. In the first place, it seems probable that isostatic adjustment of the lithosphere to the transfer of sediments would eliminate the effect. A poleward flow of material under the lithosphere would roughly equal the equatorward movement of sediment. A second objection is that there is no reason to suppose that with every position of the lithosphere the balance of sediment transfer would be toward the equator. This would require changes in the drainage systems of all the continents with each shift of the crust. The third objection is that the geomagnetic evidence suggests polar shifts were far more frequent than indicated by Bain. Bain makes no use of the continental-drift hypothesis.

6. THE CONTRIBUTION OF T. Y. H. MA

Bain has pointed out (18) that among other indications of latitude, sea crustaceans and corals may indicate latitude either by the presence or absence of evidence of seasonal variations in growth. It happens that corals have been very thoroughly investigated from precisely this point of view.

By a remarkable parallelism of development, another theory of displacement of the earth's crust took shape on the opposite side of the earth at about the same time that Mr. Campbell and I started on our project. Professor Ting Ying H. Ma, an oceanographer, then at the University of Fukien, China, came to the conclusion, after many years of study of fossil corals, that many total displacements of the earth's lithosphere must have taken place. I did not become aware of Ma's work until I was introduced to it by David B. Ericson, of the Lamont Geological Observatory, in 1954. Ericson has, in fact, taken a leading role in introducing Ma's work to American scientists.

For about twenty years previous to the time I mention, Ma had intensively pursued the study of living and fossil reef corals. He very early noticed the special characteristic of reef corals referred to by Bain but hitherto ignored by writers on corals. He saw that, at distances from the equator, there were seasonal differences in the rates of coral growth and that the evidences of these were preserved in the coral skeleton. Specifically he observed that in winter the coral cells are smaller and denser; in summer they are larger and more porous. Together these two rings make up the growth for one year.

Studying living coral reefs in various parts of the Pacific, comparing, measuring, and tabulating coral specimens of innumerable species, making photographic studies of the coral skeletons, Ma established that the rates of total annual coral growth for identical or similar species within the range of the coralline seas increased with proximity to the equator, and that seasonal variation in growth rates increased with distance from the equator.

Other writers on corals have pointed out that there are numerous individual exceptions and irregularities in coral growth rates, deriving from the fact that the coral polyps feed upon floating food, which may vary in quantity from place to place, from day to day, and even from hour to hour (125:20–21; 298:52–53). Ma, however, has guarded himself against error by a quantitative and statistical approach. In several published volumes of coral studies (285–290) he has compiled tables running into hundreds of pages, and his studies have involved thousands of measurements.

When this indefatigable oceanographer had worked out the relations of coral growth with latitude, he possessed an effective tool with which to investigate the climates of the past. He studied specimens of fossil corals from many geological periods. He devoted separate volumes to the Ordovician, Silurian, Devonian, Cretaceous, and Tertiary Periods (285–289).

As Ma assembled the coral data for these periods, it became clear that the total width of the coralline seas had not varied noticeably from the beginning of the geological record. Not only was the existence of seasons in the oldest geological periods clearly indicated; it was also indicated that the average temperatures of the respective zones were about the same as at present.

The second result of Ma's studies was to establish that the positions of the ancient coralline seas and, therefore, of the ancient equators were not the same as at present. They had evidently changed from one geological period to another. Ma first believed that this could be explained by the theory of drifting continents. Down to about 1949 he sought to fit all the evidence into that theory. By 1949, however, the accumulated evidence forced him to adopt a theory of total displacements of all the outer shells of the earth over the liquid core. By an instinct of conservatism, however, he did not abandon the theory of floating continents but combined it with the new theory.

Ma's coralline seas ran in all directions (Figs. 16, 17, 18, 19, and 20, pp. 82-86); one of his equators actually bisected the Arctic Ocean. But he had great difficulty in matching up his equators on different continents. If, for example, he traced an equator across North America, he could not match it with an equator for the same period on the other side of the earth to make a complete circle of the earth. He therefore supposed that the continents themselves had been shifting independently, and this had had the effect of throwing the ancient equators out of line. He therefore allowed, for each period, enough continental drift to bring the equators into line, and it seemed, when he did this, that in successive geological periods he did have increasing distances between the continents, as if the drift had been continuous.

Subsequently Ma developed his theory into a complete system, which is most interesting, and yet to which I think serious objections may be raised.

Corals are, according to Ma, excellent indicators of the climate for the time in which they grew, but by the nature of the case, since corals grow only in shallow water and grow upwards only as far as the surface, the period of time represented by a single fossil coral reef is of the order of a few thousand years only, as compared with the millions of years embraced by a geological period.

How short the continuous growth of a coral reef may be is indi-

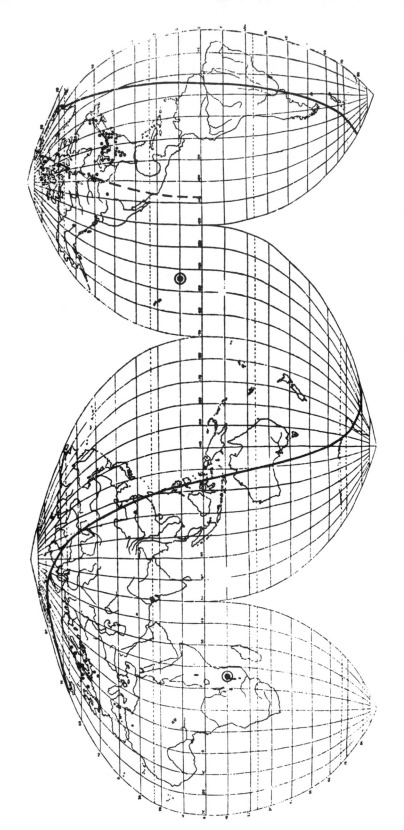

Fig. 16. The alignment of segments of the Ordovician coralline sea on the present continents, according to Ma.

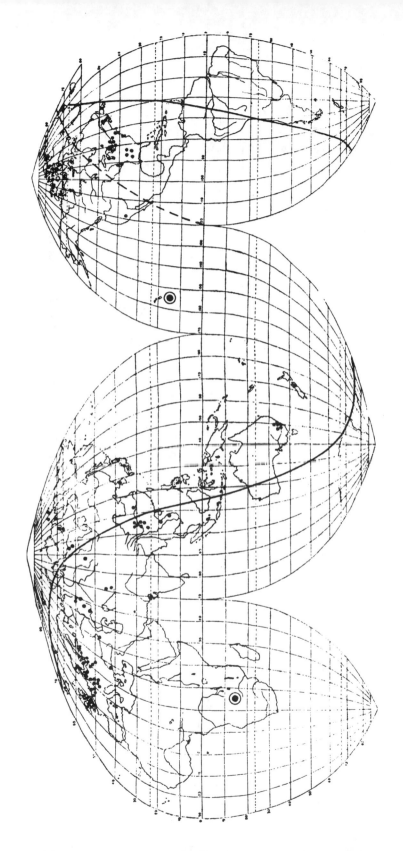

Fig. 17. The alignment of segments of the Silurian coralline sea on the present continents, according to Ma.

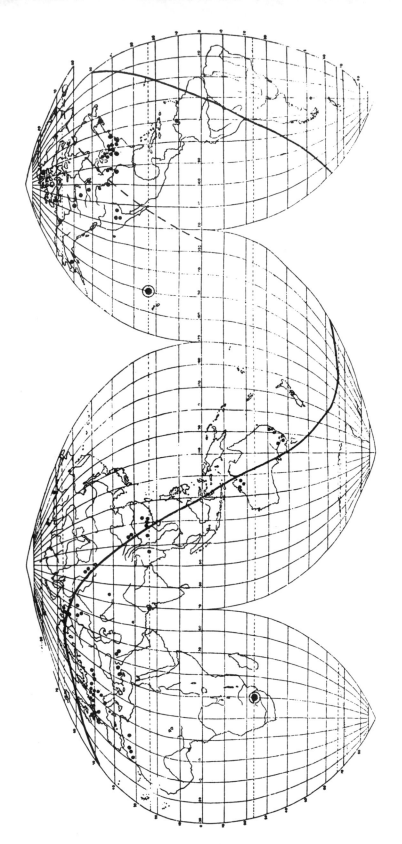

Fig. 18. The alignment of segments of the Devonian coralline sea on the present continents, according to Ma.

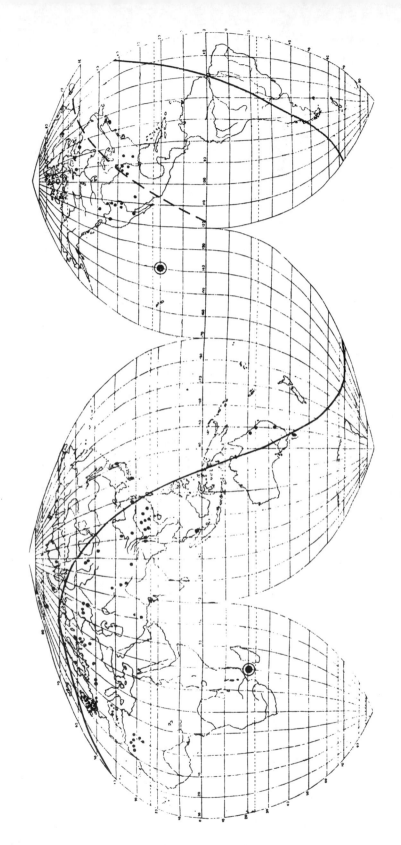

Fig. 19. The alignment of segments of the *Lower Carboniferous coralline sea* on the present continents, according to Ma.

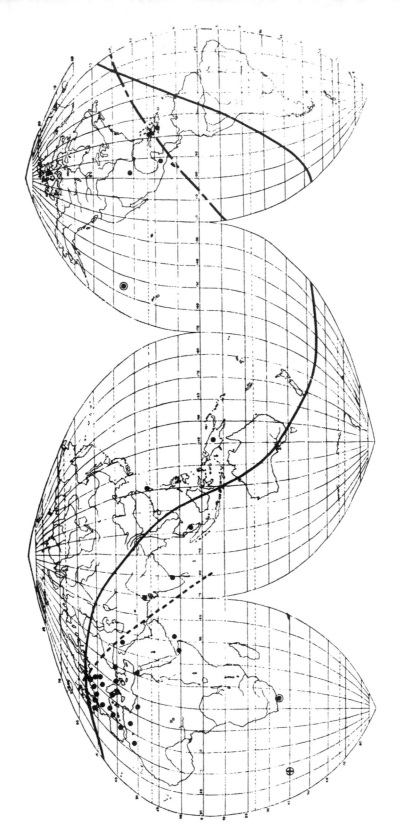

Fig. 20. *The alignment of segments of the Upper Cretaceous coralline sea on the present con-*
tinents, according to Ma.

cated by numerous studies of the coral reefs of the Pacific. A. G. Mayor, for example, says:

> . . . The modern reefs now constituting the atolls and barriers of the Pacific could readily have grown upward to sea-level from the floors of submerged platforms since the close of the last glacial epoch (298:52).

At Pago Pago harbor borings were made down to the basalt underlying the reef, and after estimates of the growth rate were arrived at, the age of the reef (Utelei) was estimated at 5,000 years. When these spans are compared with those of entire geological periods of the order of 20,000,000 or 30,000,000 years, it is clear how fragile must be any conclusions based on the assumption that a given coral reef in Europe was contemporary with another one in North America. It is quite impossible in the present state of our knowledge to decide that they were in fact contemporary.

This means that Ma's corals for a period like the Devonian may be indications of different equators that existed at different times during that period of 40,000,000 years. Therefore it is obvious that thousands of coral specimens would be required to give any certainty as to the actual climatic history of an entire geological period.

Very possibly Ma could have avoided combining the two different theories—the slipping of the shell of the earth and the drifting of continents—if he had supposed a sufficiently frequent slipping of the crust. The frequency of the displacements suggested by the theory presented in this book, which would involve many different equators in a single geological period, might remove his difficulties. As it is, he has to face all the geophysical and geological objections to the drifting-continent theory as well as difficulties with his displacement theory.

7. ON THE RATE OF CLIMATIC CHANGE

Studies appear from time to time in which attempts are made to trace climatic changes in specified areas over periods of millions of years. In one of these, for example (72), the conclusion is reached that there was a gradual cooling of the climate during a great many million years of the Tertiary Period. It is true that no cause of such a progressive cooling can be pointed to; neither is there any explanation as to why the climatic change had to be so gradual. It is simply assumed that the climatic change had to be gradual and that the cause of the change had to be such as to explain gradual changes.

It is important to define the evidence on which these conclusions are based. In the example I am considering, the facts are as follows:

a. The period of time involved is of the order of 30,000,000 years.

b. Wherever reference is made to the specific strata of rock selected for analysis of the climatic evidence (consisting of included fossils), it is clear that the time required for the deposition of any particular layer was of the order of 10,000 years.

c. It follows that during 30,000,000 years it would have been possible to have about 3,000 different layers of sedimentary rock.

d. A vast majority of these layers cannot be sampled, either because they no longer exist, or because they do not contain fossils, or simply because of the amount of work involved.

e. As a result, only spot checking is possible. Perhaps a dozen strata out of 3,000 may be studied, and from these it must be obvious that no dependable climatic record can be established.

f. Even with the unsatisfactory spot checking so far attempted, reversals of climatic trends have been observed (72).

g. Climatic conditions indicated by a layer of sediments deposited during a brief period of time in one location cannot be assumed to indicate the direction of climatic change over a great region or over the whole earth. It seems quite as reasonable to suppose that climatic change in other regions at the same time could have been in a different direction. Furthermore it cannot be assumed that two sedimentary deposits in different areas are of the same age because they both indicate climatic change in the same direction.

It may be concluded that claims for gradual climatic changes in the same direction over long periods of time and over great areas are unsupported by convincing evidence. They are supported by no reasonable hypothesis. We are left free to conclude that climatic change may have taken place in relatively short periods of time, and possibly in opposite directions at the same time, as the consequence of displacements of the lithosphere.

chapter **4**

EVIDENCE FOR THE NORTH POLE IN HUDSON BAY

In THE preceding chapters much evidence has been presented to support the assumption of the displacement of the earth's outer shells over the inner body, displacements that may have occurred rather frequently during the history of the earth. It is now time to tie down this assumption with concrete evidence that such a displacement has occurred, not in the remote past of the planet's history but in very recent time, and not once but at least three times in that recent epoch that we call the Pleistocene. I will ask the reader's indulgence for my descending to details in the examination of the evidence. This is not the sort of thing that he who runs may read, but, although the evidence is detailed, it is not particularly technical. It is the sort of evidence that murder-trial lawyer Perry Mason might present for the consideration of a jury of laymen, twelve good men and true, who may not be specialists but who are going to have to decide important issues of life and death by the use of intelligence and logic.

Several lines of solid evidence suggest that during the last ice age the North Pole was located in or near Hudson Bay. The reader will understand that if this was the case it was the result not of a change of axis of the earth but merely of a shift in the position of the crust or lithosphere relative to the earth's inner layers.

It seems desirable to have a definite point of reference rather than a vague general area to postulate as the position of the pole during the ice age. From an examination of all the evidence, which will be discussed in the following pages, the best guess for the site of the pole seems to be approximately 60° North Latitude and 83° West Longitude. We will find this definite assumption useful when we consider specific corroborating evidence from various other parts of the earth.

1. REMARKABLE FEATURES OF THE LAST NORTH AMERICAN ICE CAP

The first line of evidence that the last North American ice cap was a polar ice cap is based on the shape, size, and peculiar geographical location of the ice sheet. Two geologists, Kelly and Dachille, have pointed out that the area occupied by the ice was similar both in shape and size to the present Arctic Circle (248:39).* Many others have remarked on its unnatural location. It seems to have occupied the northeastern rather than the northern half of the continent. No one has explained why the ice cap, which extended southward as far as Ohio, did not cover some of the northern islands of the Canadian Arctic Archipelago, islands lying between Hudson Bay and the pole (87;28, note) or why it failed to cover the Yukon District of Canada or the northern part of Greenland. Later we shall examine a considerable amount of evidence indicating that the Arctic Ocean itself was warm during the ice age.

Another important problem that has long vexed geologists, and

---

* However, there is no consensus, especially as to the northern limits of the ice, as we shall see below. Available glacial maps are unsatisfactory.

for which no solution has so far even been suggested, is the distribution of the ice of the last American ice cap, usually referred to by geologists as the "Wisconsin" ice sheet. The ice appears to have been thicker and to have extended farther south on the low central plains of the Mississippi Valley, where it invaded Wisconsin and Ohio, than on the high mountain areas in the same latitudes farther west. This contradicts the basic assumption of current glacial theory, which states that the ice age involved a general lowering of world temperatures with no change in the position of the pole. If the accepted geological theory is correct, the ice should have formed first on the high mountains and it should have extended farther south on the mountainous highlands than on the lower central plains. This unsolved problem may have been one of many that led one of America's foremost geologists, Professor Reginald A. Daly of Harvard, to say that "the Pleistocene history of North America holds ten major mysteries for every one that has already been solved" (93:111).

The assumption that Hudson Bay then lay at the pole makes it possible to explain logically these hitherto unsolved problems. With that location, the northernmost Arctic islands would have been as much as 1,000 miles *south* of the pole, and the same would have been true of the western highlands. It would then have been entirely natural for the ice to be thicker and to extend farther south on the plains nearer the pole. Furthermore this assumption also explains why the European ice sheet was thinner than the North American ice sheet and did not extend as far south.

Another line of direct evidence that the Wisconsin ice sheet was a polar ice cap has been developed by the geologist Lawrence Dillon, who has examined the temperature requirements for the growth of ice sheets and compared them with average temperatures known to have prevailed around the glaciated area of North America and farther south during the ice age.

Dillon found that the essential condition governing the growth of an ice sheet is not the average year-round temperature nor the average snowfall but the average temperature during the summer (114:167). He points out that no ice sheets form at the present time in areas with average summer temperatures of 45° or higher and suggests that this situation must also have been true in the past. He cites, as a good illustration of this, the northeastern section of Siberia, which has no ice sheet despite the fact that its average annual temperature is lower than that of the North Pole itself. Furthermore northeastern Siberia has a higher average precipitation than either glaciated Greenland or Antarctica. But, as he points out, the summer temperature in Siberia is high, and this he thinks is the controlling factor.

Now, according to Dillon, the existence of the Wisconsin continental ice sheet would have required a decrease of 25° Centigrade in the average summer temperatures as they exist now (114:167). But the evidence is that in areas at a distance from the borders of the ice sheet there was no such fall in temperature. He quotes evidence produced by the geologist Antevs to the effect that, on the contrary, the average summer temperature along the 105th meridian in southern Colorado and northern New Mexico during the ice age was only 10° F. lower than it is now, while in the equatorial Andes it was only 5° or 6° F. lower. The assumption that Hudson Bay lay at the pole can explain why the apparent average fall in summer temperature should have been greatest along the southern boundary of the ice sheet, less in New Mexico, and least in the equatorial Andes. According to our assumption the equatorial Andes would, during the ice age, have been located approximately 20° of latitude to the north of the equator, but still in the tropics, and so it is to be expected that the average temperature would have been only slightly lower than now. On the other hand, Ohio, then presumably on the Arctic Circle, would have been very much colder than at present.

Dillon considers the situation that would have prevailed in the United States if during the ice age world temperatures had fallen about 10° F., as some have supposed and as might have been deduced from the findings of Antevs in Colorado and New Mexico.

By that assumption, the average July temperature around the edges of the ice sheet would have been 60° F. (as compared with 70° now). This would be similar to the average summer temperatures in England, northern Germany, or the state of Maine now. But Dillon concludes:

> Since no glaciers or permanent snowfields are known to exist today in such mild climates, it seems hardly likely that they could have done so in former times (114:168).

Dillon does not explicitly suggest a polar change, but he seems to leave no alternative.

2. THE WARM ARCTIC OF THE ICE AGE

One of the necessary corollaries of our assumption regarding the site of the North Pole during the period of the Wisconsin ice sheet would be that the point now at the North Pole, in the middle of the Arctic Ocean, would then have been 30° away from—that is, *south of*—the pole and could have been expected to have climatic conditions similar to those of Hudson Bay now, while the coast of Siberia, on the opposite side of the Arctic Ocean, would have been enjoying a warm temperate climate. Does the evidence support this corollary? I think it does.

I must again apologize to the reader for asking him to look at the details. I am sure he would not be satisfied with some general statements citing the evidence at second hand. I am sure he wants to judge matters for himself. Therefore he must bear with the details. I list below the radiocarbon tests that provide the evidence, and ask that it be critically examined.

I have already mentioned that some of the islands in the Arctic Ocean were never covered by ice during the last ice age, a fact that has been long known. The first of our radiocarbon dates relates to comparatively warm conditions on Baffin Island in the far north, 15° (about 900 miles) from the pole.

(1) Sample from the Isortoq River, Baffin Island (351a:I-731;VIII,185; 24,600±500).* ". . . Alder and birch remains [in peat] suggest a slightly warmer climate than today. Taken in conjunction with I-839 (351a:VIII,186, leafy peat taken from the same site, 30,000± 1200 yr) there is a strong implication that during the period 24,000–30,000 B.P. [before the present] the large proportion of North Central Baffin Island was ice-free or at least carried less glacier ice than today."†

It is important to note that a climate on Baffin Island *warmer than today* existed just when the Wisconsin continental glacier was going through a vast expansion (see Fig. 21, page 94). The tem-

---

* Sample numbers are made up of the initials of the dating laboratory followed by a serial number assigned by the laboratory, a reference to the source, and the date found. Where the age is followed only by a plus sign it is a minimum age rather than a finite one. Where the plus sign over a minus sign (±) is followed by a number, this gives the margin of error in years for the sample date.

† With the pole in Hudson Bay, Baffin Island would actually have been nearer the pole then, but the Arctic Ocean, much warmer than now, might have warmed the northern part of the island.

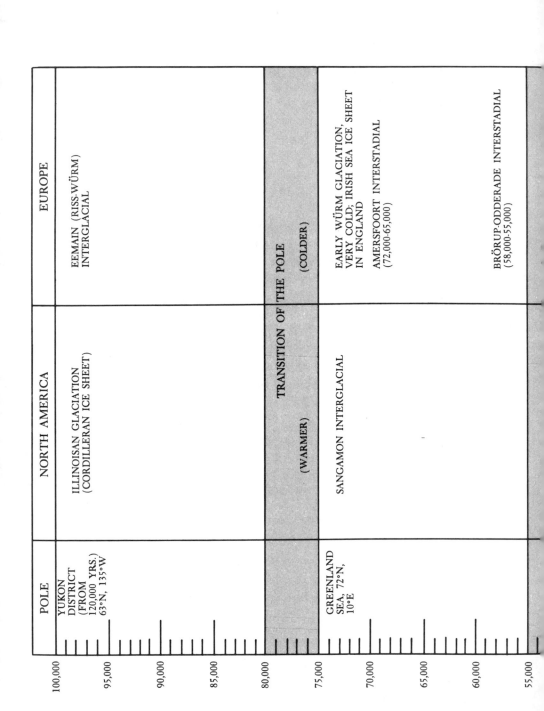

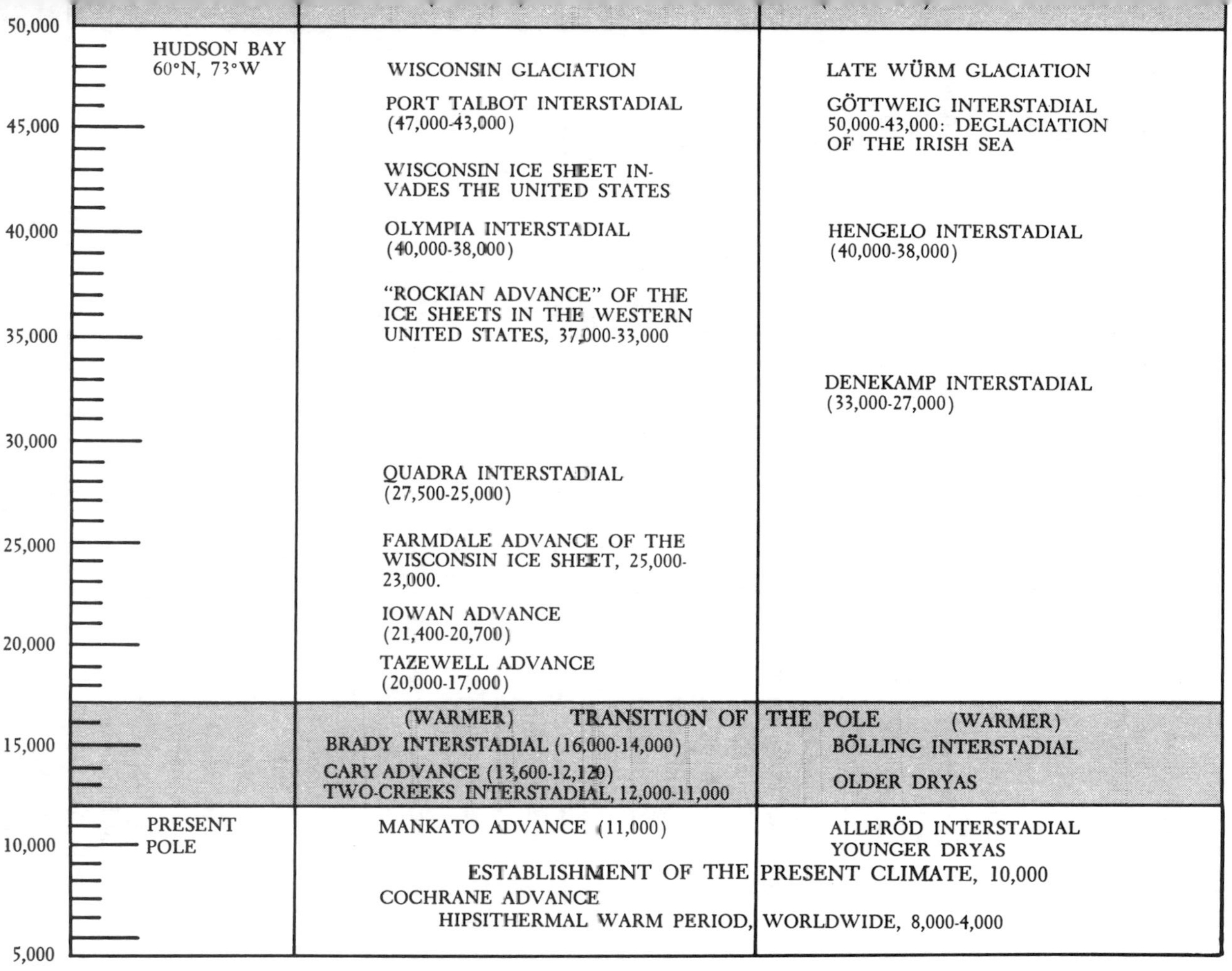

Fig. 21. One hundred thousand years of climatic change.

HUDSON BAY
60°N, 73°W

LATE WÜRM GLACIATION

GÖTTWEIG INTERSTADIAL
50,000-43,000: DEGLACIATION
OF THE IRISH SEA

HENGELO INTERSTADIAL
(40,000-38,000)

DENEKAMP INTERSTADIAL
(33,000-27,000)

WISCONSIN GLACIATION

PORT TALBOT INTERSTADIAL
(47,000-43,000)

WISCONSIN ICE SHEET IN-
VADES THE UNITED STATES

OLYMPIA INTERSTADIAL
(40,000-38,000)

"ROCKIAN ADVANCE" OF THE
ICE SHEETS IN THE WESTERN
UNITED STATES, 37,000-33,000

QUADRA INTERSTADIAL
(27,500-25,000)

FARMDALE ADVANCE OF THE
WISCONSIN ICE SHEET, 25,000-
23,000.

IOWAN ADVANCE
(21,400-20,700)

TAZEWELL ADVANCE
(20,000-17,000)

TRANSITION OF THE POLE

(WARMER)
BØLLING INTERSTADIAL

OLDER DRYAS

(WARMER)
BRADY INTERSTADIAL (16,000-14,000)

CARY ADVANCE (13,600-12,120)
TWO-CREEKS INTERSTADIAL, 12,000-11,000

ALLERÖD INTERSTADIAL
YOUNGER DRYAS
PRESENT CLIMATE, 10,000

MANKATO ADVANCE (11,000)

ESTABLISHMENT OF THE
COCHRANE ADVANCE
HIPSITHERMAL WARM PERIOD, WORLDWIDE, 8,000-4,000

PRESENT POLE

50,000
45,000
40,000
35,000
30,000
25,000
20,000
15,000
10,000
5,000

perate character of the sea around Baffin Island is indicated by
another date:

(2) Shells from Baffin Island (351a:GSC-528;IX,186;30,320±820 B.P.).
     ". . . One of the species, *Venericardia borealis* . . . indicates
     marine conditions may have been slightly warmer than today . . ."

Still another date from Baffin Island suggests that the warm conditions lasted through much, if not all, of the period of the Wisconsin glaciation:

(3) Shells from Steensby Inlet (351a:I-1242;VIII,185;19,000±1000
     B.P.). ". . . Date possibly indicates a period of open water at Foxe
     Basin at ca. 19,000 B.P. Dates compare with I-725,17,000±500
     (unpublished) and I-1314,18,700±1200 (unpublished)."

The unpublished dates, together with the published ones, suggest
that the warm conditions at Baffin Island lasted at least from 17,000
to 30,000 years ago.

Now, about as far north but to the west, we come to Banks
Island, Northwest Territories, in Lat. 73°23' N, Long. 121°54' W,
on which no evidence of glaciation has ever been found:

(4) Sample (351a:S-288;X,372;34,000+). Comment: ". . . Evidence for
     existence of Wisconsin refugium on unglaciated islands of West
     Arctic Archipelago is inconclusive . . . Date tends to substantiate
     existence of refugium."

Thus it seems that during the Wisconsin ice age there was a tem-
perate-climate refuge in the middle of the Arctic Ocean for the fauna
and flora that could not exist in Canada or the United States. There
is no suggestion that the "refugium" was limited to Banks Island.
We must acknowledge that the absence of glaciation on these islands
does not make sense according to the theory that the glacial period
was the consequence of a worldwide lowering of temperature. If
the pole had then been in its present position, the weather could
not have been warmer on Banks Island than in Ohio.

J. B. Delair has sent me an interesting passage from an old geological book entitled *Geology of Weymouth, Portland and the Coast
of Dorset (England)* by Robert Damon, published in 1860. Damon
refers to Sir Robert McClure's narrative of the discovery of the
northwest passage. McClure mentions the existence of a fossil forest
on Banks Island, where, for a depth of 40 feet, a cliff was composed

of one mass of fossil trees . . . "120 miles further north we discovered a similar kind of fossil forest." As we shall see, such fossil forests have been found not only in the Canadian Arctic Archipelago but all along the coast of Siberia.

We now go farther north to Axel Heiberg Island, in Lat. 81°3′ N, Long. 92°25′ W, only 9° from the pole:

(5) Sample $(351a:GSC-113;VI,179;36,800^{+4200}_{-2800})$. Shells of warm climate species, referred to as "interglacial."

These shells appear to extend the warm period in the Arctic back to about 40,000 years ago. A second sample from the same island $(351a:GSC-139;VI,179;36,600^{+3700}_{-2200})$ supports the same conclusion. Other samples show that the warm climatic conditions extended right down to the end of the glacial period and that down to 8,000 years ago Melville Island and Lougheed Island, in the Northwest Territories, had never been glaciated $(351a:I[GSC]-21; III,53$ and $351a:I[GSC]-24;III,53)$.

Thus we are able to say that warm conditions in the Arctic Archipelago of Canada persisted for the entire duration of the Wisconsin glaciation, from 40,000 years ago down to the establishment of modern conditions.

But we have not finished. We must get closer to the pole. We can accomplish this by examining the results of a number of sea cores taken from the middle of the Arctic Ocean. The Lamont Geological Observatory of Columbia University obtained the cores in order to explore the ancient climatic conditions of that ocean.

(6) $(351a:L-508;III,158;25,000\pm3000.)$ ". . . Foraminifera separated from . . . four cores [all from Latitude 84° N but from different longitudes]. In each case sample represents the base of zone rich in forams . . . The base of the foram-rich zone . . . apparently does not correspond to the end of the Wisconsin but to some much earlier event."

The comment reflects the view that the Foraminifera are evidence of warm climate prevailing in the Arctic Ocean about 25,000 years ago. It is worth noting that they come from the deep sea and within 6° of the pole itself. This seems to rule out the possibility that local conditions could have influenced matters. It is evidence that the

Arctic Ocean was temperate at that time, which, as the reader will have noticed, is precisely the time of a great advance of the Wisconsin continental glacier in the United States, *three thousand miles to the south* (see Fig. 21, p. 94).

The fact that the temperate conditions in the Arctic Ocean were general becomes evident when we examine evidences from the opposite shore of the ocean, along the coast of Siberia. Here, as we might expect (from the fact that we are getting farther and farther away from our pole in Hudson Bay), the evidence of warm climate during the North American ice age grows stronger still. The following dates establish the warm temperate characteristics of the climate in a vast territory which is now very cold.

TABLE 6
Radiocarbon Samples from Siberia

351a:GIN-143;X,441 .... 42,800±400
Wood from Malyy Anyuy River, gen. location Lat. 68° N, Long. 162–168° E.

351a:GIN-98;X,434 .... 36,900±400
Wood from the Yenisey Valley; date corresponds to the Rockian Advance in U.S. (see Fig. 21, p. 94).

351a:GIN-76;X,434 .... 35,800±600
Wood from the banks of the Yenisey River.

351a:GIN-155;X,441 .... 33,700±800
Wood from valley of tributary of the Lena River. (Lat. 67° 25′ N, Long. 125°20′ E).

351a:GIN-80;X,430 .... 33,000±400
Wood from Bryansk Oblast (Lat. 53° 35′ N, Long. 33°35′ E) found in floor of dwelling made of mammoth bones . . .
This indicates that the climate in this part of northern Russia was warm enough to grow food for herds of mammoths, and provide wood for fires. (See Chapter X.)

351a:GIN-99;X,434 .... 32,500±700
Wood from the Yenisey Valley.

351a:GIN-126;X,433 .... 30,700±300
Wood from Omsk Oblast, from stump in alluvium above flood plain of Irtysh River.

351a:Mo-215;VIII,320 .... 26,000±1600
Mammoth hair from the Lena River. ". . . According to the spore and pollen diagram of terrace deposits from the mam-

moth burial site, the supposed age of the find is the zero-thermic maximum of post-glacial time."

The "zerothermic maximum" refers to the period ca. 8,000–4,000 years ago (see Fig. 21, p. 94) that western geologists call the "hipsithermal," when the average temperatures were a little warmer than today. Thus the temperature when the mammoth really lived was warmer than the present temperature in the Lena River Valley.

351a:GIN-162;X,440 .... 24,800±120
Wood from the Chadobets River Valley (tributary of the Yenisey) in central western Siberia, with fossils of mammoth, horse, bison, etc.

351a:GIN-28;X,424 .... 21,350±650
Large tree trunk from banks of Yenisey River. ". . . Sample taken from accumulation of tree boles frozen for many years . . ."

Here is incontrovertible evidence of warm climate in the Yenisey River Valley when the mammoths lived there. This area, above the Arctic Circle, then evidently forested, is now barren tundra.

In summary I think I can reasonably claim with regard to this table and the preceding dates from the Arctic Ocean and Arctic Canada that they confirm an important corollary for the pole in Hudson Bay. Yes, if the pole were in Hudson Bay, the Arctic should be warm. And the Arctic was warm. However, there is much more evidence.

3. SOVIET EVIDENCE OF THE WARM ARCTIC

In recent years Soviet scientists have been very busy surveying the Arctic Ocean, in which they have a rather proprietary interest, seeing that most of its coasts belong to them. They have examined a great many cores from the ocean bottom and have dated them by the new methods of radioelement. They have come to the conclusion, just as we have, that the Arctic Ocean was warm during most of the ice age, particularly from about 32,000 to about 18,000 years ago. A report by academicians Saks, Belov, and Lapina (364) covering many phases of their oceanographic work lists the successive climatic changes in the Arctic as follows:

1. Down to 50,000 years ago the Arctic was cold.
2. From 50,000 to 45,000 years ago the Arctic Ocean water was warm.
3. From 45,000 to about 28–32,000 years ago it was cold.
4. From 28–32,000 years ago to 18–20,000 years ago it was warm.
5. From 18–20,000 years ago to 9–10,000 years ago it was cold.
6. About 9–10,000 years ago the present climate was established.

Saks, Belov, and Lapina have theories by which they explain these changes. These theories, while they do not agree with ours, may still interest the reader. They assume that the pole during this whole period was where it is now, and that consequently the Atlantic Ocean was temperate for the most part, though colder than now, especially in the higher latitudes. They correlate the warm periods in the Arctic Ocean with times when the warmer Atlantic waters could freely enter that basin, and cold periods with times when the circulation between the two oceans was cut off, possibly by a land bridge across the North Atlantic connecting Greenland, Iceland, and Scandinavia. This land bridge could not have resulted from a lowering of sea level due to the expansion of glaciers, for this would have amounted to only 300 feet at the most. It would have required an uplift of the sea bottom. The Soviet scientists cannot explain why or how the sea bottom could have risen. Never mind: The hypothesis is necessary to explain the facts. I work in the same way but with a different hypothesis.

Although my hypothesis is different, it too will include uplifts of the ocean bottom to explain some of the facts of the ice age but not the climatic changes. My explanation of the six climatic periods is as follows:

1. Down to 50,000 years ago the western Arctic Ocean (the end toward Norway—see Endpapers) was cold because of an earlier location of the pole, to be discussed later.
2. From 50,000 to 45,000 years ago the western Arctic was warm because the pole had been shifted to Hudson Bay, and warm waters from the eastern Arctic Ocean (toward eastern Siberia and Alaska) moved westward.
3. From 45,000 to about 28–32,000 years ago the waters were probably not cold except in certain places and at great depths where cold water had been trapped. Unfortunately we do not yet know the latitudes and longitudes at which the Soviet cores were taken, and we do not know how the Russian scientists interpreted their cores.
4. From 28–32,000 to 18–20,000 years ago the Soviet scientists are in complete agreement with us, and this is a positive matter of great

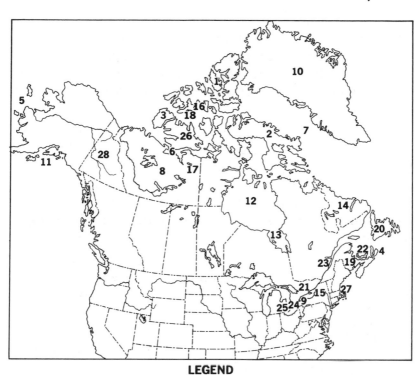

LEGEND

1. Axel Heiberg Island	10. Greenland	19. New Brunswick
2. Baffin Island	11. Gulf of Alaska	20. Newfoundland
3. Banks Island	12. Hudson Bay	21. Port Talbot
4. Cape Breton Island	13. James Bay	22. Prince Edward Island
5. Chukchi Sea, Alaska	14. Labrador	23. Quebec City
6. Coronation Gulf	15. Ledyard (Cayuga Co., N.Y.)	24. Rodney, Ontario
7. Davis Strait	16. Loughheed Island	25. Tupperville, Ontario
8. District of Mackenzie	17. MacAlpine Lake	26. Victoria Island
9. Don Valley	18. Melville Island	27. West Lynn, Mass.
		28. Yukon District

Fig. 22. *Map of North America between parallels of 35° and 85° N*

importance, in view especially of the agreement with the findings of the Lamont Oceanographic Expedition, already mentioned.

5. From 18–20,000 to 9–10,000 years ago the cold period could have resulted from the movement of the pole from Hudson Bay to its present position in the middle of the Arctic Ocean.

Leaving aside the disagreement about the period from 45,000 to 28–32,000 years ago, which cannot be settled without further analysis of the evidence, we are entitled to claim that the Soviet findings of a warm period in the Arctic from 28–32,000 to 18–20,000 years ago are in agreement with our assumption regarding the location of the pole.*

The Soviet publication *Sputnik* in its issue of November, 1968, reported the discovery of evidence of human occupation of the New Siberian Islands, as well as of Spitzbergen, during the ice age. Both archipelagoes are virtually uninhabitable now, especially the New Siberian Islands, which lie only 10°, or about 600 miles, from the pole. *Sputnik* gives the source of the information as the newspaper *Kommunist Tajikistana* and says:

> Archeologists have discovered traces of a Stone Age settlement on the Novosibirsk Islands (New Siberian Islands) . . . They have found bone implements and arrowheads, as well as needles and axes skillfully fashioned from mammoth tusks.
>
> Spitzbergen was once inhabited, too. Proof of this can be seen in the fragments of prehistoric cliff drawings found near the present-day settlement of Ny Alesund. On the rock face are well-preserved incised outlines of whales and deer . . . (391a:54).

The presence of primitive man on these islands is good evidence of warm climate in the Arctic Ocean in the Pleistocene Epoch.

4. EVIDENCE FROM JAPAN

Since a shift of the lithosphere would change the latitudes, and therefore the climate, of many countries, we must consider the evidence from as many of these as possible to see whether it agrees with our assumption. With a pole in Hudson Bay at Lat. 60° N and Long. 83° W, Japan would have been situated about twenty degrees

---

* A finding in conflict with these conclusions has recently been made public by Dr. David L. Clark of the University of Wisconsin (394a). Clark maintains that Arctic ocean bottom cores indicate no warm period in the Arctic for several million years in the past. However, while he attacks the validity of the findings of the Lamont scientists in terms that are rather tentative, he does not consider the other evidence we have discussed above.

farther south. There should, therefore, be some evidence that the climate in Japan was definitely warmer.

We do have a few radiocarbon dates that suggest the temperature during the glacial period was warmer rather than colder in Japan. One sample, of wood from Ishikawa Prefecture (351a:GaK-388;VII, 11;34,000+), carries the comment ". . . flora suggests warm climate." Another, of coral from Okinawa (351a:GaK-810;IX,43; 22,450±650), showed no signs of the glacial lowering of temperature. It was ". . . thought to be late last interglacial or early last glacial stage but the date found falls into the period of the rapid advance of the American ice cap."

Fortunately Japan is in approximately the same latitudes as the United States and therefore a comparison can be made between their conditions in the ice age. The northern Japanese island of Hokkaido is in approximately the latitude of Boston. The northern end of Honshu, the main Japanese island, is in about the latitude of New York. Tokyo is a little south of Washington, D.C. Nagasaki, in southern Japan, is on about the latitude of Atlanta, Ga. Unlike the Atlantic seaboard of the United States, Japan is highly mountainous, and the Japanese Alps compare with the other great mountain ranges of the world. They are not now glaciated, and it is interesting to note that during the ice age they were only lightly glaciated. The key matter is the elevation of the snow line. How far down did the Japanese glaciers come during the last glacial period? The answer is given by Professor Kunio Kobayashi of Shinshu University. He says ". . . the maximal expansion of glaciers in the Japan Alps took place . . . approximately 27,000 C^{14} years B. P. . . . During this phase of glaciation the snow line is supposed to have stood roughly at the level of 2500 m.a.s.l. (meters above sea level)."

From this we learn that the maximum expansion of Japanese glaciers took place at a time roughly corresponding to the main advance of the Wisconsin continental ice sheet in America (see Fig. 21, p. 94) and also that the ice did not descend farther than 2500 meters, or about a mile and a half, above sea level. Contrast the situation in America: There the continental ice cap came down to sea level all along the Atlantic coast as far south as New Jersey. The ice is thought to have stood a mile deep over New York. Surely these facts are consistent with the assumption that Japan at that time was farther south than it is now.

5. EVIDENCE FROM THE COAST OF MEXICO

According to the theory that the ice age was caused by some factor that cooled the whole earth, while the poles stayed put, the Pacific coast of Mexico should reflect the fall in temperature. But strangely enough, the indications are that the ice age made no difference in the temperature of the sea. The following dates tell the story:

TABLE 7
Radiocarbon Dates from the Pacific Coast

LJ-280;IV,231 .... 19,300±400
"Conch (*Strombus granulatus*) shells dredged at depth of 114.5 m(etres) off San Blas, Nayarit (Lat. 22°06.7′ N, Long. 106°17.8′ W) . . . the occurrence of this tropical species at ca. 22° N Lat. during a low period in Pleistocene temperature indicates that the southward displacement of the fauna then was probably not very striking at this latitude . . ."

LJ-21;II,205 .... 37,000+
"Cockle shell (*Trachycardium panamense*) from an ancient and dense inter-tidal-flat shell assemblage . . . at head of Black Warrior Lagoon, Baja California Norte . . . The molluscan species are subtropical, as at present."

LJ-13;II,203 .... 25,000+
"Clam shell (a very large and thick example of *Periglypta Multicostata*) found in place in the consolidated shell deposit of a fragment of an old lagoon barrier-island, on 'South Brushy Island' of Scammons Lagoon, Baja California Sur . . . the molluscan assemblage is subtropical, as at present . . ."

UCLA-736;VIII,491 .... 29,050±1100
Pinus radiata cones from Millerton Head, Marin County, California. Comment, "Date for flora implying relatively cool climate and corresponding to a time prior to the last glacial advance is significant because warm-water invertebrates occur in same beds . . . Whether local or due to world-wide warming, warmer sea off-shore may account for increased precipitation, with greater cloudiness and slightly lower temperature indicated by flora."

A better explanation of the discrepancy between the land and marine species is the following: The pole in Hudson Bay would actually be closer to California than it is now, and the continental ice cap would certainly lower the land temperature of the west coast as expected. However, this position of the pole involved a warm

Siberia and a warm Arctic Ocean, and warm currents might well have flowed southward from the Bering Sea along the west coast of North America. The result would certainly have been heavier rainfall. I have here substituted one speculation for another, but I insist the speculation is reasonable. All these dates unite to say that the sea was warm, and no other theory has so far offered itself to account for it.

6. EVIDENCE FROM AFRICA

There are a number of radiocarbon dates from South Africa, Northern Rhodesia, and Angola that can be understood only in terms of a change in the latitude of Africa. According to the official comment, they "confirm the wide extent of cooler and wetter climatic conditions in South Africa equating with the later part of the Würm/Wisconsin Glaciation." Cool, wet weather in low latitudes is supposed to be contemporary with periods of glacial advance in glaciated areas. However, all of the dates fall within the limits of a short warm period within the ice age called the Denekamp Interstadial (see Fig. 21, p. 94) when ice sheets were presumably retreating.

TABLE 8 *
Radiocarbon Samples from South Africa

UCLA–706;VIII,483	33,150±2500
UCLA–707;VIII,483	33,200+
UCLA–708A;VIII,483	34,000+
UCLA–708B;VIII,483	30,800±1700
UCLA–708C;VIII,483	29,800±1650
UCLA–708D;VIII,484	34,000+

How is this to be explained? It becomes understandable with a pole in Hudson Bay, because in that case South Africa would be about 10° farther south than now; that is, nearer the South Pole, and therefore cooler, despite the Denekamp Interstadial.

7. EVIDENCE FROM TASMANIA

Further support of our assumption is provided by some evidence used by Wegener to support his theory of drifting continents. He

* Reference for these dates, 351a.

quoted the glaciologist Penck as saying that the Pleistocene snow line lay about 1,500 to 1,800 feet lower in Tasmania than in New Zealand, and added, "This is very difficult to understand because of the present nearly equal latitudes of the two localities" (439:111). Wegener, of course, explained the matter by his theory of continental drift. If, however, his theory is rejected, displacement of the lithosphere may provide a solution, for if the Hudson Bay region was then located at the North Pole, as we suppose, Tasmania would have been a good many degrees nearer the South Pole than would New Zealand, as a glance at the globe will make plain.

8. CONFIRMATION FROM ANTARCTICA

Powerful confirmation of another of the corollaries of a pole located in Hudson Bay comes from Antarctica. With a North Pole at 60° N Lat. and 83° W Long., the corresponding South Pole would have been located at 60° S and 97° E in the ocean off the Mac-Robertson Coast of Queen Maud Land, Antarctica. This would place the South Pole about seven times farther away from the head of the Ross Sea in Antarctica than it is now (see Fig. 23, p. 107). We should expect, then, that the Ross Sea would not have been glaciated at that time.

We have confirmation of precisely this fact. During the Byrd Expedition of 1947–1948 Dr. Jack Hough, then of the University of Illinois, took three cores from the bottom of the ocean off the Ross Sea, and these were dated by the ionium method of radioactive dating at the Carnegie Institution in Washington by Dr. W. D. Urry, who had been one of those to develop the method.

The cores showed alternations of types of sediment (see Fig. 24, p. 108). There was a coarse glacial sediment, as was to be expected, and finer sediment of semiglacial type, but there were also layers of fine sediment typical of temperate climates. It was the sort of sediment that is carried down by rivers from ice-free continents. Here was a first surprise, then. Temperate conditions had evidently prevailed in Antarctica in the not distant past. The sediment indicated that no fewer than three times during the Pleistocene Epoch a temperate climate had prevailed in the Ross Sea.

Then, when this material was dated by Urry, it was revealed that the most recent temperate period had been very recent indeed. In fact it had ended only about 6,000 years ago. Hough wrote:

The log of core N–5 shows glacial marine sediment from the present to 6,000 years ago. From 6,000 to 15,000 years ago the sediment is fine-grained with the exception of one granule at about 12,000 years ago. This suggests an absence of ice from the area during that period, except perhaps for a stray iceberg 12,000 years ago. Glacial marine sediment occurs from 15,000 to 29,500 years ago; then there is a zone of fine-grained sediment from 30,000 to 40,000 years ago, again suggesting an absence of ice from the sea. From 40,000 to 133,500 years ago there is glacial marine material, divided into two zones of coarse- and two zones of medium-grained texture.

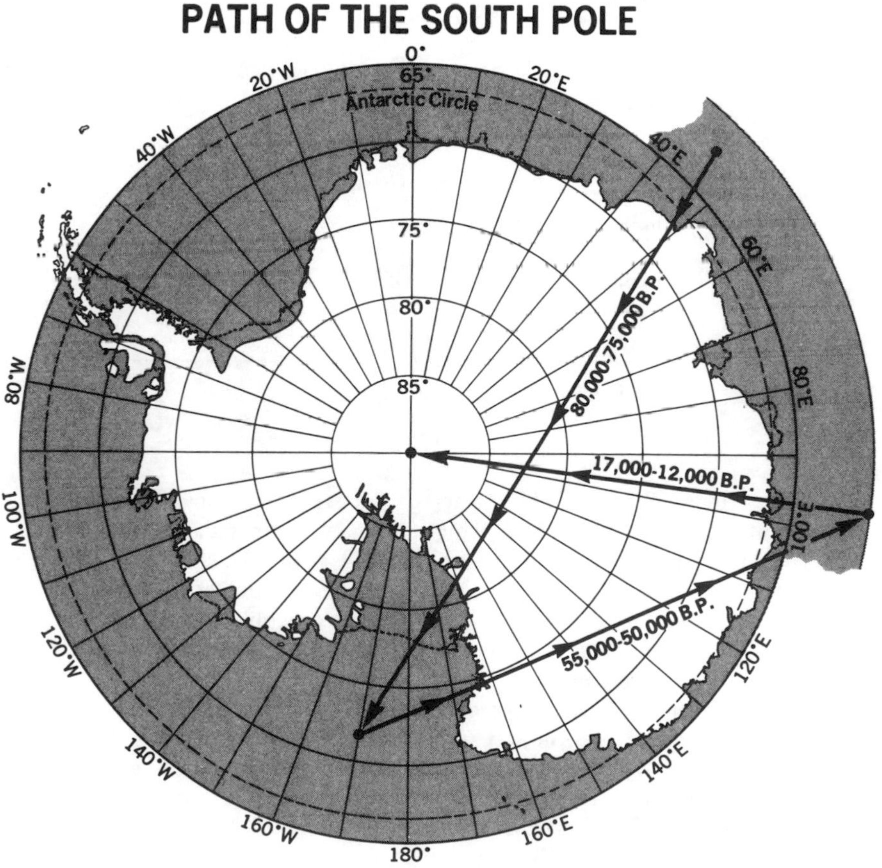

PATH OF THE SOUTH POLE

Fig. 23. Map of the Antarctic, with three former positions of the South Pole.

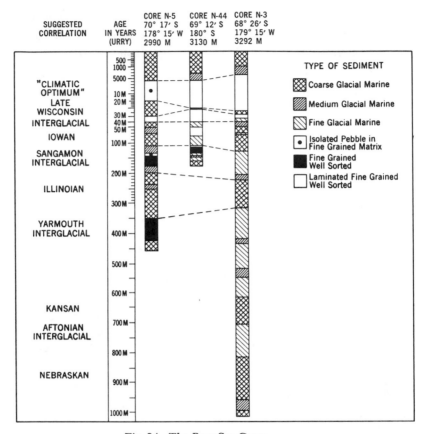

Fig. 24. The Ross Sea Cores.

The period 133,000–173,000 years ago is represented by fine-grained sediment, approximately half of which is finely laminated. Isolated pebbles occur at 140,000, 147,000 and 156,000 years. This zone is interpreted as recording a time during which the sea at this station was ice-free, except for a few stray bergs, when the three pebbles were deposited. The laminated sediment may represent seasonal outwash from glacial ice on the Antarctic continent.

Glacial marine sediment is present from 173,000 to 350,000 years ago, with some variation in the texture. Laminated fine-grained sediment from 350,000 to 420,000 years ago may again represent rhythmic deposition of outwash from Antarctica in an ice free sea. The bottom part of the core contains glacial marine sediment dated from 420,000 to 460,000 years by extrapolation of the time scale from the younger part of the core (225:257–59).

A comparison of the three cores shows that the end of the temperate period and the beginning of the most recent glacial period is well defined; that is, the cores are in close agreement. In Core N-5 and in Core N-4 the change takes place 6,000 years ago. In Core N-3 it is shown 1000 years later.

The next previous time when the three cores are in such close agreement is 40,000 years ago, when the first two cores show change from glacial to temperate sediment. Core N-3 is again a little later, showing the change about 2,000 years afterward.

Between the dates 40,000 and 6,000 years ago the sediment shows temperate climate in the Ross Sea except for a period of 10,000 years of glacial deposition in Core N-5, a very short period of glacial deposition in N-4, and a somewhat longer glacial interval in Core N-3. Here, as we see, the cores are in disagreement, and it is possible that the differences are due to some disturbances of the bottom sediments. At any rate, it will be agreed that the evidence of the cores is more impressive when they agree than when they disagree.

All three cores were taken in deep water at depths of between 6,000 and 12,000 feet. Core N-3, however, was taken farther off shore, about twice as far from the edge of the continental shelf. Whether this has anything to do with the slight time lag shown in it I do not know. There seems at first nothing to show why Core N-5 should have a longer period of glacial sediment interrupting the temperate phase than do the other cores. The cores are close enough together so that the same climatic conditions should have affected all three at the same time.

One geologist has suggested that perhaps the sediments in these cores do not represent regular deposition at all but merely indicate the action of "turbidity currents." Such currents are set in motion in the sea by slumping of sediments from the steep slope where the continental shelf ends. The sediments accumulate until they become unbalanced or an earthquake or other disturbance starts a landslide (in this case, of course, a seaslide) very much like an avalanche in the high mountains. Then the sediments from the shelf plunge down into the depths and are deposited on the deep ocean bottom. Turbidity currents appear to have been very active in arranging and rearranging the sediments on many areas of the ocean floors of the world.

However, if a turbidity current has influenced the deposition in the core the fact is shown by the arrangement or displacement of the materials. It has therefore been possible to check the question. I asked the specialists of the United States Geological Survey and also Doctor Hough himself for their judgments on the cores. Their answer was that they saw no evidence of the action of turbidity currents.

A suggestion has been made that the warm period in the Ross Sea may represent the "hipsithermal" period, when temperatures all over the world were a little warmer than now. This suggestion is not satisfactory, because the hipsithermal phase began about 8,000 years ago and ended about 4,000 years ago. It lasted only about four thousand years, while the warm period indicated in the cores lasted (though with interruptions) for about 34,000 years and ended when the hipsithermal was beginning.

A question has been raised about the reliability of the dating method used by Urry. However, the Soviet scientists have used the method with success, and Ericson has found that the results obtained by the use of this method in deep-sea cores agree very well with comparable radiocarbon datings. We have observed the same thing in the cores already discussed. I had the opportunity to discuss the ionium method with Albert Einstein before his death, and he said he had studied the method and thought it reliable.

A possible explanation of the glacial deposition interrupting the long warm period in Core N-5 is that the site of this core was nearest to the east coast of the Ross Sea, which is at that point about 200 miles distant. The massive Admiralty Range, with peaks up to two miles high, parallels this coast, and even at a time when the Ross Sea area might have been generally unglaciated, it is possible that the mountains would have had large glaciers descending to the sea and throwing off icebergs. The area would have been about 2000 miles from the hypothetical South Pole. The mountains, then, would have been approximately in the latitude of 60° S, similar to the north latitude of Hudson Bay. A mountain range two miles high at that latitude would probably be glaciated, and even if their glaciers did not reach the sea, rivers during spring thaws might have carried much ice.

Accepting the cores at face value, we must realize that the date found by Urry for the beginning of the present glacial conditions in the Ross Sea is not the date of the beginning of the last change of climate in Antarctica. Rather, it was the end of the change, the establishment of present conditions. If our hypothesis is correct, the change took the same length of time as the whole period of the melting of the Wisconsin ice sheet in the northern hemisphere (see Chapter VI).

9. THE EVIDENCE OF AN ANCIENT MAP

It is rare that geological investigations receive important confirmation from archeology; yet, in this case, it seems that this matter of the deglaciation of the Ross Sea can be confirmed by an old map that has somehow survived for many thousands of years. A group of ancient maps, including this one, was the subject of an entire book which I published in 1966 under the title *Maps of the Ancient Sea Kings* (199a).

In some way or other which is still not, and may never be, entirely clear, this extraordinary deglacial map of Antarctica has come down to us. Apparently originated by some ancient people unknown, preserved perhaps by Minoans, Phoenicians, and Greeks, it was discovered and published in 1531 by the French geographer Oronce Finé, and is part of his Map of the World (see Fig. 25, p. 112).* "Impossible!" That would be the opinion of practical people of intelligence and learning. "Utterly impossible!" But sometimes truth is really strange. It has been possible to establish the authenticity of this ancient map. In several years of research, the projection of this ancient map was worked out. It was found to have been drawn on a sophisticated map projection, with the use of spherical trigonometry, and to be so scientific that over fifty locations on the Antarctic continent have been found to be located on it with an accuracy that was not attained by modern cartographic science until the nineteenth century. And of course, when this map was first published, in 1531, nothing at all was known of Antarctica. The continent was not discovered in modern times until about 1818 and was not fully mapped until after 1920. With the reader's permission I reproduce here a table from the above-mentioned work. For further details the reader should refer to that book.

---

* The original engraving is now in the Library of Congress.

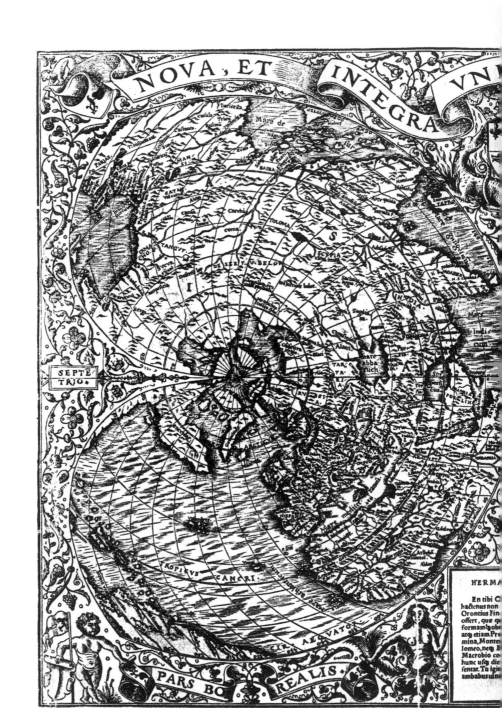

Fig. 25. The Oronteus Finaeus map of Antarctica.

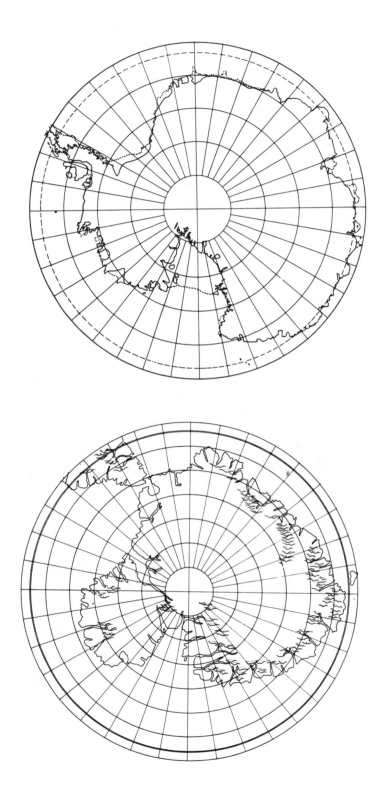

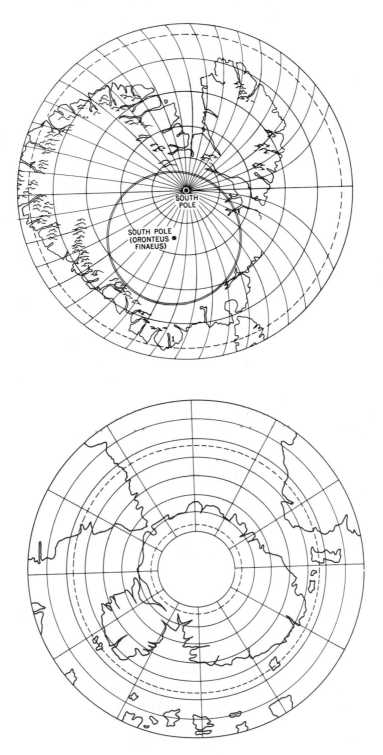

Fig. 26. Four maps of Antarctica compared.

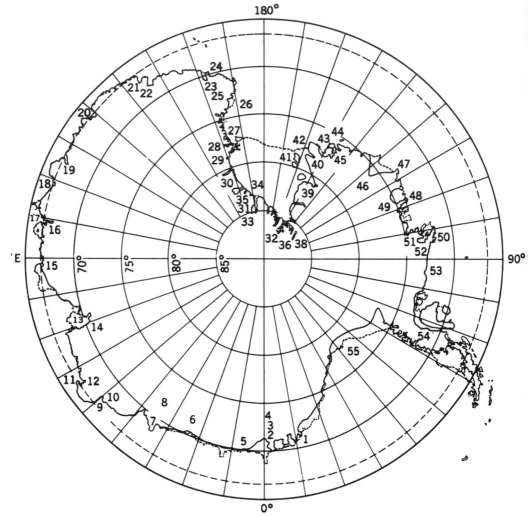

Fig. 27. Modern map of Antarctica with numbers corresponding to the table.

The table should be compared with the map in Figure 28 (p. 121), which is the Oronteus Finaeus Map with a modern grid worked out by my students and myself at Keene State College in about seven years of research. The serious student should obtain a large modern map of Antarctica, either that produced by the National Geographic Society or the more elaborate and up-to-date map produced by the American Geographic Society. With the large map he should follow the coast, comparing it with the Oronteus Finaeus

Map and this table. He will conclude, I am sure, that the agreement of the ancient and modern maps is entirely beyond the probabilities of coincidence.

In evaluating the errors on the map, the reader should allow about 60 miles per degree of latitude, but he should remember that the degree of longitude is very short in the high latitudes, diminishing to zero at the poles. As a result, longitudes on this map average out as accurately as latitudes.

THE ORONTEUS FINAEUS WORLD MAP OF 1531

Antarctica

Geographical Localities	True Position	Oronteus Finaeus Position	Errors	
(a) QUEEN MAUD LAND				
1. Cape Norvegia	71.5 S	66.5 S		5.0 N
	12.0 W	6.0 W		6.0 E
2. Regula Range	72–73 S	68–69 S		4.0 N
	2–5 W	0–3 E	c.	5.0 E
3. Penck Trough	71–74 S	66.5–69 S	c.	3.0 N
	3.0 W	4.0 E		7.0 E
4. Neumeyer Escarpment	73.0 E	68–69 S		4.0 N
	0–4 W	0.5 E	c.	4.0 E
5. Muhlig-Hofmann and	71–73 S	70–71 S		0.0
Wohlthat Mountains	0–15 E	10–15 E		0.0
6. Sor Rondanne and	72.0 S	72–73 S		0.0
Belgica Mts.	22–33 E	20–30 E		0.0
7. Prince Harold Coast,	69–70 S	70.0 S		0.0
Lützow-Holm Bay,	35–40 E	35–37 E		0.0
Shirase Glacier				
8. Queen Fabiola	76–77 S	73.0 S	c.	3.0 N
Mountains	35–36 E	30–40 E		0.0
(b) ENDERBY LAND				
9. Casey Bay (Lena Bay)	67.0 S	70.0 S		3.0 N
or Amundsen Bay	48–50 E	48.0 E		0.0
10. Nye Mountains, Sander-	73.0 S	72–73 S		0.0
cook Nunataks	49.0 E	50.0 E		1.0 E
11. Edward VIII Bay	66–67 S	69–70 S		3.0 S
(Kemp Coast)	58–60 E	55.0 E		4.0 W

Geographical Localities	True Position	Oronteus Finaeus Position	Errors	
12. Schwartz Range, Rayner Peak, Dismal Mountains, Leckie Range, Knuckley Peaks, etc.	71–74 S 54–57 E	71–73 S 60–75 E		0.0 12.0 E
13. Amery Ice Shelf, MacKenzie-Prydz Bays	73–78 S 70–75 E	67.0 S 73.0 E	c.	8.5 N 0.0
14. Prince Charles Mountains and adjacent peaks	72–74 S 60–69 E	68–70 S 70–75 E	c. c.	4.0 N 5.0 E

(c) WILKES LAND

15. Philippi Glacier, Posadowsky Bay	67.0 S 88–89 E	66.0 S 77.0 E		1.0 N 11.5 W
16. Denman-Scott Glaciers (Shackleton Ice Shelf)	66–67 S 99–101 E	66.0 S 85.0 E		0.0 15.0 W
18. Vincennes Bay	66–67 S 109 E	65.0 S 105.0 E		1.5 N 4.0 W
19. Totten Glacier	67.0 S 115–117 E	66.0 S 112.0 E		1.0 N 4.0 W
20. Porpoise Bay	67.0 S 128–130 E	67.0 S 122.0 E		0.0 7.0 W
21. Merz Glacier	68.0 S 144–145 E	70.0 S 140.0 E		2.0 S 4.5 W
22. McLean, Carroll Nunataks, Aurora Peak, Medigan Nunatak, etc.	67–68 S 143–145 E	70–71 S 132–134 E		3.0 S 11.0 W
23. Pennell Glacier, Lauritzen Bay	69.0 S 157–158 E	70.0 S 150.0 E		1.0 S 7.5 W

(d) VICTORIA LAND and THE ROSS SEA

24. Rennick Bay	70.5 S 162.0 E	72.5 S 155.0 E		2.0 S 7.0 W
25. Arctic Institute Range	70–73 S 161–162 E	74–75 S 150–155 E		3.5 S 7.0 W
26. Newnes Iceshelf and Glacier	73.5 S 167.0 E	74.5 S 170.0 E		1.0 S 3.0 E

Geographical Localities	True Position	Oronteus Finaeus Position	Errors
27. Ross I. (Mount Erebus)	77.5 S 168.0 E	76.5 S 172.0 E	1.0 S 4.0 E
28. Ferrer Taylor Glacier	77–78 S 159–163 E	77.0 S 160–170 E	0.0 0.0
29. Boomerang Range and adjacent peaks (Escalade Peak, Portal Mt., Mt. Harmsworth, etc.)	77–78 S 159–163 E	77–79 S 142–152 E	0.0 14.0 E
30. Mountain group: Mt. Christmas, Mt. Nares, Mt. Albert Markham, Pyramid Mountains, Mt. Wharton, Mt. Field, Mt. Hamilton	80–82 S 157–160 E	79–80 S 140–150 E	1.5 N c. 13.5 W
31. Queen Alexandra Range	84–85 S 160–165 E	84–87 S 145–155 E	0.0 c. 10.0 W
32. Queen Maud Range	87.0 S 140 W–180 E/W	81–82 S 140–160 E	5.5 N c. 20–40 E
33. Nimrod Glacier	82.5 S 157–163 E	81.0 S 160–175 E	1.5 S c. 0.0
34. Beardmore Glacier	84–85 S 170.0 E	82–84 S 170 E–170 W	0.0 c. 0.0
36. Leverett Glacier	85.5 S 150.0 W	81–82 S 140–160 W	4.0 N c. 0.0
37. Supporting Party Mountain, or Mt. Gould	85.3 S 150.0 W	80.0 S 160.0 W	5.3 N 10.0 W
38. Thiel and Horlich Mts.	86.0 S 90–130 W	83–85 S 130–150 W	3.0 N c. 30.0 W
39. Bay, unnamed, west coast of Ross Sea	80–81 S 150.0 W	77.5 S 150–160 W	3.0 N c. 0.0
40. Prestrude Inlet, Kiel Glacier	78.0 S 157–159 W	74.0 S 160.0 W	4.0 N 2.0 W
42. Edward VII Peninsula	77–78 S 155–158 W	72–73 S 158–165 W	5.0 N c. 4.0 W

Geographical Localities	True Position	Oronteus Finaeus Position	Errors
43. Sulzberger Bay	77.0 S	73.5 S	3.5 N
	146–154 W	150–155 W c.	0.0
44. Land now submerged?			

(e) *MARIE BYRD LAND*

45. Edsel Ford Range	76–78 S	73–74 S	3.5 N
	142.0 W	135–143 W	0.0
46. Executive Committee Range	77.0 S	73.0 S	4.0 N
	125–130 W	130–135 W c.	5.0 W
47. Cape Dart, Wrigley Gulf, and Getz Ice Shelf	75.0 S	70.5 S	4.5 N
	130.0 W	130.0 W	0.0
48. Cape Herlacher, Martin Peninsula	74.0 S	72.0 S	2.0 N
	114.0 W	108.0 W	6.0 E
49. Kohler Range and Crary Mountains	76–77 S	74–75 S	2.5 N
	111–118 W	108–115 W	0.0
50. Canisteo Peninsula	74.0 S	74.0 S	0.0
	102.0 W	98.0 W	4.0 E
51. Hudson Mountains	74–75 S	72–74 S	1.0 N
	99.0 W	105–110 W c.	8.0 W

(f) *ELLSWORTH LAND*

52. Jones Mountains	73.5 S	74.0 S	0.5 S
	94.0 W	100.0 W	6.0 W
53. Inlet in Ellsworth Land, indicated to exist under the present ice cap	73–80 ?S 80–95 W	72–74 S 85–95 W	0.0 0.0

54. Base of the Antarctic (Palmer) Peninsula?

55. The Weddell Sea, shown almost connected with the Ross Sea?

56. Duplicated coastline of Ellsworth Land?

57. Duplicated base of Antarctic Peninsula?

58. Berkner Island in the Weddell Sea shown extending over the continental shelf to the north?

In final comment on this extraordinary evidence, I will say that though this map is proved to have existed as far back as 1531, no

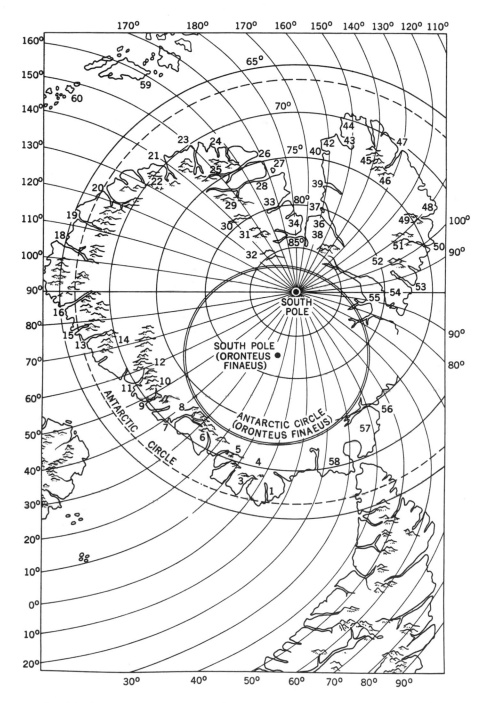

Fig. 28. Oronteus Finaeus map of Antarctica; tracing with projection and numbers corresponding to the table.

map of this accuracy could have been drawn in modern times until the invention of the chronometer in the reign of George III about the year 1780. This instrument first made possible the accurate determination of longitude.

If this were not all so completely proved it would, of course, sound like a fairy story. Statistics, however, do not lie. The mathematical probabilities of anyone's accidentally getting so many points right on a map are fewer than one in a hundred million. This is good evidence for the displacement of the lithosphere.*

10. CONFIRMATION FROM SOUTH AMERICA

Support for the assumption of the pole in Hudson Bay comes also from South America. That assumption involves the southward movement of both North and South America at the end of the ice age. The southward movement of southern South America would result in its climate becoming not warmer but colder at the end of the northern ice age. Evidence of this has unexpectedly been furnished by Dr. Calvin Heusser, of the American Geographical Society, in the course of presenting a paper to an international symposium on world climate (210b). During the discussion after the presentation of the paper, Heusser remarked that the first known Argentine glacial advance had been dated about 350 B.C. (210b:141). He further stated, however, that there were indications of cooling of the climate in Argentina between 6,500 and 4,500 years ago, just when the warmer hipsithermal climatic phase was setting in in Europe and other parts of the world (see Fig. 21, page 94). He thought that this cooling might have caused glacial expansion at that time in Argentina.

The contrast between this story and the events in Europe and America will be apparent to the reader. Let us recapitulate: We have two areas lying at similar distances from their respective poles. In one, the northern, we have many evidences of heavy glaciation, extending over a period of perhaps 40,000 years, but ending about 14,000 years ago, to give way to the present climate about 10,000 years ago.

---

* It also has shattering implications for ancient history, but those are not my concern here. They are dealt with in my *Maps of the Ancient Sea Kings*.

In Chile and Argentina, on the other hand, in the same relative latitude just as close, presumably, to a pole, we have *no glaciation* until after the climate has become normal for the present temperate zone in the north. It appears that in Argentina a cool period set in just as the hipsithermal phase with higher temperatures set in all over the northern hemisphere! Clearly, then, there was no similarity in climatic trends, but rather the opposite. Argentina and Chile grew colder when they should have grown warmer. This supports the assumption of a crustal shift moving the western hemisphere southward toward the Antarctic Circle.*

To sum up, the evidence for the proposition that the North Pole really was located in Hudson Bay during the Wisconsin glaciation is very strong indeed. A great deal of additional evidence will be found scattered through other chapters in this book, especially in Chapter X, which goes into the question of the extinction of the mammoths.

11. THE RELATIONSHIP OF THE WISCONSIN ICE CAP WITH THE GLACIATIONS IN EUROPE AND IN ALASKA

Our assumption of a pole in Hudson Bay confronts us with the problem of explaining why, with that location of the pole, there was a glaciation in Europe, the northern part of which would have been farther from the pole than it is now, and the same problem arises with Alaska. Why, during the ice age, did Alaska have many mountain glaciers but no continuous ice sheet?

---

* Additional confirmation is indicated in this note, received from Oppé and Delair: "Dr. H. von Ihering concluded from a study of a collection of marine molluscs obtained from the so-called 'Pampas Formation' near the La Plata estuary, that they were of recent (that is, present) species, though adapted to a climate warmer than that now prevailing on the coast of Argentina. He recognized almost all of them as still living on the southern Brazilian shores. Sir Arthur Woodward, who discussed these shells, observed that of the species represented sixteen still live on the Argentine coast, one (*Purpura haemastoma*) does not survive to the south of the Rio Grande de Sud, and that two more (*Littorina flava* and *Nassa polygona*) live only in Santa Catalina, San Pablo and more to the north." ("Notes on the Geology and Paleontology of Argentina," *Geological Magazine*, Vol. IV, p. 19, footnote 2.)

"The past and present distribution of these shells, as well as their association in the same deposits as the reptilian remains, numerous fragments of a sub-tropical flora, and warm-temperate species of fresh-water molluscs suggested to Woodward that the ocean waters which appear to have covered the Pampas plains in late Pleistocene time were of somewhat higher temperature than that of the ocean in the same latitude today." (*Ibid.*)

The explanation of the glaciation of northwestern Europe is, I think, as follows. First, the heaviest glaciation of Europe was not contemporary with the Wisconsin ice sheet but was the consequence of an earlier polar position, which will be discussed farther on. Secondly, the comparatively thin European ice sheet of Wisconsin time (which in Britain really consisted only of discontinuous mountain glaciers) was made possible by a very special combination of meteorological conditions. In North America a vast ice cap extended eastward from its center near Hudson Bay. Much of the continental shelf in this whole area was then above sea level, and this was covered by ice. Then anticyclonic winds, blowing outward in all directions from the ice cap, had only to cross the narrow North Atlantic, raising moisture from the sea and depositing it upon Scandinavia and Britain.*

So far as the glaciation of Alaska is concerned, again, the climate there was colder than it is now because of the vast refrigerating effect of the ice cap that covered 4,000,000 square miles of the continent. Just as at present the Antarctic ice cap makes the South Polar region colder than the Arctic (because it is a perfect reflector of the sun's radiant energy back into space), so then the great Wisconsin ice cap meant that the prevailing temperatures at the center of the ice sheet (presumably the pole) were much lower than the temperatures prevailing at the present North Pole, where no great ice cap exists. But although the intensely cold anticyclonic winds blowing off the Wisconsin ice cap made Alaska colder than it is now and thereby produced larger glaciers than exist at present, still these winds blew only over continuous land, not over the sea, and so they could not pick up the moisture required to produce a continuous ice sheet. This explains why Alaska warmed up at the end of the North American ice age even though it actually may have moved closer to the pole.

A displacement of the earth's surface layers would necessarily have implied a great deal of disturbance on the earth's surface in addition to climatic change. The turbulence accompanying the polar shift is the subject of the next chapter.

* At present in Antarctica winds blow toward the pole at high elevations. As they cool they sink to the surface of the ice cap and then blow outward in all directions, greatly affecting the wind patterns and climate of the southern hemisphere. A similar wind pattern must have existed in North America during the Wisconsin glaciation.

chapter **5**

THE VIOLENT LIFE
OF THE LAST
GREAT ICE SHEET

1. THE ACCELERATION OF EROSION
AND ITS CAUSE

IT goes without saying that the geological effects of a polar shift would be many and varied. Some of them will be hard to imagine, much less reconstruct. Others may be suggested in a general way, and for still others definite evidence may be presented.

There is good evidence that some geological processes were accelerated during the ice age. There was greater precipitation of rain or snow; there were more violent or rapid processes of erosion and deposition of sediment. For such a deviation from natural conditions it is important to pinpoint a cause if we can. First, however, it is important to present the evidence. It is all the more important to do this since many geologists have tended to ignore it, and those who have observed the evidence have been unable to account for it.

David B. Ericson and other oceanographers have repeatedly pointed to evidence of this accelerated rate of sedimentation in the ice age. On one occasion Ericson wrote:

Decrease of turbidity-current activity with amelioration of climate and rise of sea level toward the close of the Wisconsin glacial stage is indicated by the decrease in the number and thickness of graded layers in the upper parts of most [marine sedimentary] cores. It is inferred that the Pleistocene Epoch and the glacial stages in particular were times of exceptionally rapid sediment accumulation (141).

In another paper Ericson wrote,

Radiocarbon age determinations show that the rate of sediment accumulation in the equatorial region varies from 2.2 cm per thousand years to 4.3 cm per thousand years. It is at least suggestive that the rapid rate of accumulation took place during the latter part of Wisconsin time. A similar relationship, though in less degree, is found in one core (A-179-4) from the Caribbean . . . (142:124).

Two radiocarbon dates of samples from cores taken in the Chuckchi Sea, Alaska (ML-160;351a;VI,213,3960±110 and ML-159;351a;VI, 213,13,700±150) suggested the same thing:

. . . This series permits estimate of the rate of sediment accumulation in this . . . area. If derived ages can be considered valid, then data show a remarkable decrease of sediment accumulation during the past 10,000-12,000 years.

From France we have the following statement relating to a series of radiocarbon dates of the glacial period, made on samples from a core taken in the Mediterranean:

. . . Ages found for sections 43 to 51 cm and 51 to 59 cm are too old because they contain much reworked carbonate, which probably indicates that after the end of the last glacial period a very important erosion period occurred in SE France . . . (351a:232).

From Australia we have evidence that the turbulence of climate was worldwide. A radiocarbon date from the Goulburn River, Victoria, (ANU-29;351a;X,181,13,500±700) produced the comment:

Channel geometry of streams which deposited silts . . . indicates higher discharges than in present hydrological regime. . . .

From California we have the suggestion of much heavier rainfall than at present:

Sample UCLA-728;351a;VIII,492,38,000±2500, indicates ". . . Flora represents a pinyon-juniper woodland like that now in San Rafael Mts., a few tens of miles to SW, and indicates 10 to 15 in. more precipitation than (at site) today . . ."

One writer, Leland Horberg (222:281), assembled much evidence detailing the same thing by comparing the rates of retreat of the ice sheets as shown by radiocarbon dates with present known rates of ice movement. In the following table radiocarbon dates of the glacial phases are compared with estimates of their ages based on "normal" rates of retreat of glaciers:

TABLE 9
Rates of Ice Movement

	C 14 Dates	Normal Retreat
Postglacial	8,000	8,000
Mankato Advance	11,000	19,400
Two Creeks Interstadial	12,000	20,000
Cary Advance	14,000	32,250
Brady Interstadial	17,000	35,250
Tazewell Advance	20,000	?
Iowan-Tazewell Interstadial	?	49,500
Iowan Advance	22,000	63,350
Farmdale-Iowan Interstadial	25,000	69,550
Farmdale Advance	?	?

Horberg believed that the radiocarbon dating method must be wrong, because it led to impossibly rapid processes of erosion and sedimentation in the ice age. The radiocarbon method, however, has been very well tested and is now considered highly accurate, although mistakes are made now and then with this method as with all others.

Horberg has collected and studied the radiocarbon dates bearing on the history of the Wisconsin ice cap. He summarizes its short and violent history as follows (222:281):

a. The first appearance of the ice sheet in Ohio is dated about 25,000 years ago, although it had probably been advancing from its center in the Hudson Bay area for 20,000 or 25,000 years before that. This advance of the ice sheet into Ohio is called the Farmdale Advance (see Fig. 21, pg. 94). It was formerly thought to have occurred as much as 100,000 or even 150,000 years ago.* This date, then, cuts the time for the later history of the ice sheet by about three quarters. Six different radiocarbon dates, all of the Farmdale Advance, show that the expansion continued until at least 22,900 years ago, or for about 3,000 years. Then there was an unexplained interval of warm climate, called the "Farmdale-Iowan Interstadial." This warm period lasted about 1,500 years, during which the ice withdrew a certain distance.

b. Following the recession, a new advance of the ice cap occurred. This is referred to as the "Iowan Advance." It began about 21,400 years ago, lasted about 700 years, and was interrupted by a new recession about 20,700 years ago.

c. This second recession, after less than a thousand years, was succeeded by an extremely massive advance during the period from 19,980 to 18,050 years ago. These dates must not be taken as exact; there is always a small margin of error. This new expansion, called the "Tazewell Advance," apparently carried the Wisconsin ice cap to its maximum extension. †

d. The Tazewell Advance was interrupted by a prolonged period of warmth and recession called the "Brady interval" or "Brady Interstadial." This lasted about three thousand years. It began before 16,720 years ago and ended sometime after 14,042 years ago. The ice retreated a long way.

e. A fourth advance of the ice sheet beginning about 13,600 years ago and continuing to about 12,120 years ago (called the "Cary Advance"), was followed by the "Two Creeks Interstadial," an interval of warmth and recession about 11,404 years ago.

f. A fifth advance of the ice, referred to as the "Mankato Advance," appears to have taken place between 10,856 and 8,200 years ago. The high point of this advance is called the "Mankato Maximum." Another writer, Emiliani, finds that a sixth expansion of the ice sheet, the "Cochrane Advance," took place less than 7,000 years ago (132).

* Since this was written, earlier advances have become known, as we shall see below, but this does not reduce the acceleration of those phases discussed by Horberg, compared with the previous estimates.

† This date has since been revised. (See Fig. 21, p. 94).

g. There was a sudden, virtually complete disappearance of the ice sheet (which had, however, according to Flint, been getting thinner ever since the Tazewell Advance) (375:177).

h. Shortly after the disappearance of the last vestiges of the ice, there was a period of warm climate, which was actually warmer than the climate today. It is called the "hipsithermal" (or sometimes "climatic optimum"). It lasted from 8,000 years ago to about 4,000 years ago. According to Brooks (52:296), the temperature then averaged about 5° Centigrade warmer than now. There is at present no accepted explanation of this warm period.

Compared with the usual geological time concepts, even the period of 10,000 years for the decline of the ice sheet from the end of the Tazewell Advance is incredibly rapid. Horberg, as I have mentioned, has pointed out that if the radiocarbon method is valid, the rate at which the ice must have advanced and retreated indicates that geological processes (that is, meteorological processes like rainfall) must have been greatly accelerated during the ice age. Now it is easy to show that these processes inevitably would have been accelerated by a movement of the crust; we shall return to this matter below.

Another line of evidence suggesting an acceleration of the rate of geological change is presented by Emiliani, who has applied a technique of determining ancient temperatures of seawater developed by Harold C. Urey. Urey's method is based on the use of an isotope of oxygen. Emiliani has noted many important temperature changes in a comparatively short period during the latter part of the Pleistocene; he has reached the conclusion that the four known Pleistocene ice ages all occurred in the last 300,000 years. He agrees essentially with Horberg as to the date of the beginning of the Wisconsin glaciation (132).

Assuming the radiocarbon dates to be correct, then, we find that at the end of the Tazewell Advance there was a recession and that despite the readvances the ice gradually thinned until the ice sheet disappeared. This can be accounted for by the assumption that the crust was in motion and that it continued to move slowly during all or most of the 10,000 years during which the ice cap was in intermittent decline. As I have already pointed out, there is no other reasonable explanation for the disappearance of the ice sheet. But the assumption is strengthened by a most remarkable fact. It would have to be considered probable, as following naturally from the theory, that as the crust moved there would be a period, possibly prolonged, when the melting on the equatorward side of the ice cap would be balanced and even more than balanced by further build-up of the ice cap on the poleward side. Thus, as the Wisconsin ice cap moved southward, build-up of the ice would continue on its northern side. The result would be that the ice center, the center of maximum thickness, from which the ice sheet would move out by gravity in all directions, would be displaced to the north. And this is exactly what happened. Coleman writes:

Two important facts have been established by Low, who worked over the central parts of the Labrador sheet; first, that the center of the glaciation shifted its position, at one time being in Lat. 51 or 52, later in Lat. 54, and finally in Lat. 55 or 56. Instead of beginning in the north and growing southward it reversed this direction; second, that the central area shows few signs of glaciation, so that the pre-glacial debris due to ages of weathering are almost undisturbed. A broad circle around it is scoured clean to the solid rock . . . (87:117).

This is good evidence that the lithosphere was in motion. In addition, it provides a suggestion that the initial phase, the beginning of the Wisconsin glaciation, was rather sudden. A vast unglaciated area was covered all at once with snow which did not melt.

2. THE CAUSE OF OSCILLATIONS OF THE ICE SHEET AND THE CAUSE OF THE CLIMATIC OPTIMUM: VOLCANISM

But what about the alternating phases of retreat and readvance of the ice sheet? The retreats can be explained, of course, by the assumption that the ice cap was moving slowly into lower latitudes with the displacement of the crust. But how are the readvances to be explained? Up to the present there has been no explanation for these.

I think we can assume that a corollary of any crust displacement would be an increase of volcanic activity because of the resulting strains within the earth. There have been times in the past when the quantity of volcanic action has been extraordinary (231:629). As an example of this, there appears to be evidence that in a small area of only 300 square miles in Scandinavia during Tertiary times there may have been as many as 70 active volcanoes at about the same time. Bergquist, who cites the evidence, remarks, "Volcanic activity on this scale, erupting through about 70 channels, and concentrated in a relatively short period, must have been very impressive" (31:194).

Because the earth is a flattened sphere, it is highly probable that the movement of its outer shell or shells over its interior would produce intense strains of various kinds. In any such movement some parts of the surface would be moved toward the equator and others toward the poles. Any sector moved toward the equator would have to be stretched to pass over the equatorial bulge; any sectors moved toward the poles would have to undergo compression because of the reduction of the surface. Stretching of the crust, as I have already suggested (Chapter I), could create fissures, great fractures and rifts, while the compression of the crust in areas moved poleward could create great numbers of volcanic eruptions. Unusual volcanism, then, would be expected.

A special phase of this volcanism must now attract our attention. Most volcanoes produce dust, sometimes in vast quantities (87:271), that is rapidly distributed through the atmosphere. The effects of volcanic dust on the climate have been the subject of intensive studies. We must stop for a moment to summarize the results of these studies.

The fundamental work on the relationship of volcanic dust to climate is *The Physics of the Air*, by Humphreys (231). Humphreys shows that volcanic dust can have a remarkable effect in lowering temperature. He points out that the effect of the particles depends upon whether they happen to be more efficient in intercepting the sun's light and reflecting it back into space than they are in preventing the radiation of the earth's heat into outer space. What is important is the size and shape of the dust particles as compared with the wave lengths of the radiation. Particles of a given length will have great reflecting and scattering effect on sunlight, and none on the radiation of heat from the earth (which, of course, is not in the form of light). Humphreys concludes that it is necessary to determine the approximate average size of the individual grains of floating volcanic dust as well as the wave lengths of the radiation involved. He accomplishes this satisfactorily. After mathematical treatment of the various factors he concludes: ". . . the shell of volcanic dust, the particles all being of the size given, is some thirtyfold more effective in shutting out solar radiation than it is in keeping terrestrial radiation in . . . " (231:580). He also points out:

> . . . The total quantity of dust sufficient . . . to cut down the intensity of solar radiation by 20% . . . is astonishingly small—only the 174th part of a cubic kilometer, or the 727th part of a cubic mile . . . (231: 583).

This, of course, means that the sun's radiation is reduced by 20% over the whole surface of the earth. It requires only a few days for volcanic dust projected into the upper atmosphere to be distributed around the world. The amount of dust produced by the eruption of Mt. Katmai in Alaska in 1912 was sufficient to cause a slight lowering of the temperature of the whole earth's surface for a period of two or three years (87:270;231:569). For long-range effects a continual series of explosions would be necessary, because volcanic dust settles out of the atmosphere in periods of about three years. Humphreys presents a great deal of evidence correlating variations in average annual global temperatures, through the nineteenth century, with specific volcanic eruptions. He establishes the fact that the eruptions certainly had an important influence on temperature.

If this is true of our times, what results should we expect from the activation of very great numbers of volcanoes during a displacement of the crust? Not only would the temperature fall, and perhaps very drastically, but continuing volcanic outbursts would keep it low. At the same time, the alternation of periods of massive outbursts with periods of quiet would produce violent fluctuations of the climate between extremes of cold and warmth.

Here we have our explanation of the five or six major readvances of the Wisconsin ice sheet (there were, apparently, many more minor ones). They may well have resulted from the long continuation of massive outbursts of volcanism which are explained by the displacement of the crust.

It is not necessary, however, for us merely to assume without evidence that there must have been unusual volcanic activity at the end of the ice age. On the contrary, there is a rather remarkable amount of evidence of excessive volcanism during the decline of the Wisconsin ice cap. It comes from many parts of the earth. For North America it is particularly rich. From radiocarbon dating we have learned that during the last part of the ice age there were active volcanoes in our northwestern states. One of the greatest eruptions was that of Mt. Newberry in southern Oregon less than 9,000 years ago (242:23). Other late glacial or early postglacial volcanic activity in Oregon was reported by Hansen (199). Farther south the story is the same:

In Arizona, New Mexico, and southern California there are very fresh looking volcanic formations. The lava flow in the valley of the San Jose River in New Mexico is so fresh that it lends support to Indian traditions of a "river of fire" in this locality (235:113).

Volcanic disturbances in South America about 9,000 years ago have been dated by radiocarbon (242:45). Huntington reported "lava flows of the glacial period interstratified with piedmont gravel" in Central Asia (232:168). Ebba Hult de Geer quoted Franz Firbas as follows: "The volcanic eruptions that produced the Laacher marine volcanic ash are about 11,000 years old, or a little older . . ." (108:515). Hibben suggested that the extinctions of animals in Alaska at the end of the ice age may have been due to terrific volcanic eruptions there, of which the evidence is plentiful (212). We will return to his account later.

Volcanic dust is not the only important product of volcanic eruptions. They also produce vast quantities of carbon-dioxide gas. Tazieff, for example, estimated that in one eruption he observed in Africa, the volcano emitted, along with about seventy-eight million tons of lava, twenty billion cubic yards of gas (417:217), not all of which, of course, was carbon dioxide.

The carbon dioxide emitted by volcanoes has an important effect on global temperature but one quite different from the effect of the volcanic dust. Being a translucent gas, it does not interfere with the entrance of sunlight, of radiant heat, into the atmosphere. But it is opaque to the radiation of the earth's heat into outer space. A small quantity of the gas will act effectively to prevent loss of heat from the earth's surface. A considerable increase in this small percentage will tend to raise the average temperatures of the earth's surface.

Carbon dioxide differs from volcanic dust also in the fact that because it is a gas it will not settle out of the atmosphere. It will remain until, in the course of time, it is absorbed by the vegetation or by chemical processes in the rock surfaces exposed to the weather. Therefore, as compared with volcanic dust, carbon dioxide is a long-range factor, and its effect is opposite to that of the dust.

In any displacement of the crust it follows that massive outbursts of volcanism must have added to the supply of carbon dioxide in the air. Its proportion in the atmosphere must have finally been raised far above normal. In consequence, it is likely that whenever volcanic activity declined sufficiently to permit a warming of the climate, the high proportion of carbon dioxide in the air may have acted to intensify the upward swing of the temperature. This would have increased the violence of the oscillations of the climate and would have accelerated many geological processes.

Evidence that the proportion of carbon dioxide in the air was, in fact, higher toward the end of the ice age than it is now is provided by recent studies of gases contained in icebergs. Scholander and Kanwisher, writing in *Science*, reported that air frozen into these bergs, presumably dating from the ice age, showed lower oxygen content than air has at the present time, and theorized:

Possibly this ice was formed as far back as Pleistocene time, when cold climates may have curbed the photosynthetic activity of green plants over large parts of the earth, resulting in a slight lowering of the oxygen content of the air (368:104-05).

The weight of a great deal of evidence presented in this book is opposed to this particular speculation; we must suppose, on the contrary, that the earth's surface as a whole was then no colder than it is now, and that just as many plants were absorbing carbon dioxide and releasing oxygen into the air then as now. But the same fact—the lower proportion of oxygen—may perhaps be explained by supposing a higher proportion of carbon dioxide.

Another consideration that greatly strengthens this line of thinking about the carbon dioxide is that the assumption of a cumulative increase in the proportion of this gas in the air, during the movement of the crust and the waning of the ice sheet, helps to explain not only the extraordinarily rapid final melting of the ice but also the succeeding hipsithermal.

The hipsithermal is the most important climatic episode since the end of the ice age; the fact of the occurrence is well attested, but it is unexplained. Scientists have been aware that this 4,000-year warming of the climate could have resulted from an increase in carbon dioxide content of the air, but this has not been helpful, since hitherto no way has been found by which to account for an increase of the required magnitude. No other possible cause of the warm phase (such as an increase in the quantity of the sun's radiant heat) has been supported by tangible evidence. It seems that the assumption of a displacement of the crust furnishes the first possibility of a solution.

To return for a moment to the question of the several readvances of the ice, it may be asked, Why did the volcanism occur in massive outbursts separated by quieter periods? Why was it not continuous through the whole movement of the lithosphere? I have a suggestion to make, knowing that its validation will have to depend upon the efforts of others in the future. The evidence to be presented in detail below appears to point to an extraordinary conclusion: that the movements of the outer shells, once started, gather speed rapidly and achieve a terrific tempo, completing the main part of their displacement within a millennium or two. Because of the oblate shape of the earth, as already mentioned, this must mean the production of millions of strains and pressures which, because of the resistance of the crystalline lithosphere, would have to work themselves out over a comparatively long period. Periodically, then, parts of the crystalline shell might be expected to yield to pressures by fracturing, and whole families of volcanoes could result. Not individual eruption but long-lasting episodes of volcanicity could be expected, and the dust from these might initiate the glacial readvances.

The following table gives radiocarbon dates for volcanic eruptions in late glacial times, during a part of which the lithosphere was presumably in motion.

TABLE 10

Volcanism in the Glacial Period

Date	Sample	Place	Phase*
8,620±350	Sa-243;VII,244	Japan	Postglacial
11,520±400	Sa-244;VII,244	Japan	Mankato
11,720±220	GaK-870;IX,48-49	Japan	End of Two Creeks Interstadial
12,750±350	W-1644;IX,517	Montana, U.S.	Cary Advance
13,800±300	W-1548;IX,526	Costa Rica	Cary Advance
13,800±250	N-355-1;X,233	Japan	Cary Advance
13,900±250	GaK-521;IX,48-49	Japan	Cary Advance
14,260±160	GX-279;VIII,147	Costa Rica	End of Brady Interstadial

* Phase is assigned on assumption that the volcanism either falls within the period of glacial advance or immediately precedes and contributes to causing the advance.

Date	Sample	Place	Phase*
15,000±500	A-636B;VIII,10	Chile	Brady Interstadial
15,350±320	GaK-494;IX,44	Japan	Brady Interstadial
16,350±350	GaK-558;VIII,57	Japan	Tazewell Advance
16,400±300	GaK-868;IX,48-49	Japan	Tazewell Advance
17,710±750	N-93;VI,112-113	Japan	Tazewell Advance
17,900±450	GaK-457;VII,12	Japan	Tazewell Advance
19,600±600	GaK-409;VII,12	Japan	Tazewell Advance
20,200±800	GaK-456;VII,12	Japan	Tazewell Advance
21,230±720	GaK-476;VII,14	Japan	Iowan Advance
21,470±1130	N-96-2;VI,112	Japan	Iowan Advance
22,720±800	N-97-1;VI,112	Japan	Iowan-Farmdale Interstadial
22,970±800	N-97-2;VI,113	Japan	Iowan-Farmdale Interstadial
23,200±800	GaK-513;IX,44	Japan	Farmdale Advance
23,400±800	GaK-734;VIII,57	Japan	Farmdale Advance
23,400±800	GaK-558;VIII,56	Japan	Farmdale Advance
23,800±700	GaK-514;IX,44	Japan	Farmdale Advance
24,000±650	GaK-725;IX,46	Japan	Farmdale Advance
24,500±900	GaK-472;VIII,56	Japan	Farmdale Advance
25,900±1000	GaK-381;VII,11	Japan	Farmdale Advance
27,700±1500	GaK-469;VIII,56	Japan	End of Denekamp Interstadial
28,000±1000	NZ-217;V,144	New Zealand	End of Denekamp Interstadial
29,400±1800*	GaK-867;IX,48	Japan	Denekamp Interstadial
29,800±1200	GaK-489;VIII,57	Japan	Denekamp Interstadial
(8–10,000-year gap)			
37,400±2000	LJ-959;VII,90	California	Rockian Advance
39,050$^{+1450}_{-1230}$	NPL-79;VII,160	West Indies	Hengelo Interstadial
39,160$^{+6090}_{-3410}$	NPL-49;VI,29	Tristan da Cunha	Hengelo Interstadial
39,600+2000	NZ-373;V,160	New Zealand	Hengelo Interstadial

From such a brief list as the above it is, of course, impossible to come to any conclusion as to the amount of volcanism throughout the world at any one time during the glacial period; perhaps 5,000 radiocarbon dates would be necessary for this. Nevertheless there is evidence of considerable volcanism in areas now comparatively quiet. One has the impression that Japan was much more active volcanically then than now. One cannot help feeling, too, that there was some significance in the curious intermission of about 10,000 years in this volcanism—an interruption that took place just after the previous presumed shift of the pole, from the sea off northern Norway to Hudson Bay. That shift apparently produced no volcanism in Japan (unless we are being deceived by lack of evidence). Could it be because, in that particular shift, Japan moved into slightly lower latitude? It would have moved equatorward but not far, perhaps about 10°.

Possibly it is significant that North America, a continent that hypothetically was moved toward the equator in the last displacement, is volcanically quiet, while South America, which then moved toward the South Pole, is seismically active. Again, India and the East Indies, which moved toward the North Pole, are seismically active, while Europe, which theoretically moved southward, is now quiet.

To carry this idea a step further, Africa, which underwent very little movement in the last displacement, is relatively free of earthquakes and volcanoes.

I would like, in concluding this chapter, to urge the relevance and importance of this question of the high turbulence toward the end of the ice age: the increased rates of sedimentation and the indication of violent fluctuations of climate during the numerous advances and retreats of the ice sheets. The suggestion advanced in this book is not an alternative to any existing theory. It may be difficult for some to believe that up to now there has been no theory advanced to account for these facts. However, nobody has ventured to hazard even a guess, so far as I know. It appears that a displacement of the lithosphere at the end of the glacial period, because of the volcanism that was probably (and almost inevitably) connected with it, is the first suggestion of a possible explanation.

As I have already pointed out, geologists have long been committed to the assumption of modern geology that basic geological processes have always proceeded at a uniform rate, the rate at which they proceed today. But here we see that there was indeed an exceptional situation at the end of the ice age. Some factor was disturbing the usual course of geological processes. Continental drift, being so very slow, could hardly have accomplished this. But the displacement of the lithosphere at a considerable speed could have done so. This matter will be discussed in greater detail in Chapter IX.

chapter **6**

THE SUDDEN
MELTING OF
THE ICE SHEET

1. THE TEMPO OF DEGLACIATION

OUR present objective is to establish, by marshaling all the necessary facts, the rate at which the Wisconsin ice cap melted. If it can be shown, as I believe it can, that the ice sheet melted at a rate that is entirely inexplicable in the light of presently accepted theories of geology, we shall have established our right to look at new ideas. We shall have established that in this instance the concepts of uniformitarianism do not apply, that an exceptional cause must be found, but one that fundamentally reflects the dynamic realities of the strange planet on which we live.

We shall, first, by examining numerous radiocarbon dates and other kinds of evidence, determine the approximate time when the major part of the melting of the ice sheet had been completed, and then go back and find out just when the process of melting started.

There is little trouble about the time of the establishment of the postglacial climate. Evidence from most parts of the world is in agreement. In England a large number of radiocarbon dates average out to about 10,000 years ago, or about 8,000 B.C. (351a;I:65–67). From Chile and from the American Northwest similar dates are reported (210b:124).

However, there was a considerable time lag between the disappearance of the ice and the establishment of a climate about like that of the present. It appears, indeed, that this time lag amounted to several thousand years, so that when the present climate was established the ice had long since gone from most areas. We have a number of radiocarbon samples that indicate the extent of the time lag.

Sample GSC-614 (351a;II:216) is of great interest. It dates mastodon bones found at the Ferguson Farm, Tupperville, Ontario. The date is 8910±150 years ago. Tupperville is located at Lat. 42°33′ N, Long. 82°17′ W, on the north side of Lake Erie. This, of course, is toward the central part of the area glaciated by the Wisconsin ice sheet. The ice is supposed to have been at least a mile thick over the Great Lakes. But a mastodon died there about 9,000 years ago. It must be remembered that while mammoths are alleged to have been arctic animals (wrongly, I believe—see Chapter X), mastodons have never been so classed. The mastodon did not have a hairy coat, which is supposed to prove the arctic adaptation of the mammoth. However, like the mammoth, he did require great quantities of vegetation for his food. He is supposed to have lived on bushes and trees as well as grass. One would suppose, then, a considerable time to have elapsed for the development of this vegetation.

But now a mystery appears. The contents of the mastodon's skull consisting of humus material, was dated, and turned out to be only 6230±240 years old (Sample S-16, *ibid.*). Therefore the reporter of the date concludes, ". . . . date, which agrees with pollen diagram from site, suggests nonsedimentation and exposure of bones for about 2,500 years. . . . " In other words, the mastodon died and its bones were exposed to the presumably rather cold air for 2,500 years before the bog developed over them. Now this is an extraordinary thing. Where did the mastodon come from? He had to come from a place where there was food for him. In any case we seem to have evidence here of a gap of 2,500 years between deglaciation and the development of vegetation cover.

In connection with this case, we have another interesting date from very nearly the same place. It is Sample GSC-620 from Walker Pond, Ontario, in Lat. 42°57′ N, Long. 81°13′ W (ibid.) Taking a minute of latitude to be about a mile, Walker Pond is about twenty-one miles north of Tupperville and about four miles farther east. Yet the humus material from this pond dates 12,190±230 years ago. This proves two things: that the minimum time since the deglaciation of this region is about 12,000 years and that, despite the exposure of the mastodon's bones for 2,500 years, there was food for him in the vicinity. We are going to have to allow plenty of time for the development of his food supply, which was evidently there a long time before he was. In fact this evidence points to the deglaciation's having taken place perhaps 13,000 to 14,000 years ago.

Support for this view is produced by another date from Canada. This is Sample GSC-419 (ibid.), plant materials found twenty-nine feet below the surface of a bog near St. Hilaire Station, Quebec, in Lat. 45°33′30″ N, and Long. 73°08′30″ W, somewhat north and west of our mastodon. This sample is dated 12,570±220 years ago and is considered by the reporter a "minimum for deglaciation of the SW part of St. Lawrence Lowland." Again, we must remember that considerable time was required for the development of the materials here dated: at least one or two thousand years. The area is very close to the central portion of the former continental ice sheet. We would suppose that deglaciation here probably occurred about 14,000 years ago.

To return for a moment to our Tupperville mastodon, if his bones lay uncovered for about 2,500 years, which really doesn't seem very likely, it must have been because he happened to die in a dry (and cold) place that much later developed into a bog. But let us consider the case of another mastodon, this time from New York, where they are known to have lived in great numbers (see Chapter X). Sample Y-460 (351a:147) consists of spruce wood, associated with the remains of a mastodon, taken from the Miles Colgan Farm, near Ledyard, Cayuga County, in Lat. 42°40′ N, Long. 76°36′ W. This is the same general region from which the previous samples were taken. The age of the wood was 11,410±410. There is no question but that here we have an actual forest growing, not just postglacial scrub. The comment reads, ". . . The fossil flora appears to record a boreal coniferous forest with mosses. . . The implication of closed forest in south-central New York during Two-Creeks time [see Fig. 21, pg. 94] is also of interest." It seems probable that at least two or three thousand years would have been involved in the development of this forest, and yet the wood itself is between 11,000 and 12,000 years old. The deglaciation, then, can quite reasonably be pushed back to the neighborhood of 14,000 years ago. A few other dates confirming the existence of spruce forests in Connecticut are the following. I give only the sample numbers and the dates:

Y-285	13,550±460
Y-477d	13,290±120
W-46	12,700±280

These dates support the existence of forests in this region about 14,000 years ago and push the probable date of deglaciation back even further (351a;I:146).

From Rodney, Ontario, in Lat. 42°34′ N, Long. 81°4′ W, again in the same region, we have a sample of vegetable muck associated with parts of a mastodon skeleton, which dates 12,000±500 years ago (S-30:351a;II,74). This indicates that mastodons (not arctic animals) were present, probably in large numbers, in the forests of the United States and Canada as early as 12,000 years ago. Deglaciation was probably at least 2,000 years earlier.

A date that seems to push deglaciation in the eastern part of North America to at least 14,000 years ago comes from West Lynn, Massachusetts. The sample consists of barnacles, and the age found is 14,250±250 (W-736;351a:II,133). Barnacles, we might suppose, could have appeared along the beaches of Massachusetts almost as soon as the ice left, and they would not have required the long prior development of a soil. Therefore we may, with these barnacles, be getting pretty close to the period of actual deglaciation.

For this same reason shells of various species may be more reliable guides to the time of deglaciation than peat, wood, or land animals. In the following table nine shell samples are arranged in the order of increasing latitude (from south to north):

The first two shell samples, from New Brunswick and Prince Edward Island, are supposed to have originated during or shortly after the glacial retreat from those places, which therefore can be dated about 13,000 years ago. It is natural enough that the glacial retreat here might have been somewhat later than in the more southerly latitudes of New York and Connecticut. It has also been considered that the last ice center of the continental ice cap may have been in eastern Canada, in Quebec or Labrador, in which case it would be here that the ice may have made its last stand.

TABLE 11
Deglaciation Dated by Shell Samples

	Sample	Reference	Latitude	Age
1.	I(GSC)–7	III,50	45°13′	13,324±500
2.	GSC–160	VI,168	46°58′	12,670±340
3.	GSC–87	V,41	49°35′	11,880±190
4.	I(GSC)–14	III,49	50°13′	7,875±200
5.	GSC–110	VI,173	66°49′	8,160±140
6.	I(GSC)–17	III,51	67°30′	10,215±220
7.	I(GSC)–16	III,51	67°39′	9,100±180
8.	I(GSC)–25	III,51	68°47′	10,530±260
9.	I(GSC)–18	III,52	73°18′	12,400±320

A little farther north and east, in an area more exposed to oceanic climatic control, and also nearer to the presumed ice center in Labrador, we have a date for Newfoundland (our No. 3): 11,880±190. This seems reasonable. With regard to these first three dates the reporter wrote: "These three dates seem to indicate that the ice withdrew from the north coast of Newfoundland about 12,000 yrs. ago" (351a:V,41).

Our fourth date is supposed to record the deglaciation of James Bay, far to the west of the Atlantic Coast region above, and a part of Hudson Bay. The dated shells are supposed to have grown "immediately following the deglaciation of James Bay." The age is 7,875±200. It is notable that this is much later than deglaciation on the Atlantic Coast or in New York or New England. This suggests that the last stand of the ice sheet was in Hudson Bay.

For our next date, No. 5, we move north of Hudson Bay, to the Northwest Territories. Here we find the ice retreat from the Mac-Alpine Lake area, Lat. 66°49′ N, Long 103°28′ W, dated at 8,160±140 years ago, a little earlier than the retreat from James Bay—a surprise, perhaps, except for a remarkable fact that has been observed by numerous geologists. It appears that the ice cap did not melt from south to north, as might have been expected, but from all sides inward toward the central area, about Hudson Bay. The southern edge of the ice retreated northward, the northern edge southward, the western edge eastward exactly as if, indeed, it had been a polar ice cap. A member of the Canadian Geological Survey wrote:

> . . . If it is assumed that the marine limit* at any locality relates to the time of glacial retreat, this group of dates supports the following (conclusion): . . . retreat of the glacial margin within the zone proceeded from W to E and occurred at about the same time as the retreat of the southern margin of the ice. (351a:IV,22-24)

Our next date, No. 6, is from Coronation Gulf, above the Arctic Circle, in the District of Mackenzie, about as far west of Hudson Bay as Newfoundland is east of it (see Fig. 22, p. 101). The age of the dated shells here is 10,215±220. Our date No. 7, also from Coronation Gulf, is dated a little later, 9,100±180. No. 8, from somewhat north and west of these two, is dated 10,530±260. It seems that here, in the far north, deglaciation preceded the deglaciation of James Bay.

Our last sample, from farthest north, from Victoria Island, a thousand miles north of Hudson Bay, is dated 12,400±320, and therefore it seems that deglaciation was proceeding simultaneously at about the same rate at equal distances north and south of the center of Hudson Bay (for New Brunswick lies about the same distance south of that point).

This, of course, leads us back to our assumption that the North Pole was located in Hudson Bay during the last ice age. As we have seen, the ice sheet seems to have melted from west to east, and from north to south, as well as from south to north. Dr. Fred Earll, the author of the foreword of this book, has raised a question about this. He asks† how this can be reconciled with our assumption that the pole was moving at this time from Hudson Bay to its present site in the Arctic Ocean. If this was the case he suggests that the ice should not have melted from north to south. The pole was

---

* "Marine limit" means highest stand of the sea, assumed to mark greatest depression of the land under the ice load.

† Personal correspondence.

moving north, and the snow at the northern edge of the ice sheet should have been the last to disappear. In fact, if Doctor Earll is right, glaciation should have developed at this time on the islands of the "refugium" that were not glaciated earlier.

Considerable thought on this matter had led me to the conclusion that the evidence points to a very rapid transit of the pole from its old to its new home. It must have completed its transition in a matter of centuries rather than millennia. We shall see further evidence for this conclusion very shortly. If the transit of the pole was very sudden, then the pattern of melting indicated by the evidence becomes understandable. The ice sheet was thickest toward its center. If it melted all over at about the same rate (probably a little faster on its southern side) then the thin edges would disappear first, while the core would remain to the last. A factor that would work to this end would be the ice cap itself, which, by reflecting the sun's light back into space, would tend to minimize melting in the central parts until the edges drew in.

In summary it seems that the principal part of the continental ice cap had melted prior to 14,000 years ago. In order to determine the speed at which it melted we must next examine evidence bearing on the question of when the melting started.

2. THE LAST MAJOR ADVANCE OF THE ICE SHEET

It is safe to assume that if we can fix the time when the last major advance of the ice sheet ground to a halt and gave way to withdrawal, we shall have determined the time of the polar shift; that is, the time when the movement of the lithosphere had proceeded far enough to have an important effect on the climate. We can suppose that when the lithosphere started to move it moved slowly for a time, gathering momentum. It would, however, have gathered momentum very rapidly, for the centrifugal effect of unbalanced masses within it would be multiplied geometrically by its movement. (See Note 4, p. 343).

The last major advance of the Wisconsin ice sheet, which carried that ice sheet to its maximum volume, is called the Tazewell Advance (see Fig. 21, page 94). A number of radiocarbon dates enable us to fix the time of this advance rather exactly.

It is important here to distinguish between two sorts of samples. Sometimes a sample is included in the glacial materials. Thus we may have a log buried in a mass of debris left by the glacier. We can say that the age of this log dates the time that the tree it came from was killed by the advancing glacier (unless in advancing, the glacier has picked up material buried by an earlier glacier). On the other hand, the sample may come from the soil underlying the glacial material. In that case it dates the warmer period before the glacier came. We have a few samples from the soils underlying the Tazewell glacial deposits, and these will help us decide when the Tazewell Advance began. These do not necessarily all agree, because the advancing glacier reached different areas at different times. Table 12 gives these dates, together with the locations from which the samples came:

TABLE 12
Samples Predating the Tazewell Advance

Indiana: OWU-8;351a:VI,432. Spruce wood . . . . 19,906±691
"Sample came from 30 cm beneath contact with overlying Wisconsin till." (30 centimeters of accum. indicates a considerable lapse of time.)

Indiana: W-577;351a:II,139. Wood . . . . 20,500±800
"Wood from peaty mollusk-rich silt at base of till which overlies a paleosol . . ."

Iowa: W-1687;351a:IX,507 . . . . 18,300±500
Organic carbon from buried soil immediately below material indicating glacial climate.

Iowa: W-879;351a:II,145 . . . . 19,059±300
"Spruce wood from gray silt loam under Tazewell-age loess . . ." (Probably dates arrival of ice.)

Kentucky: W-521;351a:II,147 . . . . 18,530±500
"Wood from highly fossiliferous gray sandy silt, related to the Tazewell-age valley train along the Ohio River."

Ohio: W-738;351a:II,152 . . . . .18,750±300
Carbonaceous matter from soil.

Washington: W-1186;351a:VI,58 . . . . 18,100±700
Plant fragments overlain by till. (This was not the main ice sheet, which did not reach Washington, but a simultaneous advance of a local glacier.)

Table 12 suggests that the Tazewell Advance of the ice did not begin much before 20,000 years ago. Table 13 gives dates directly associated with the advance:

TABLE 13
The Tazewell Advance*

Indiana: W-598;II,140 .... 20,100±800
         W-597;II,140 .... 20,300±800
         W-580;II,140 .... 20,900±800
    Wood samples included in Tazewell till.

Illinois: W-524;II,136 .... 18,460±500
    Wood from till.

Ohio: W-724;II,152 .... 19,100±300
    Log in till.

Ohio: Y-449;I,148 .... 18,500±420
    Log in till.

Ohio: OWU-76;VII,166 .... 17,290±436
    Spruce wood . . . from silts in till.

Ohio: OWU-52;VI,345 .... 17,880±224
    Spruce log projecting from till bank.

California: UCLA-269;IV,227 .... 19,500±500
    Highest level of Lake Manix shoreline—a "pluvial period" associated with last major ice advance.

California: UCLA-121;IV,113 .... 19,300±400
    Encrusting tufa from high-stand of Glacial Lake Manix.

North Carolina: I-1750;X,263 .... 16,200±290
    Singletary Lake . . . "date supports pollen data indicating full glacial correlation."

Table 13 gives us some idea of the duration of the Tazewell Advance. It chronicles its expansion in the Middle West, and the correlated pluvial conditions in California and Texas, down to about 17,000 years ago. It is clear that if a shift of the earth's crust stopped the advance of the ice, the shift must have had its first climatic effect between 17,000 and 16,000 years ago. It may have been moving imperceptibly for some time before that; it would have had to move the area of the Middle West at least two or three degrees, one would think, to arrest the advance of the ice. The advance had been

* Reference for all dates 351a.

continuing for about 3,000 years, and the ice was now at its maximum; it was now covering about 4,000,000 square miles of territory, and over the Great Lakes it has been estimated to have been at least a mile, perhaps two miles, deep.

3. THE RETREAT OF THE ICE

Now the recession began. Table 14 tells the story of the retreat that reduced the ice cap to a fraction of its size and wholly deglaciated New England and New York and Europe. This retreat has various names. In the United States it is called the "Brady Interstadial" or "Brady Retreat." In Europe it is referred to as the "Bölling" phase of warm climate and is considered postglacial.

No matter what part of the world we turn to, Table 14 appears to indicate that the Brady Interstadial set in with astonishing suddenness and that the deglaciation proceeded at a very rapid pace. By comparison with Table 13, we see that there was apparently very little lapse of time between the interruption of the glacial advance 16,000 years ago and the establishment of an advanced state of plant and animal life. There is no doubt that considerable local differences in the beginning of the ice retreat would occur because of differences in elevation and differences in the wind patterns in various areas. Nevertheless there is evidence from all over the world that something very drastic indeed occurred.

TABLE 14
The Brady Interstadial*

Saskatchewan: S-241;X,367 .... 15,200±260
    Date is minimum for deglaciation.

Ohio: OWU-83;VII,167 .... 14,780±192
    Spruce wood . . . "may be early postglacial for this area . . ."

Iowa: I-1024;X,256-257 .... 16,000±500
    Spruce wood from loess interbedded between tills.

Idaho: M-1409;VIII,280 .... 14,500±500
    M-1410;VIII,280 .... 15,000±800
    Humus and fossils, incl. horse, camel, sloth.

Washington: W-1227;VII,386 .... 15,000±400
    Wood fragments in non-glacial silt and sand.

* Reference for all dates 351a.

Massachusetts: W-1187;VII,373 .... 15,300±800
"Leaves, needles, fruits, etc., from thin-bedded silty clay . . ."

New Jersey: W-1151;VI,41 .... 16,700±420
Bog material overlying gravel.

Alaska: CX-250;VIII,145 (65°5' N, 147°45' W) .... 14,760±850
Rodent nest. "Dates a ground squirrel living in thawed layer
above frozen ground."

Alaska: LJ-631;VII,85 .... 15,500±600
Sea cores from Gulf of Alaska. Sample "close to end" of glacial
period.

USSR: GIN-97;X,436 .... 14,570±120
Irkutsk Oblast. "Fossil bones . . . in loess-like loam: last interstadial of Zyranka [Wisconsin] glaciation."

USSR: MO-302;VIII,303 .... 16,260±640
Lithuania: Moss from River Ula: ". . . On basis of pollen
analysis Lithuanian scientists assign the lacustrine-bog stratum
to an interstadial between the Pomeranian [Tazewell] and
Brandenburg [Cary] stages . . ."

USSR: TA-137;X,380 .... 13,970±115
      TA-138;X,380 .... 14,725±260
Estonia. Peat with vegetable remains.

USSR: TA-50;VIII,437 .... 15,500±575
Estonia. Sandy loam with plant remains "from intermoraine
deposits."

USSR: RUL-197;VII,225 .... 15,700±300
      RUL-161;VII,225 .... 15,080±270
The first date from "interglacial deposits"; the second based
on wood sample from near Moscow, "from lower horizon of
Holocene [post-glacial] deposits."

Sweden: St—449;II,186-187 .... 16,040±200
        St—431;II,186-187 .... 15,740±290
Marine clay from near Goteborg, classed as "post-glacial."

Monaco: MC-53;VIII,287 .... 16,600±700
Worm *Sabellaria alveolata* "may date a sea-level of the last
interstadial Wurmian period . . ."

France: GrN-1671 ....16,240±120
Pollen diagram. ". . . indicates the beginning of the Late-
glacial period."

A few of the samples listed above are particularly interesting. We
may note that, though the maximum phase of the continental ice

cap has been shown to have occurred as late as 16,000 years ago, nevertheless a minimum date for deglaciation of Saskatchewan is about 15,000 years ago. In Ohio where the ice cap was at its thickest, we have a postglacial sample dated about 14,000 years ago. And that was spruce wood, suggesting a forest that must have taken (as we have already pointed out) a few thousand years, by conservative estimate, to get established. What, indeed, does this mean? Does it not clearly suggest that the ice cap, estimated to have been at its maximum at least a mile thick in Ohio, disappeared completely from Delaware County in that state within only a few centuries?

The same thing is suggested by many of the other samples. In Massachusetts we find "leaves, needles and fruits" flourishing about 15,300 years ago. In New Jersey we have a bog which developed over glacial material at least 16,280 years ago, immediately after the interruption of the ice advance. In the Soviet Union, in the Irkutsk area, deglaciation was complete and postglacial life fully established by 14,500 years ago. In Lithuania another bog developed as early as 15,620 years ago. These two dates taken together are rather suggestive. A bog can develop much faster than a forest. First, however, the ice must disappear. And let us not forget that there was a great deal of ice. It continued, at least in the United States, to increase in depth and area until at least 17,000 years ago: then a millennium or two later it was gone! And furthermore life was already reestablished on the vacated tracts, which, we must understand, were devoid of life when the ice left.

The sea samples may give us a reliable measure of the rate of the climatic change. We may note that marine clay from Sweden, about 16,000 years old, is classified as "postglacial." A comparison of samples suggests that the age of the clay falls between 15,500 and 16,250 years, agreeing very closely with the end of the Tazewell Advance in the United States. In other words the temperature of the seawater in Sweden (and in the Mediterranean) was changed just at the time that the growth of the ice cap stopped. One may say that the change must have been drastic to have made an observable difference in sea conditions.

There has recently been an important change in the estimates made of the time when the sea level during the ice age stood at its lowest point. Theoretically this would signify the greatest expansion of glaciers throughout the world, the greatest amount of water withdrawn from the oceans. Until last year (1968) we had the impression that the lowest sea level was reached about 19,000 years ago. Now, basing their conclusions on much more data than have been available before, Drs. J. D. Milliman and K. O. Emery, of the Woods Hole Oceanographic Institution, announce that the time of minimum sea level was 15,000 years ago.*

This finding is of the greatest importance. It means that for the end of the glacial period we have a remarkably tight fit of dates. In fact there is quite a mystery. It does not seem possible that the Wisconsin glacier and the Scandinavian glaciers could have retained so much of their bulk so late. In fact all the dates we have cited argue strongly on the other side. By 15,000 years ago deglaciation was indeed very far advanced. How shall we explain this mystery? Where did the water go?

The answer is quite obvious: If the crust was moving; if North America was moving southward; if South America was moving toward the South Pole; and if Antarctica, which had not been within the Antarctic Circle at all, was now moving to the pole, then surely the Antarctic glaciers were expanding fast and were taking up the water released by the melting glaciers in the north.

Let us summarize the situation. The maximum extension of the ice was not earlier than 17,000 years ago and may have been considerably later. Deglaciation was completed in some places by 16,000 years ago and in many places by 15,000 years ago, while vegetable and animal life had been re-established by 14,000 years ago even in areas close to central parts of the former ice sheet. Much-diminished glaciers continued to exist, and even re-expanded several times after this (see Fig. 21, pg. 94), but they were confined to a very small part of their former territory. In effect a vast ice cap covering half of a continent, and as deep as the Antarctic ice cap is today, disappeared, it would seem, in little more than a millennium—perhaps two. In any case it was, geologically speaking, a sort of miracle. There was nothing in this to suggest the painfully slow pace of usual geological history. To be blunt about it, it was a catastrophe, a cataclysm; it was a *revolution*.

There are some deep-sea cores that may help us make a sound

* New York *Times*, Sept. 5, 1968, and personal communication from Doctor Milliman, Oct. 7, 1968.

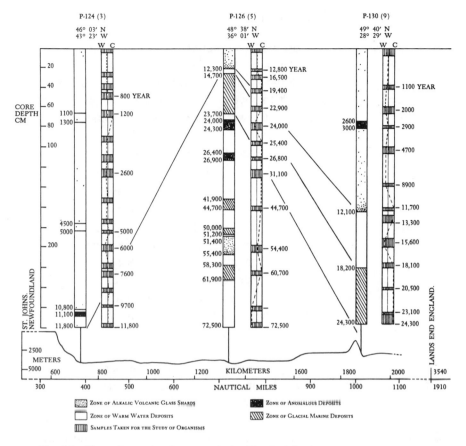

Fig. 29. Chronology of sediments of the North Atlantic cores, after Piggott and Urry.

guess as to when the crust must have started to move, to precipitate this geological revolution.

The first of these comes from the North Atlantic (Fig. 29, above)(344). Core P-126–5, taken from the bottom of the mid-Atlantic, indicates that the ocean started warming up about 16,400 years ago, deposition of glacial sediment stopped 14,700 years ago, and the water reached its present temperature about 12,800 years ago. This agrees very well with the samples we have had from the land. It suggests that, regardless of when the crust actually started moving, it had moved far enough to begin to affect the sea temperature by 16,400 years ago. It would seem, too, that the movement might have continued for four thousand years, but this is probably not the case. If the crust had completed its shift quickly, in one millennium, let us say, it would still have taken thousands of years to heat up the waters of the oceans. In fact, in a way, it is almost necessary for the polar shift to have been completed very quickly, for otherwise the warming might have gone on several thousand years more.

Cores P-126–5 and P-130–9 are in agreement as to the sudden warming that took place between 16,000 and 14,000 years ago, although Core P-130–9 would have the warming start somewhat earlier.

In three other cores, from the Caribbean and the equatorial Atlantic (Fig. 30, p. 153) (409), we have evidence for the climatic change. Core A-172–7 shows a sharp change at about 16,500 years ago, Core A-179–4 shows it about 14,000 years ago, and Core 180–73 indicates it at 12,000 years ago. The differences among the cores are related, perhaps, to varying conditions in the different parts of the ocean from which they were taken, including possible variances in ocean currents.

However, we are in a position to reach some tentative conclusions on this evidence. The rapidity of the deglaciation suggests some extraordinary factor was affecting the climate. The dates suggest that this factor first made itself felt about 16,500 years ago, that it had destroyed most (perhaps three quarters) of the glaciers by 2,000 years later, and that, *if it was a crust displacement,* the main part of the polar shift may have been completed in a millennium or less.

Our assumption is that the pole moved from a point approximately in Lat. 60° N and Long. 83° W to its present position. This means a displacement through 30° of latitude, or 1800 miles, and if the shift was accomplished in a millennium and a half, it would mean an average travel rate of about a mile a year; but of course the start and the finish would necessarily be very slow; therefore the speed of travel during the principal phase of the displacement must have been dizzy indeed.

It is no wonder that, with circumstances like these, there are evidences of extreme disturbances on the earth's surface. It is not particularly popular to talk about these things, because catastrophes, so to speak, went out with the Flood. Yet facts are facts, and come what may, we shall have to face some quite remarkable ones as we

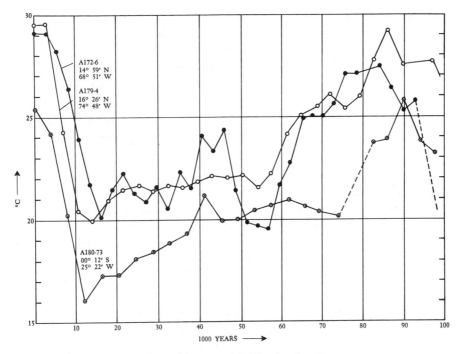

Fig. 30. *Three cores from the equatorial Atlantic, after Suess.*

proceed. These will be dealt with especially in Chapters X and XI, but there are some we should discuss now.

Various scientists have noted with puzzlement, from time to time, cases where animals of cold and warm climates or animals and plants of different climates, are found assembled in caves or other places where they have all apparently suffered death under violent conditions. One such case is reported from the Puy de Dôme, France. A radiocarbon date (Sa-103;351a:VII,239) of 13,500±450 years was obtained on a peat bed containing fossils of warm-climate animals but of cold-climate flora. The reporter remarks, ". . . Simultaneity of fauna and flora of cold and warm climates is not yet explained." A solution occurred to me as I looked at the date. It is the date of the Cary Advance of the diminished Wisconsin ice sheet. The Cary Advance was a short but sharp cold spell, probably brought on, as I imagine, by particularly violent outbreaks of volcanism. As it was apparently worldwide, it affected France. It may have killed the animals, and their bodies may have lain on the ground until they were covered by peat containing cold-climate plants.

A similar case is reported from Japan. From Hanaizumi, a wood sample from a conifer bed was dated 15,850±360 years ago (Y-594, 351a;II,55). The reporter comments, ". . . Plant fossils imply a climate colder than today's; accompanying vertebrate remains . . . belong to several extinct species, including fossil elephant." This suggests an event accompanying the polar shift. As I have already explained, if Hudson Bay moved southward 30° along a meridian of 83° W, then Japan would have been moved northward about 20°, and there would have been, we may well suppose, some pretty violent storms during this period. Even though this was the time of the Brady Interstadial in the United States, it was a time of reverse trends in Japan, of falling, not rising, temperature, and the animals died, no doubt, under the most unpleasant conditions. As in France, cold-climate plants came to bury them.

C. P. Brooks, in his *Climate Through the Ages* (52:241), mentions a similar problem, apparent contradictions between fossil indications of a warm climate combined with glacial evidences, occurring in what geologists refer to as "erratic blocks" carried by glaciers. Here the answer may be: animals killed by the sudden advent of cold, and a glacier developing shortly afterward.

To sum up the evidence, the continental glacier of the last ice age expanded, with unexplained interruptions, until it covered an area of four million square miles in North America. About 17,000 years ago it stopped growing, and two thousand years later most of it was gone. Its later history consisted of occasional readvances followed by further dwindling. The suddenness of the change cannot be explained by present geological theory. If, however, it is to be explained by the assumption of a displacement of the entire outer shell of the earth, then other questions arise: What factor initiated the glaciation? What caused it? If it was ended by the movement of the pole from Hudson Bay to the center of the Arctic Ocean, we shall have to assume it was started by a movement of the pole to Hudson Bay from somewhere else. From where? And when? Those are the questions that face us and that will be considered in the next chapter.

chapter **7**

EVIDENCE FOR THE NORTH POLE IN THE GREENLAND SEA

Part I.

THE RADIOCARBON EVIDENCE

1. THE WEAKNESSES OF THE ACCEPTED GLACIAL CHRONOLOGY

GEOLOGISTS are used to thinking of four major glaciations during the Pleistocene Epoch. They have assumed that each ice age affected the earth as a whole simultaneously, causing ice sheets in both northern and southern hemispheres and lowering temperatures generally. Some geologists have questioned this concept of four glaciations; it is at least necessary to recognize several successive phases of advance and retreat for the older glaciations. Whether these interruptions were merely interstadials, like those of the Wisconsin glaciation, or were true interglacials it is increasingly hard to decide the further back in time one goes. According to the accompanying chart of the glacial periods (see Table 15, p. 159), it is evident that the intervals between the different stages of the earlier glacial periods are in some cases longer than the entire duration of the Wisconsin glaciation. It does not seem reasonable, therefore, to insist that they were merely interstadials, or, consequently, to insist upon the number of just four glaciations during the Pleistocene.

This becomes more apparent when we consider the implications of a recently discovered Eurasian continental glaciation of late Pleistocene time. Attention was called to it by Hobbs in 1946 (219). Although known to Russian geologists since the nineteenth century, it seems still unknown in the west, and this in spite of the fact that the evidences are apparently spread widely over two continents. In view of this we must ask how many other glaciations in various parts of the world may have been overlooked. Flint has pointed out how easily glacial evidence can be destroyed (156:171). Coleman also emphasized the same thing:

> It might be supposed that so important a change would leave behind it evidence that no one could dispute, and that there should be no room for doubt as to what happened in so recent a time of the earth's history. In reality the proof of the complete disappearance of the ice and its return at a later time is, in the nature of things, a matter of great difficulty and it is not surprising that there are differences of opinion (87:20).

Croll pointed out the ephemeral character of glacial evidence eighty years ago in books that are still eminently readable. After first discussing the accumulations of strata containing plant and animal remains during a period of temperate climate, he comments thus on their subsequent destruction:

> . . . We need not wonder that not a single vestige of [these strata] remains; for when the ice sheet again crept over the island [Britain] everything animate and inanimate would be ground down to powder. We are certain that prior to the glacial epoch our island must have been covered with life and vegetation. But not a single vestige of these is now to be found; no, not even of the very soil on which the vegetation grew. The solid rock itself upon which the soil lay has been ground down to mud by the ice sheet, and, to a large extent, as Professor Geikie remarks, swept away into the adjoining seas (91:257).

It is obvious, of course, that whatever could destroy all the surface deposits of a temperate period might, at the same time, destroy any evidence of former glaciations. Croll goes on to say:

> It is on a land surface that the principal traces of the action of ice during a glacial period are left, for it is there that the stones are chiefly striated, the rocks ground down, and the boulder clay formed. But where are all our ancient land surfaces? They are not to be found. The total thickness of the stratified rocks of Great Britain is, according to Professor Ramsay, nearly fourteen miles. But from the bottom to the top of this enormous pile of deposits there is hardly a single land surface to be detected. True, patches of old land surfaces of a local character exist, such, for example, as the dirt beds of Portland; but, with the exception of coal seams, every general formation from top to bottom has been accumulated under water, and none but the under-clays ever existed as a land surface. And it is here, in such a formation, that the geologist has to collect all his information regarding the existence of former glacial periods. . . .
>
> If we examine the matter fully we shall be led to conclude that the transformation of a land surface into a sea-bottom (by erosion and deposition of the sediments) will probably completely obliterate every trace of glaciation which the land surface may once have presented. . . .
>
> The only evidence of the existence of land ice during former periods which we can reasonably expect to meet with in the stratified rocks, consists of erratic blocks which may have been transported by icebergs and dropped into the sea. But unless the glaciers of such periods reached the sea, we could not possibly possess even this evidence. Traces in the stratified rocks of the effects of land-ice during former epochs must, in the nature of things, be rare indeed (91:267-69).

Croll was interested in pointing out the impermanence of glacial evidence. He continued, therefore, as follows:

> The reason why we now have, comparatively speaking, so little direct evidence of former glacial periods will be more forcibly impressed upon the mind, if we reflect on how difficult it would be in a million or so of years hence to find any trace of what we now call the glacial epoch. The striated stones would be that time be all, or nearly all, disintegrated, and the till washed away and deposited in the bottom of the sea as stratified sands and clays. . . . (91:270).

In view of the facts presented by Croll, it would appear to be most unreasonable to insist on any fixed number of Pleistocene glaciations simply because hitherto it has been possible to group, in a very rough way, the comparatively few evidences we have into four "glacial periods."

It is a well-known fact that the chronology of four Pleistocene glaciations has been built on the assumption that all glacial epochs were the result of lowered world temperatures. Thus the European glaciations were declared to have been contemporary with the glaciations in America, although, as a matter of fact, no evidence of this existed. The assumption was based solely on astronomical and other theories of the causes of glaciation that we have shown to be inadequate. If the grouping of all European glacial evidences into only four major glaciations is questionable, and if, in addition, there is no good evidence that these glaciations were really contemporary with those in America, then the possibility of a large number of different glaciations in America and Europe during the Pleistocene must be taken seriously.

If the number of the alleged major Pleistocene glaciations is not satisfactorily established, the attempts at dating them leave even more to be desired. A review of the past and current literature on the subject reveals lack of agreement. Estimates vary widely, and none of them has convincing support. To make this plain, it is necessary only to compare the various estimates. Table 15 , on page 159, shows the estimates made by Penck and Bruckner, considered the leading European experts (whose work, however, was done before the development of nuclear techniques of dating), and by Zeuner, whose estimates were endorsed by the climatologist Brooks. The reader will note that Zeuner divides each of the older glaciations into a number of substages, some of which are longer than the entire period covered by the Wisconsin glaciation. The reader will recall that the interstadials and the successive advances of the Wisconsin glaciation had durations of the order of two or three thousand years; he may also note in·the various cores shown later on that all the cores show brief climatic changes of the same magnitude. It is therefore impossible to concede that the earlier glaciations could have had interstadials 40,000 or more years long. The only explanation ever advanced for the oscillations of the Wisconsin ice sheet that I know of is the one advanced in this book: massive volcanism caused by displacement of the crust. This explanation cannot, however, be reasonably applied to oscillations 40,000 years in length.

We see that the estimates of Zeuner (52:107) and of Penck and Bruckner (52:107) are in profound disagreement. It would be easily possible to multiply the number of such contradictory estimates, or, if the reader pleases, he may accumulate authorities who will support one of them; but is it not obvious that if leading professional geologists can differ to such an extent, no real reliance can be placed upon any of their very approximate and very speculative estimates? And when, in addition, we find they have all been wrong as to the number of Pleistocene glaciations—since a fifth one has just turned up—are we not justified in dismissing all these estimates as speculations that are no longer worth discussing?

TABLE 15
The Pleistocene Glaciations

Zeuner (European)		Penck & Bruckner (European)
Würm Glaciation		40-18,000
Stage III	25,000	
Stage II	72,000	
Stage I	115,000	
Riss Glaciation		130-100,000
Stage II	187,000	
Stage I	230,000	
Mindel Glaciation		430-370,000
Stage II	435,000	
Stage I	476,000	
Gunz Glaciation		520-490,000
Stage II	550,000	
Stage I	590,000	

If there is any doubt as to the reasonableness of dismissing these estimates, it should be put to rest by an entirely new estimate of the glacial chronology produced by Emiliani. Emiliani, working with marine cores and applying some of the new techniques of dating, has found that the earliest Pleistocene glaciation occurred only 300,000 years ago and that all the four recognized European glaciations, and their alleged American counterparts, have to be compressed into that comparatively short period (132). This finding, which there is no good reason to reject, completes our picture; it disposes finally, it seems to me, of the traditional glacial chronology of the Pleistocene.

As a consequence of this breakdown of the old theory, it seems to me that we must now start from the beginning and build a new glacial chronology of the Pleistocene. Our method can only be the tested method of science; to proceed from the known to the un-

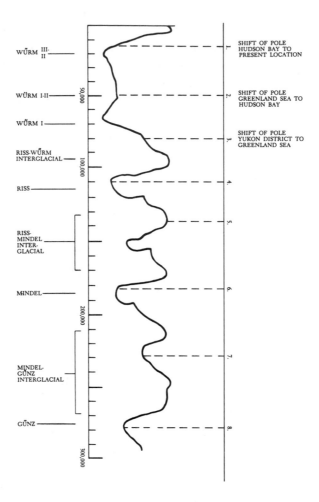

Fig. 31. *Temperature variations of the glacial Pleistocene, according to C. Emiliani, "*SCIENCE*," Vol. 123, p. 925, 1956, correlated with displacements of the lithosphere.*

known; from the Wisconsin glaciation, where our information is most ample, backward.

2. THE BEGINNING OF THE WISCONSIN ICE AGE

First of all, I shall have to jump the gun by suggesting that the North Pole was in the ocean north of Norway, approximately in Lat. 72° N, Long. 10° E, from about 75,000 until about 50,000 years ago. A good deal of the research work which has indicated this location for the pole at this time was carried out by Professor T. Y. H. Ma, of the University of Taiwan (289), to whom I am much indebted. Some general reasons for suggesting this location are:

1. The last ice age appears to have started a great deal earlier in Europe than in America. Some of the European geologists call their later ice age, contemporary with the Wisconsin glaciation, "Würm II," and the earlier, heavier glaciation "Würm I."

2. While this earlier European glaciation, Würm I, was in progress America seems to have enjoyed a long warm period called the "Sangamon Interglacial." A glacial period in Europe simultaneous with a very warm period in America can be well explained by a pole in this position. The evidence for all this will be presented in detail in the following pages. It is all the more important to discuss the evidence fully, because American geologists are at present all convinced on theoretical grounds that the American glaciation extends back as far as the European. They are, in fact, so convinced of this *that they have not looked for the evidence bearing on this question.** They expect that, when they get around to it, they will find evidence of glaciation in America between 40,000 and 80,000 years ago. I have news for them, however. They will not find it.

3. The transition between Würm I and Würm II in Europe was a much greater change than those fluctuations of climate that are called "interstadials." The evidence of radiocarbon tests shows that at the end of Würm I a vast ice cap of great thickness retreated from England and the Irish Sea and never returned. In some respects its history parallels that of the Wisconsin glacier. We shall return to this question a little later.

---

* I wrote a letter to one of America's leading glacial geologists asking him why none of the radiocarbon laboratories in America had tried to date materials as old as those dated in Europe showing the earlier phases of the European glaciation. I was much surprised, I told him, that no American laboratory had attempted dates in the range of 55,000–65,000 years ago, as had been done in Europe. I did not tell him my opinion that if they had done so they would have discovered that the Sangamon Interglacial had lasted well down toward 45,000 years ago, but suggested that perhaps they (or rather the geologists selecting and submitting samples) had been too busy with other things, especially since they felt justified in operating on the assumption that the European dates of the early glaciation must also apply in America. I received a very courteous reply, in which the writer agreed that the most probable reason the earlier dating had not been attempted was that the laboratories (and the geologists) were too busy with other (and by inference more important) matters.

3. THE BIRTH OF THE ICE CAP

I have already cited (Chapter V, page 130) a statement of Coleman's, quoting Low to the effect that the central area occupied by the Wisconsin ice sheet now shows few signs of having been glaciated at all. The preglacial debris in the central area is largely undisturbed, while a broad circle around it is "scoured clean to the solid rock." This fact is actually of very great significance; it is, or might be, one of the great unsolved mysteries of the Pleistocene glacial period mentioned by Daly. It seems that the ice cap did not start in a small area and expand outward, but rather it started all at once over a great area. It deepened until it became thick enough to move by gravity, and then the edges of the ice cap, in contact with the ground, were set in motion by the pressure of the ice behind them and began their scouring action, striating the bed rock of the continent over which the ice cap expanded. This suggests, indeed, a rapid transit of the pole from its (assumed) position in the Greenland Sea. It was fast enough so that the ice cap did not start growing at the coast and move inland; it *started* in the Hudson Bay region, and after it had grown thick enough to move, it spread outward in all directions.

A very remarkable evidence of the suddenness with which the ice cap was born is the fact that it contained thousands (and perhaps millions) of animals of a temperate climate, many of them frozen entire into the ice, including mastodons, mammoths, bear, elk, beaver, and so forth.

When the ice cap melted, many of these animals were dropped into bogs, which preserved their bodies and sometimes even the contents of their stomachs. It is evident enough, from the assemblage of species, that the snow overwhelmed them while they were living in temperate conditions. What this may mean is actually rather frightening to contemplate.

The ice sheet had to grow to a thickness of several hundred feet before it began to move. Then it began its long journey, advancing century after century until finally it reached the limit of its expansion in the Ohio valley to the south. It seems to have entered Ohio for the first time about 43,000 years ago. Simultaneously we have radiocarbon evidence of the expansion of mountain glaciers on the Pacific Coast. After this the ice cap seems to have gone on growing until its climax with the Tazewell Advance about 17,000 years ago. The whole development of the ice cap, then, seems to have required between 30,000 and 40,000 years. This is exceedingly interesting considering its very abrupt departure. An ice cap that may have taken as much as 40,000 years to develop disappears, for the most part, in 2,000. It must be obvious that this *could not* have been the result of gradually acting climatic factors usually called upon to explain ice ages. With such factors (see Chapter II) the ice retreat would be at about the same tempo as the advance. But what factor could have produced melting at a rate twenty times faster than accumulation, except such a basic factor as a polar shift? Considering the indicated speeds of the polar shifts, is it surprising that there was an acceleration of the rate of erosion and accumulation of sediments at the end of the ice age?

4. DATING THE POLAR SHIFT

Radiocarbon evidence has established the occurrence of a warm period in Europe between the two phases of the ice age, Würm I and Würm II. This is called the Göttweig Interstadial (see Fig 21 page 94), and lasted from about 50,000 to about 43,000 years ago. In our view this warming of the climate marked the departure of the pole from the Greenland Sea. In a way it was not an interstadial at all. Its length itself suggests that it had a different cause. In any case we interpret it as a climatic improvement resulting from the fact that the pole had moved from just north of Norway to distant America. Table 16 gives the radiocarbon evidence for this warm period.

TABLE 16

The Göttweig Interstadial*

GrN-4252;IX,68:Hengelo, Netherlands .... $50,000 \begin{smallmatrix} +4000 \\ -2400 \end{smallmatrix}$

GrN-4289;IX,68:Hengelo, Netherlands .... 51,600+
Reporter's comment: "Pollen diagrams for layers dated . . . indicate tundra vegetation almost devoid of trees. Existence of organic remains, however, may suggest a slight improvement in the climate during this period."

* Reference for all dates, 351a.

Note: It is worth noting that these two dates, taken together, suggest that the warming of the climate began at *least* 51,600 years ago, but *no more than* 54,000.

GrN-2677;IX,88:Brühl, Germany .... $50,400^{+2600}_{-1900}$

Reporter's comment: ". . . soil referred to as Göttweig."

GrN-2675;IX,87:Erkelenz, Germany .... $49,000^{+2000}_{-1700}$

Reporter's comment: ". . . Humus extracted from fossil soil . . . considered to have originated from Göttweig soil."

Ta-100;X,128:Karakula series, Estonia .... $48,100\pm1700$

Ta-101,X,128:Karakula series, Estonia .... $48,100\pm1650$

Reporter's comment: These two dates on samples from intermorainic deposits dated a warm period, according to the pollen diagram.

Hv-527;IX,232:Niedersachsen, Germany .... $45,000\pm800$

Reporter's comment: According to pollen analysis, ". . . bed belongs to Eem Interglacial. Two samples of same trunk (of tree) . . . furnished identical results."

Note: Since the Eem Interglacial is now considered to have ended about 80,000 years ago, and the bed is of Göttweig age, this sample shows that by this date the climate had warmed up beyond the tundra stage. See below.

GrN-1715;IX,73:Moerschooft series, The Netherlands... $43,500\pm1000$

GrN-1718;IX,73:Moerschooft series, The Netherlands... $46,250\pm1500$

Reporter's comment: "The above dates . . . support the view that tundra phase of Lower Pleniglacial started sometime before 50,000 B.P. and lasted until ca. 43,000 B.P."

Note: It is possible that conditions in The Netherlands reported as "tundra" may have differed from those at Niedersachsen, in Germany (Hv-527, above) at the same time. Niedersachsen, in Holstein, is on the eastern side of the Danish Peninsula, facing eastward toward the Baltic. Holland, on the other hand, faces the North Sea. According to our assumptions, the pole was now in Hudson Bay, so that The Netherlands were north of Denmark, while the Baltic, lying then on the *south side* of the Danish Peninsula, would have been warm. In addition, the expansion of the North American ice cap by this time would have been having a refrigerating influence on the North Atlantic.

GrN-1063;IX,80:Upton Warren, Great Britain ....$42,100\pm800$

Reporter's comment, on samples of organic material: ". . . It is thought that deposition of gravel was connected with the retreat of the Irish Sea Glacier . . . Floral and faunal remains

indicate treeless landscape, but climate prevailing was comparable to southern Sweden [today] . . ."
Note: Upton Warren, in Lat. 52°19′ N., Long. 2°5′ W., is about in the center of England. It appears that, sometime before, the glacier had withdrawn from its farthest south position about 20 miles north of Upton Warren. It was perhaps quite a while before, as by now the climate was very much improved.

It may be recalled that, at the last presumed displacement of the crust, at the end of the ice age, the polar shift was indicated by the Brady Interstadial, which evidently was, like the Göttweig phase, not really an interstadial. The two periods were similar in being caused, it would seem, not by intervals between volcanic upheavals but by two different polar shifts. They were similar in that they were comparatively long, the Göttweig lasting about 7,000 years and the Brady about 3,000. They were followed by colder periods, the Göttweig by a cold phase that lasted about 3,000 years and the Brady by the Cary Advance, that lasted about 1,500 years. It seems that the proportions were maintained. The cold phase after the Göttweig was followed by the Hengelo Interstadial, which lasted 2,000 years, while the Cary Advance was followed by the Two Creeks Interstadial, that lasted about 1,000 years. There seems to be a suggestion here of a repetitive pattern, but the evidence is as yet insufficient for definite conclusions. Some important Hengelo dates are presented in the table below:

TABLE 17
The Hengelo Interstadial *

GIN-149;IX,441:Yakut, USSR . . . . 40,760±580.
Reporter's comment: "Wood from the Niimingde River, a branch of a tributary of the Lena River, in Lat. 67°25′ N., Long. 125°20′ E., in peat layer overlain by redeposited glacial sediments from the Verkhoyansk Mts. Cold pollen spectrum."
Note: The wood dates a nonglacial phase, but does not date the glaciation of the mountains, which may be related to the period of the pole in the Greenland Sea, which would have rendered the Siberian coast glacial. The shift of the pole to Hudson Bay may account for the deglaciation and the growth of the tree. Later the glacial material may have been redeposited by river action on the peat bed.

U-181;II,120:Spitzbergen . . . . $40,300 {+4100 \atop -2900}$

* Reference for all dates, 351a.

U-118;II,120:Spitzbergen .... 37,000$^{+3000}_{-2000}$

U-89;NN,119:Spitzbergen .... 39,700$^{+1500}_{-1300}$

The above three samples were shells from the highest beach levels on Spitzbergen, thought to mark the high stand of the sea, supposedly marking the minimum amount of ice locked up in the glaciers.

Note: If this deduction is correct, then this period, the Hengelo Interstadial, should mark a time after the major deglaciation in Europe following the polar shift to Hudson Bay, and before the major growth of the Wisconsin continental ice sheet, when the latter was still comparatively thin.

TA-136;X,380:Estonia .... 39,180±1960

Large wood fragments from inter-morainic stratum. Evidence of alder, fir, spruce, pine, white beech, elm.

Note: Here is good evidence of our warm Baltic after the polar shift to Hudson Bay.

GrN-4828;IX,68:Hengelo, The Netherlands .... 30,350±550

GrN-1763;IX,68:Hengelo, The Netherlands .... 37,500±650

Reporter's comment: ". . . the above dates suggest that the Hengelo Interstadial began about 38,000 to 39,000 B.P."

UCLA-957;VIII,479:Niah Cave, Sarawak .... 37,500±1600

Reporter's comment: Oyster shells . . . date highest ocean level at Niah in late Pleistocene and confirm the 40,000 yr (minimum) age of homo sapiens skull, as shell layer stops just above skull . . .

Note: As with the Spitzbergen shells, these oysters may date a time of minimum ice after the previous polar shift.

GrN-2438;VI,352:Radosina, Czechoslovakia .... 38,400$^{+2800}_{-2100}$

Reporter's comment: [on charcoal] ". . . geological age would be optimum of Würm I/Würm II Interstadial."

Note: This means that the charcoal dates the warmest time of the Hengelo Interstadial. This warm phase is explainable through the polar shift to Hudson Bay, as in the case of the Baltic samples. Radosina, with the pole in Hudson Bay, would have lain nearly 400 miles *south* of The Netherlands, which is roughly half the latitude difference between New York and Jacksonville, Florida.

NPL-97;VIII,344:North Pembrokeshire, Wales .... 37,310$^{+1515}_{-1275}$

Marine shell from gravel pit. Reporter's comment: "relates . . . to outwash gravels laid down during melting of the Irish

Sea ice sheet . . . evidence suggests that interstadial boreal [non-glacial] conditions persisted in Irish Sea 38,000 yr B.P."

Note: We may here note that the Irish Sea ice sheet, apparently beginning its retreat from central England sometime before 42,000 years ago, had left the Irish Sea itself by 5,000 years later. Here, then, was a major deglaciation, not a mere oscillation of the ice sheet. The Irish Sea ice sheet never returned. These dates are minimal: we cannot tell as yet how long the deglaciation took. However, we do have the following date for the advance of the glacier, one of the oldest radiocarbon datings ever made.

GrN-1475;IX,80:Chelford, England . . . . 60,800±1500
Reporter's comment: ". . . Sample from upright stump of *Picea*. According to Simpson and West . . . stratigraphic and pollen analytic evidence points to early Weichselian [Würm I] Interstadial—mud band presumably correlated with Brørup Interstadial of Denmark. Above bed is boulder clay thought to be associated with first main advance of last glaciation, the Irish Sea Glaciation, which therefore postdates Chelford Interstadial . . ."

If the advance of the Irish Sea ice sheet into England occurred after the Brørup Interstadial (see Fig. 21, page 94), which ended 55,000 or 56,000 years ago, and the withdrawal began before 42,000 years ago, as suggested by the radiocarbon date from Upton Warren (Table 16, GrN–1063), then we have about 10,000 years for the maximal stage of that glaciation. This duration compares very favorably with the maximal stage of the Wisconsin glaciation, from its main advance about 25,000 years ago to the beginning of the Brady Interstadial about 16,000 years ago.

The Irish Sea ice sheet that invaded England about 55,000 years ago was very thick. It filled the Irish Sea basin and was thick enough to sweep across the ridges of the Welsh mountains. Forrest describes the glacial evidence of this ice sheet that entered England from the northwest (164). He feels that so massive an ice sheet had to originate on a land mass larger than Ireland and therefore supposes that there was then such a land mass in the North Atlantic. His views have been opposed by geologists, but it now appears that Soviet geologists have been led by other evidence to suppose the same thing (see Chapter IV).

To reconstruct the chronology of the previous shift of the pole it is necessary to go back a stage earlier to the time when the pole arrived in the Greenland Sea. Evidence of this still earlier shift will be presented below. Here I want to mention merely the fact that it occurred, so far as we can see, between 80,000 and 75,000 years ago. If we are correct, the chronology, then, is as follows:

By 75,000 years ago glacial conditions had begun in Europe, and a great ice sheet formed, perhaps on a land bridge connecting the British Isles with Iceland and Greenland. After a few thousand years it had grown thick enough to move out from the center of thickest accumulation by gravity. It covered Ireland, filled the Irish Sea basin and swept across the Welsh mountains into England about 55,000 years ago. Its advance, it appears, required a period of 20,000 to 25,000 years, comparable to the time that was apparently required for the growth of the Wisconsin ice sheet. We have just noted evidence suggesting that a polar shift occurred at about the time this ice sheet reached its maximum: the shift from the Greenland Sea to Hudson Bay. It is easy to connect the two events: The shift brought the advance to an end and initiated the glacial retreat. This retreat probably started long before 42,000 years ago, and by about 37,000 years ago the ice sheet had left the Irish Sea basin. The decline of the Irish Sea ice sheet seems to have required a period of time similar to the time required for the decline of the Wisconsin ice sheet. The parallel is extremely interesting.

It cannot be claimed that events of this magnitude are comparable to the minor fluctuations we have termed interstadials.

5. WHAT HAPPENED IN AMERICA

We have just seen that, between 55,000 and 43,000 years ago, there was a vast improvement in the climate in Europe with the disappearance of the Irish Sea ice sheet. Exactly the opposite thing happened in America, as the Wisconsin ice sheet entered on its vast expansion. Table 18 lists radiocarbon samples that have dated its advance:

TABLE 18*
The Advance of the Wisconsin Ice Sheet

1. GSC-217;VII,29:Ontario, Canada .... 47,700±1200
   Wood from the so-called "Port Talbot Interstadial beds" on

---

* Reference for all dates, 351a

the north shore of Lake Erie. Other samples from these beds have been dated as follows:

> Gro-2570 ...... 47,000±2500 †
> Gro-2597 ...... 47,500±250 †
> Gro-2619 ...... 44,900±1000 †

Note: These dates establish the fact that the Wisconsin ice sheet, probably advancing from the area of Hudson Bay, had not yet reached the Great Lakes. Lake Erie is about 22° or 1,300 miles south of our assumed position for the pole. Since forests were still growing on the north shore of Lake Erie it is probable that the ice cap was still in a very early stage of development. This in turn suggests that the polar shift may not have been completed much before 50,000 years ago.

2. L-409;I,9(a) and (b) :Ontario, Canada (a) .... 40,760±580
    (b) .... 43,100±3000

These two dates are from one sample. (a) is from wood cellulose; (b) from untreated wood. (a) is presumably more dependable. The sample is from the Don Valley, Lat. 43°41′ N, Long. 79°22′ W. The Don beds contain warm climate fossils. There is a possibility that the sample was from a tree that grew at that locality just before the ice arrived, while the warm climate fossils could date from a considerably older (and warmer) period. The dates indicate that as late as 40,000 years ago southern Ontario was still ice-free.

3. Y-1165;V,312:Quebec, Canada .... 42,000+

The sample was peat from Seven Mile Island, Quebec. Reporter's comment: ". . . the sample . . . gives [a minimum date for] marine inundation of James Bay. The rhythmites [in the sample] are correlated with the Missinaibi beds of Ontario, with an age of 53,000 (Gro-1453, unpublished). This is the first indication of the possibility that James Bay was ice-free during a Wisconsin Interstadial."

Note: A Gröningen date of 53,000 years for an ice-free James Bay is of great importance for us. James Bay is a part of Hudson Bay and therefore very close to the center of the area to be glaciated by the Wisconsin continental ice cap. The date suggests that the pole had not completed its transit to Hudson Bay, for the freezing of James Bay would have been the first result of the shift. Unfortunately the margin of error for the date is not given, and it may have been of the order of two or three thousand years. It is therefore possible that the ice cap did not begin to form until about 50,000 years ago.

† See references, 351a:VII, 29.

4. L-477B;III,144:Fayette County, Indiana .... 41,000+
   Wood from a silt layer underlying till. Reporter's comment:
   ". . . appears to represent pre-glacial sediments laid down before the first advancing Wisconsin glacier."
   Note: This suggests that the ice cap reached Indiana after
   41,000 years ago, but we do not know how long after.

5. W-869;II,139:Alexander County, Illinois .... 37,000±1500
   Shell from depth of 39 feet in 52 ft.-thick layer of fossiliferous
   loess overlying Sangamon age soil.
   Note: This sample suggests that in this local area 52 feet
   of loess accumulated over soil formed during the Sangamon
   Interglacial, before the ice sheet entered Illinois. Since 39 feet
   of this loess accumulated after the sample date, and only 13
   feet before, there might be a ratio of 3:1 in the two lapses of
   time. If glacial conditions producing the loess began 3,000
   years before the date, the ice sheet would have entered Illinois
   about 27,000 years ago. The date seems reasonable.

6. W-729;II,137:Madison County, Illinois .... 35,200±1000
   Snail shells from thick loess section. Sample comes from 4th
   layer from bottom. The lowest layer overlies "Sangamon age
   soil, with typical soil horizons." Since no glacial materials un-
   derlie this loess deposit it must be taken as representing the
   period of cold climate preceding the first advance of the Wisconsin ice sheet into Illinois. The considerable thickness of loess
   over the shell suggests a considerable delay after this for the
   arrival of the ice, so this sample is in good agreement with the
   previous sample (W-869).

7. W-526:II,136:Virginia, Illinois .... 29,000±1200
   Wood from a depth of 20 ft. in rusty-gray silt, resembling the
   Farmdale loess.

8. W-638;II,153:Lake Geneva, Wisconsin .... 31,800±1200
   Spruce wood, found in glacial till, apparently from a tree destroyed by the "Rockian" advance of the ice sheet.

9. W-901;III,89:Waukesha, Wisconsin .... 30,800±1000
   Spruce log from depth of 60 ft. in gravelly sand outwash considered to date the "Rockian" advance of the ice sheet.

10. UW-31;VII,498:Des Moines, Washington .... 36,800±1800
    Wood from deposit of peat, silt and clay ". . . beneath about
    300 feet of sand and gravel formed by advancing Vashon
    glacier."
    Note: The Wisconsin ice sheet did not reach the Pacific
    coast states, but simultaneously with its expansion local glaciers
    expanded in the state of Washington. It is worth noting that
    the expansion of the glaciers on the Pacific coast was coinci-

dent with the retreat of the Irish Sea Glacier. There is no satisfactory explanation for this in the geological literature. This was the time of the "Rockian Advance" in the United States.

11. M-1256;V,230:Klickitat, Washington .... 30,000±2500
Wood from 241 ft. below the surface in drill hole. It occurred in a layer of sand and gravel below 210 ft. of olivine basalt, in turn overlain by 40 feet of silt, clay and gravel. Comment by E.E.C.: ". . . in view of stratigraphic position of wood beneath massive lavas, it seems doubtful that the C_{14} dating is correct."

Note: The wood appears to date the advance of the glaciers in Washington. The views of the commentator illustrate a tendency of scientific workers to discount otherwise acceptable evidence when it conflicts with expectations. It seems probable that the thick stratum of lava merely reflects a volcanically more active period in the state of Washington.

12. L-163a;I,5:Kenai Lowland, Alaska .... 39,000±2000
Iron-stained, lignitized wood collected from silt, sand and gravel unit from Kenai Lowland, Lat. 60°40′ N, Long. 151°23′ W. Reporter's comment: ". . . apparently represents sedimentation during advancing phases of an ice tongue from Alaska Range sources . . ."

13. The Meadow Creek series, British Columbia:
GSC-740;X,224 .... 43,800±800
GSC-720,*ibid* .... 42,300±700
GSC-733,*ibid* .... 41,900±600
GSC-716,*ibid* .... 41,800±600
GSC-542,*ibid* .... 33,700±300
GSC-493,*ibid* .... 32,710±800
GSC-715;X,225 .... 25,840±320
General comment: ". . . sequence of dates from separately conformable series of sediments indicates that this part of British Columbia was not occupied by ice from at least ca. 44,000 B.P. until after 26,000 yr B.P. . . ."

Note: This fact of the ice age in North America is one of many that are incomprehensible according to traditional methods of interpretation. This northern region (in Lat. 50°15′ N, Long. 116°59′ W) remained unglaciated during an ice age assumed to have started 80,000 years ago, until only 26,000 years ago. Why? Presently accepted ideas provide no solution. On the other hand, with our assumptions, the matter is quite simple. We suppose that when the ice cap developed in the area of Hudson Bay, 2,600 miles to the east of British Columbia, that area then lay that same distance *south of the pole*, or about 7° farther south than it does with regard to the present pole, which at present gives British Columbia a temperate climate. It was, then, only at the climax of the glaciation, when the continental ice cap covered about 4,000,000 square miles and had approached the foothills of the Canadian Rockies, and was having a marked effect in lowering the temperatures, that mountain glaciers finally descended into the valleys of British Columbia.

Let me summarize the conclusions to be reached from the dates reviewed in Table 18. It is believed by most geologists that the ice age began at the same time in Europe and America and followed the same course in both continents down to the final establishment of the present climate. Since in Europe an early part of the ice age apparently began between 80,000 and 70,000 years ago, the same is assumed for America. But, as this table shows, there is not the slightest evidence that glacial conditions existed in America prior to about 55,000 years ago. On the contrary, the evidence is that James Bay, practically in the very center of the area that was to be glaciated by the Wisconsin ice cap, was ice-free 53,000 years ago.

Secondly, the dates suggest that the climate changed in opposite directions in Europe and in America. In Europe, between 50,000 and 37,000 years ago, a gigantic ice sheet that had covered Ireland, the Irish Sea, Wales and half of England melted away, while in America the Wisconsin ice sheet was expanding rapidly. By 30,000 years ago the latter had invaded the United States and stimulated the growth of glaciers on the Pacific coast. It continued to expand rapidly through the Farmdale and Tazewell Advances, while, on the other hand, in Europe the Scandinavian ice sheet was comparatively thin, and in England glaciation was mostly confined to mountain glaciers.

It is time that these discrepancies were taken account of. I present a theory that can reconcile the facts. It is unfortunate that the theory makes so complete a break with accepted ideas. Nevertheless the truth, if it is the truth, must eventually be faced.

There are two radiocarbon dates that can be added to the impressive evidence for the location of the pole in the Greenland Sea.

One of them comes from South Africa, and it is especially interesting because the scientists who examined it frankly admitted themselves stumped in their efforts to explain it.

Y-468;351a:V,329:South Africa .... 27,700±2000
    Shells from Sedgefield, Knysna Division, South Africa (34°1′
    S, 22°48′ E) of species not now living at this latitude, but
    reflecting a water temperature about 5° C. higher.

This would be explainable by a pole located in the Greenland Sea
(at Lat. 72° N, Long. 10° E) because that position of the pole
would put Sedgefield 20° nearer the equator. The date of 27,000
years is too young to fit our assumption, but this may be the result
of some contamination of the sample, as the following "comment"
suggests:

". . . shells include *Calliostoma fultoni*, not now living S of Delagoa
Bay 8° of latitude farther N, and *Cerithium kochi*, an Indo-Pacific
species of similar implication . . . The possibility of slight contamination of the shells by modern (atmospheric) C_{14}, always present when
very old carbonate is used for dating, makes it impossible to be sure that
the Sedgefield deposit is not beyond the limit of C_{14} dating, and therefore
probably of last interglacial age; but the shells were not exposed to air
until collection, and contamination is not particularly likely."

A very small amount of contamination could account for an error
of 30,000 years in the date. However, I would not have cited this case
if the scientists involved had not considered it so puzzling.

GSC-370;351a:VIII,100:Cape Breton Island .... 51,000+
    Reporter's comment: ". . . pollen study by R. J. Mott indi-
    cates a northern boreal forest cover rather than that of the
    region today. . . ."

The minimum date suggests that the forest flourished when the
pole was, according to our hypothesis, north of Norway, in the
Greenland Sea. With this location of the pole Cape Breton Island,
Nova Scotia, would have lain slightly north of its present latitude and
much closer to the then arctic North Atlantic than it does to the
Arctic Ocean today, so that a more northern type of forest would be
expected to grow there. On the other hand, according to presently
accepted ideas, if the American ice age extended back as far as the
European, the forest could not have grown there during Wisconsin
time. The significance of the date is increased when we compare
it with Sample Y-1165 (Table 18), which indicates the deglaciation
of James Bay at possibly the same time.

6. THE YUKON DISTRICT AT THE POLE

The logic of the facts we have marshaled to support the assumptions of two different positions of the North Pole prior to its present location leads us to postulate a third. It is known that, prior to about 75,000 years ago, the North Atlantic was warm. The warm Eemian Interglacial climate prevailed in Europe. For this there has to be a reason, though none has been convincingly presented for it. We are compelled by logic to assume there must have been a still earlier shift of the lithosphere about that time. We have been compelled to look for the most logical site the pole may have occupied before it was located in the Greenland Sea. The site we have selected as an assumption for further scientific testing is the Yukon District of Canada, as indicated on our climatic chart, Fig. 21, p. 94.

The two greatest advantages of this assumption are that it can explain both the Eemian Interglacial in Europe and a preceding ice age in North America, which geologists have named the Illinoisan Glaciation. We note that here again Europe and America would have opposite climates. We assume that during the period of the Greenland Sea pole Europe was very cold while America enjoyed the Sangamon Interglacial. With the Yukon pole there was a reversal, and while lions and hippopotami enjoyed a warm European climate, the bitter cold of the Illinoisan ice age gripped the United States.

We will now examine various long-range climatic studies, covering the last 100,000 years and more, that have been made possible by the datings of deep-sea cores, to see how closely they confirm the conclusions we have reached for the whole period from radiocarbon evidence.

Part II.

THE EVIDENCE FROM DEEP-SEA CORES

1. THE ROSS SEA CORES

Let us first return to the Ross Sea cores (Fig. 24, page 108) already discussed (Chapter IV) and review their evidence for the whole period of the last 120,000 years (for the assumed pole in the Yukon carries us back that far).

We noted that a temperate period seems to have begun in the Ross Sea area of Antarctica about 40,000 years ago, the glaciers of a preceding cold period having by this time melted back from the coast. We assumed that such melting may have begun 10,000 years or more before that time and therefore have coincided with the previous polar shift.

How do the cores reflect the previous site of the pole, in the Greenland Sea? We have assumed by an educated guess that the North Pole was located then at Lat. 72° N, and Long. 10° E. The corresponding South Pole would have been located just off the Ross Sea (see Fig. 23, page 107). What have the cores to say about this? For the period in question, from 40,000 to roughly 80,000 years ago, the cores differ, and so we shall take them separately.

Core N-3 shows glacial sediment for the whole period but with changes that are rather significant. Proceeding backward (that is, downward in the core), we find a layer of fine glacial sediment, suggesting the tapering off of glacial conditions, between 36,000 and 40,000 years ago; then a layer of medium glacial material from 40,000 to 50,000 years ago; and then coarse glacial material back as far as 120,000 years ago.

Core N-4 shows fine glacial material between 40,000 and 50,000 years ago, and nonglacial material from 50,000 to 80,000 years ago. The nonglacial material is in disagreement with our assumption; however, it may be explained. There is always the possibility that bottom currents in the ocean may have eroded temperate-type sediment from a nearby part of the bottom, where it may have been deposited in one of the earlier temperate phases of climate indicated by the core, and redeposited it here. This possibility is favored by the fact that both Core N-3 and Core N-5 show glacial sediment during the whole period.

Core N-5 shows coarse glacial sediment from 40,000 to 50,000 years ago and medium glacial sediment from 50,000 to about 70,000 years ago. It then shows coarse glacial sediment back to 110,000 years ago and medium glacial sediment back to 140,000 years.

It is clear that there are several instances in the cores of sediment being displaced by bottom currents, but they appear to be exceptional. Despite them, the agreement of the cores is rather good. In the case of our assumed Greenland Sea pole, we have a 2:1 agreement in support of the assumption.

Going farther back and considering the assumed Yukon North Pole and the corresponding South Pole (Fig. 23, page 107), which would have been located in Lat. 63° S and Long. 45° E, we note that this South Pole would have been much farther from the Ross Sea than the present pole. The shores of the Ross Sea would have been deglacial. We find there is temperate-type sediment in cores N-4 and N-5, but the date for the suggested warm period is older than expected. The cores suggest a warm period in the Ross Sea between 180,000 and 110,000 years ago, which is perhaps not too much out of the way.

For this earlier period we note that Cores N-3 and N-5 show coarse glacial material, while Core N-4 shows fine glacial deposition, all three cores agreeing on the essential point of glacial deposition.

2. THE PIGGOTT-URRY CORES FROM THE NORTH ATLANTIC

Figure 29, page 151, shows three cores taken from the North Atlantic Ocean and dated by the ionium method of radioelement dating. The first of these, P-124–3, extends back only to 11,800 years ago and therefore can be disregarded. The second, P-126–5, taken from the middle of the Atlantic, goes back to 72,500 years. Going downward in the core, we find the following:

Back to 12,300 years ago the material is volcanic glass, indicating volcanic eruption somewhere at no great distance, a matter of some interest in itself, since the site of the core, taken in Lat. 48°38′ N and Long. 36°1′ W, is hundreds of miles from the nearest land. The volcanic glass itself was not dated by the ionium method; the date applies to a layer of warm-water sediments deposited on it and extending to 14,700 years ago. From 14,700 years ago to 23,700 years ago the core contains glacial materials. The fact that glacial deposition ceased 14,700 years ago agrees very well with dates of the deglaciation of North America as found by radiocarbon (see Chapter VI).

From 23,700 years to 41,900 years ago Core P-126–5 shows warmwater deposits except for two brief periods, when the sediments are "anomalous," meaning (I suppose) confused so that they cannot be classified.

What is the meaning of warm-water deposition in the mid-Atlantic for most of the Wisconsin glacial period? Here is an astonishing thing. I have not seen any reference to this contradictory fact in the geological literature. Nobody tries to explain it, yet it obviously has tremendous importance.

Perhaps it cannot be explained in terms of the present position of the pole, but, if we suppose that pole in Hudson Bay, an explanation offers itself. We have shown that, for the whole duration of the Wisconsin ice sheet, the Arctic Ocean was warm. Therefore, while the site of the core would have been closer to the pole then than now, it was nevertheless in a location to be influenced by currents of warm water from the Arctic Ocean. The core indicates this happened.

From 41,900 years ago back to 44,700 years the core shows glacial sediment. This can have resulted from a diversion of the warm Arctic current. Then, from 44,700 years back to 50,000, we again have the warm water.

At 50,000 years ago a major change of climate is indicated by the core. There are glacial deposition and volcanic glass back to 61,900 years ago, except for a brief period of warm water for 2,900 years, between 55,400 and 58,300 years ago. It would seem that the end of glacial deposition 50,000 years ago agrees very well with our assumption of a shift of the pole at that time from the Greenland Sea to Hudson Bay. It is not that the site of the core was any farther from the Hudson Bay pole than it had been from the Greenland Sea pole. The difference lay in the fact that the pole in Hudson Bay was cut off from the Atlantic by land masses, while there was nothing to interfere with the refrigeration of the North Atlantic by having the pole in the Greenland Sea. And of course, the Arctic Ocean itself was much colder when the pole was in the Greenland Sea. (The reader should follow this discussion with a handy globe.)

From 61,900 years to the bottom of the core, at 72,500 years ago, the core shows warm-water sediments. This is in agreement with our assumption of a pole in the Yukon District of western Canada. We explain the end of the warm interglacial in Europe by the shift of the pole from the Yukon to the Greenland Sea about 75,000 years ago. It would have taken some time to refrigerate the central Atlantic after this change.

As the reader will note, temperature studies based on organisms included in the core show that the temperature of the ocean fell steadily from about 70,000 years to 60,000 years ago, even while the deposition of warm-climate sediments continued. The polar shift, then, began to affect the ocean temperature about 70,000 years ago, and this is in close agreement with the other evidence we have discussed.

The temperature curve shown for Core P-126-5 contains another possibly very significant detail. Between 24,000 and 24,300 years ago there was "anomalous" deposition. At this very time the temperature of the ocean suddenly and briefly warmed up. The curve shows that the ocean took 600 years to warm up and 1,100 years to cool off again to the same level. It is probable that this event is connected with the "anomalous" character of the deposits. To me it suggests turbulence, some sort of upheaval of a geological nature. It does not seem to be reflected in the evidence from the land.

It is strange, too, that while the sediments deposited during most of the Wisconsin glacial period were warm-water sediments, the species of organisms found in the sediments suggested cold climate. We have noted similar contradictions on land, the mixtures of cold-climate flora with warm-climate fauna (See Chap. IV, pp. 104, 153, 169). This may be explainable as a result of the fact that, during Wisconsin time (50,000 to about 16,000 years ago), the bottom of the North Atlantic at the site of this core lay, by our hypothesis, about 12° farther north than it does now. It would have been about 750 miles nearer the pole in Hudson Bay than it is to the present pole. This would naturally have involved colder bottom waters than now. At the same time, the sediments may have been borne from the warm Arctic Ocean by warm currents traveling nearer the surface, as warm currents would naturally do. In this way we can explain the appearance of cold-climate organisms in warm-climate sediments.

There is good correlation between this core and the succession of climatic changes we have followed in Europe and in America. In order to point out the connections I will summarize the changes in the core chronologically, starting with 72,000 years ago.

Core P-126-5

(a) 72,500 to 61,900: nonglacial sediment in the core corresponding to the Amersfoort Interstadial in Europe.

(b) 61,900–58,300: glacial sediment resulting from the advance of the Irish Sea ice sheet and general advance of glaciers in Europe.

(c) 58,300–55,400: nonglacial sediment, corresponding to the Brørup Interstadial in Europe.

(d) 55,400–51,200: volcanic glass shards, indicating heavy volcanism at the very time of the assumed polar shift from the Greenland Sea to Hudson Bay.

(e) 51,200–50,000: glacial deposition, probably the last part of a cold period represented by the layer of volcanic shards, following the Brørup Interstadial.

(f) 50,000–44,700: nonglacial sediment, contemporary with the Göttweig Interstadial in Europe.

(g) 44,700–41,900: glacial sediment, corresponding with the time of the major advance of the Wisconsin glaciation in North America.

(h) 41,900–23,700: with two interruptions by anomalous deposits: warm-water deposits, discussed above.

(i) 23,700–14,700: glacial deposits, correlating with the period from the beginning of the Farmdale Advance to the Brady Interstadial.*

We may now turn for a moment to Core P-130-9, taken from the ocean bottom on the eastern side of the Mid-Atlantic Ridge. This core extends only to 24,300 years ago, but it differs from Core P-126-5 on some important points. For one thing, it does not indicate the warming of the ocean that took place in the mid-Atlantic. This suggests that what happened in the mid-Atlantic may have been due to the temporary diversion of some warm ocean current, perhaps from the Arctic.

Core P-130-9 indicates the end of glacial deposition 19,200 years ago. However, the organisms in the sediment indicate that warming of the climate started about 18,100 years ago, reached a peak 13,600 years ago and then cooled off again by about 12,000 years ago. This seems to reflect the Brady Interstadial, followed by the Cary Advance. However, it is surprising to find the end of glacial deposition in the eastern Atlantic apparently occurring at the beginning of the massive Tazewell Advance of the Wisconsin ice sheet. Of course, any dating method, however sound, is capable of minor errors, and

---

* It is worth noting that there is evidence here of apparently good agreement between the ionium and the radiocarbon findings.

therefore perhaps we should be satisfied with the degree of agreement that we have.

Another rather interesting point about Core P-126–5 is that the deposition of glacial sediment ceased about 14,000 years ago. It may be noticed, however, that in another of the Urry cores, P-130–9, which was taken much farther to the east, the deposition of glacial sediment ceased 18,000 years ago. One might at first be inclined to pass this over as an unimportant detail, until one realizes that with Hudson Bay at the pole the difference is completely explained. Under those circumstances the second core, which now lies to the east, would have been due south of the first core, and it would be entirely natural that the warming of the climate at the end of the ice age would be felt first in the more southerly region. This, then, constitutes additional evidence for the location of the Hudson Bay region at the pole during the period of the Wisconsin glaciation.

3. CORES FROM THE EQUATORIAL ATLANTIC

We now have to consider three cores taken in low latitudes, two from the Caribbean and one from the eastern Equatorial Atlantic (Fig. 30, p. 153).

To begin with, we note that according to these cores the temperature of the Atlantic Ocean was at a peak about 75,000 years ago. They are in substantial agreement on this matter with Core P-126–5 from the North Atlantic and are in accord with our assumption of a pole in the Yukon at this time.

We note, too, that the temperature in the Atlantic during this warm period was not, at the sites of these cores, as high as the temperatures now prevailing there, except for two very brief spurts in Cores A-180–75 and A-179–4. Yet we have seen that Core P-126–5 indicates clearly that the water was warmer at that time than it is at present. How is this conflict to be resolved? Shall we be forced to discredit the reliability of one finding or the other? Not at all. Our theory offers the possibility of eliminating this apparent conflict in the evidence.

Let us consider the present and past latitudes of these cores. Core P-126–5 was taken in Lat. 48°38′ N; the others were taken in very low latitudes. If we assume that the Yukon was at the pole during this warm period in the Atlantic, the site of this core would then have been farther from the pole than it is now; that is, it would have been south of its present latitude. Quite naturally the water would have been warmer. A glance at the globe will suffice to make this plain.

On the other hand, if we consider Core A-180–75, taken in the eastern Equatorial Atlantic, nearly on the equator, the opposite situation is revealed. A pole in the Yukon would displace this core southward from the equator, possibly as far as the 15th parallel of South Latitude (depending, of course, on the precise position of the pole, which is uncertain). Quite probably the water would be colder than it is now, other things being equal.

The two Caribbean cores would, with the Yukon pole, have approximately the same latitude as now; the uncertainty as to the precise location of that pole makes it impossible to draw any reliable conclusions from them.

Both the Caribbean cores indicate a temperature minimum about 55,000 years ago, which would be about the time of the maximum expansion of the ice sheets of the early ice age in Europe, when we suppose the pole to have been in the Greenland Sea. They show the following warm period which we interpret as the result of the polar shift to Hudson Bay between 55,000 and 50,000 years ago. They then show a gradual temperature decline from about 40,000 to about 11,000 years ago, which may correspond first to the growth of the Wisconsin ice sheet and then to the movement of the center of that ice sheet to Labrador during the declining phases of the glaciation.

At this point it is important to consider a contradiction between the Caribbean cores and the core from the eastern Equatorial Atlantic. It appears that the ocean temperature reached its minimum, after·the early warm period about 70,000 years ago, in the eastern Atlantic about 20,000 years before it did in the Caribbean. How is this to be explained?

It appears that this seemingly anomalous fact may constitute, in itself, one of the most impressive confirmations of the whole theory of displacements of the crust, for, if you look at a globe and visualize the crustal shift that moved the Yukon from, and Greenland to, the pole, and if you use a tape measure to determine the distances from each polar position to the Caribbean and to the equator off the bulge of Africa, you will see that that particular polar shift would make only a comparatively slight change in the latitude of the Caribbean but a very radical change indeed in the latitude of the eastern Equatorial Atlantic. And since the movement would take the same period of time in both cases, the rate of movement would necessarily be much more rapid in the eastern Atlantic, and this both agrees with and explains the core evidence.

A final point should be emphasized. All three cores show the sudden upswing of the sea temperature at the end of the ice age. However, they date the upswing at different times, and the time differences may be very significant. I have arranged the cores, below, in the order of the longitudes of their sites:

	Core	Longitude	Temperature Change
1.	A-179-4	74°48′ W	14,000 B.P.
2.	A-172-6	68°51′ W	16,000 B.P.
3.	A-180-75	25°22′ W	12,000 B.P.

Now, to begin with, according to our hypothesis, North America was shifted southward about 30° when the pole moved from Hudson Bay to its present site. The two Caribbean cores would accordingly have been shifted that distance southward, and the sudden rise in the temperature can be thus explained.

It is not satisfactory to retort to this idea with the answer that everybody knows there was a sharp rise of temperature in the ocean when the ice age came to an end. What brought the ice age to an end? *That is the question.* My contention is that if the lithosphere had not moved, the temperatures never would have risen. They would be just where they were. The rise indicated in these cores is exactly what would be expected with a displacement of the lithosphere in the direction and to the extent we have assumed.

Let me remind the reader that no one has ever explained the cause of any ice age or the cause of the end of this ice age. It certainly did not just happen by itself.*

But we still have to explain the differences of the dates shown in the cores for the sudden rise of temperature. Let us first consider Core A-180–75 from the eastern Atlantic. While the other two cores

---

* I have not discussed the Ewing-Donn Theory by which the Pleistocene glaciations of North America caused alternating periods of a frozen and an open Arctic Ocean. (See the Bibliography.) Besides being in conflict with the Lamont evidence cited above, this theory fails to help us with ice ages in general.

lay near the meridian of maximum southward displacement (which we suppose to be the 83rd meridian of West Longitude),* Core A-180–75 lay far from it and accordingly was displaced through less latitude. The western side of the Atlantic would then naturally warm first, and the eastern side would warm later.

So far as the two Caribbean cores are concerned, the differences of their longitudes (only 6° or about 360 miles) was probably less important than their geographical positions. Core A-172–6 was closer to the open Atlantic, while core A-179–4 was protected from the open ocean by the Greater and Lesser Antilles. It is natural, then, that the smaller body of water should have warmed up more quickly, protected from the influence of the great cold-water masses of the Atlantic Ocean.

4. THE GEOMAGNETIC EVIDENCE FOR THE POLAR SUCCESSION

Having concluded the presentation of the empirical evidence bearing on the assumed polar succession, the successive poles in the Yukon, the Greenland Sea and Hudson Bay, we may now briefly analyze the implications of the geomagnetic evidence of the Japanese lavas presented in Table 4 of Chapter I, page 14. In the following table I have rearranged the first six polar positions of that list in chronological order, the oldest first.

TABLE 19
The Geomagnetic Evidence from Japan

(1) Pole	(2) Comment
6. 50° N, 134° E (19° certainty oval)	Pole in the North Pacific, near coast of Siberia (Maritime Provinces). Could have been the cause of the Cordilleran ice sheet in North America and the Eemian Interglacial in Europe.

---

* Since the Hudson Bay pole is assumed to be at 60°N and 83° W, then its movement to the present site of the pole would be along the 83rd meridian, and a great circle around the earth would represent the meridian of the maximum movement. Two points 90° away on each side of the earth would represent the "pivot points" on which the crust or lithosphere would turn, and these points would not move at all. All points from the meridian of maximum travel to these pivot points would move precisely in ratio to their distance from the meridian.

5. 75° N, 7° E (20° certainty oval)

Magnetic pole moved great distance (about 50°) to vicinity of Greenland Sea, could have coincided with heavy early Würm I glaciation in Europe and Sangamon Interglacial in America.

4. 75° N, 22° W (23° certainty oval)

Magnetic pole moved 30° westward toward North America; geographic pole may have been en route to Hudson Bay from the Greenland Sea.

3. 76° N, 14° W (6° certainty oval)

Indicated 8° eastward motion of magnetic pole doubtful because of wide certainty oval of 4. The motion could still have been westward as the theory requires.

2. 80° N, 0 Long. (7° certainty oval)

Magnetic pole apparently moved eastward some distance, inconsistent with hypothesis.

1. 78° N, 80° W (14° certainty oval)

Magnetic pole moved a great distance westward (80°) suggesting movement of the geographic pole to North America, consistent with the Hudson Bay pole or with the present position of the geographic pole.

0. 75.5° N, 100.5° W

Present position of the magnetic north pole (as of 1965).

Despite the uncertainties in the present state of this geomagnetic evidence (and its small amount), I think the findings can be considered significant. It can hardly be maintained that enormous shifts of the position of the magnetic pole (up to 80°!), such as are indicated here, could have taken place while the position of the geographic pole was unchanged. Despite some contradictions, the evidence is in good agreement with the assumptions of our polar succession. Indeed the degree of agreement is a surprise to me at this early stage in the collection and analysis of the geomagnetic evidence.

chapter **8**

POLAR CHANGE IN
THE REMOTER PAST

IN THE previous chapters I have reviewed the evidence for three displacements of the earth's crust during the last 110,000 years. They seem to have occurred at intervals of 30,000 to 40,000 years. There are indications that they may have occurred at this rate through much of the Pleistocene Epoch. From the evidence we now have it seems futile to try to determine the locations of the poles in the more remote cases. With every step backward in time the evidence naturally becomes thinner. Eventually perhaps we may know more. At present, however, there are other matters to be considered.

We must consider whether the rate of geological change suggested for the last 110,000 years by the evidence presented in this book can be typical for the entire history of the earth. It is plain from the cores that rapid change has characterized the record for the Pleisto-cene. Radiocarbon dating has established the fact that all the geological processes of glacial growth and decay, precipitation and sedimentation, were accelerated during the Wisconsin ice age. Emiliani has argued, as already mentioned, that all the ice ages of the Pleistocene occurred in the last 300,000 years, which implies a threefold increase in the velocity of geological change as compared with the older views. Studies of the delta of the Mississippi River suggest numerous important changes at short intervals (165, 276, 349). Blanchard has shown that there were at least twelve major climatic changes in the valley of the Somme since the first glacia-tion, accompanied by changes in sea levels, fauna and flora, and human cultures. He argues that only polar change can explain this record (38).

For the older geological periods, there are a number of lines of evidence that suggest rapid change. So insistently, indeed, does this theme occur in the strata that Brooks, in his *Climate Through the Ages*, refers to a 21,000-year cycle of climatic change which he be-lieves operated through the whole Eocene Period, or for about 15,000,000 years. His figure, of course, is only a rough average, and the intervals may have been very unequal in length. With reference to a still older period he remarks, "Alternations in the Cretaceous of U.S.A. suggest a cycle that is estimated at 21,000 years, but there are no annual layers" (52:108).

Irregularities in the cycle are indicated by another study of Eocene beds covering about 5,000,000 to 8,000,000 years. In this case annual varves were present, and they indicated long-term changes at 23,000 and 50,000 years (52:108). Some scientists have attempted to explain these cycles as the result of the earth's astronomical precession, but in view of the above-mentioned irregularities the phenomena seem better explained in terms of crust displacements.

Naturally such frequent changes in climate have had profound effects on the formation of sedimentary rocks, one consequence, perhaps, being the thinness of the individual strata. Very seldom can deposits be found that indicate with any certainty the uninterrupted deposition of more than a few thousand years. On the other hand, innumerable cases of conditions interrupted after a few thousand years can be proved. In addition to the evidence mentioned above,

Brooks, for example, refers to a great salt lake or inland sea that existed in Europe in the Permian Period, and says:

The number of annual layers indicates that the salt lake existed for some 10,000 years, after which the salt deposits were covered by a layer of desert sand (52:25).

Wallace, too, refers to the evidence of sudden changes in climate at short intervals in his *Island Life*: ". . . the numerous changes in the fossil remains from bed to bed only a few feet and sometimes a few inches apart" (435:204).

Some of the best evidence is provided by coal seams, which are ordinarily thin and interlayered with rock indicating very different climatic conditions. There has developed a considerable literature on the rate of coal formation, and some recent experimentation has thrown light upon it.

Croll devoted considerable attention to the problem. He estimated that it would take about 5,000 years for the formation of one yard (or about a meter) of coal (91:429), and came to the conclusion that the periods of coal formation between changes in climate were about 10,000 years long. It is obvious that any changes that replaced conditions required for coal formation by conditions suitable for the formation of sedimentary deposits beneath the sea (for Croll points out that rock strata between the coal strata are usually of marine origin) (91:424) were indeed radical changes, taking place in short periods of time. Another writer, Otto Stutzer, after very careful calculation, concluded that a Pittsburgh coal bed seven feet thick could have been formed in no more than 2,100 years (397).

In view of all this evidence we must not be too much impressed by the very thick layers of rock that are occasionally found. Croll, who was a sound geologist even if his theory about ice ages was not accepted, pointed out that

. . . The thickness of a deposit will depend upon a great many circumstances, such as whether the deposition took place near to land or far away in the deep recesses of the ocean, whether it occurred at the mouth of a great river or along the sea-shore, or at a time when the sea-bottom was rising, subsiding or remaining stationary. Stratified formation 10,000 feet in thickness, for example, may under some conditions, have been formed in as many years, while under other conditions it may have required as many centuries (91:338).

It is worth noting that at a number of points the evidence for great and frequent changes in the earth's climatic conditions is linked with evidence of structural changes in the earth's crust; that is, with changes in the elevation of lands and in the distribution of land and sea.* Croll remarked:

. . . It is worthy of notice that the stratified beds between the coal seams are of marine and not of lacustrine origin . . . If, for example, there are six coal seams, one above another, this proves that the land must have been at least six times below and six times above sea-level (91:424).

Coleman has emphasized the frequent association of abrupt breaks in the continuity of the strata with extreme changes of elevation above or below sea level. In discussing the Permo-Carboniferous period in India, he says:

There are the usual cold climate fern leaves in these beds, and above them, without an apparent break, come the Productus limestones with marine fossils (87:102).

Now, it seems altogether reasonable to suppose that if changes of climate were associated with changes of elevation in these different kinds of cases, then the two may have occurred at the same tempo and have proceeded from the same cause. The hypothesis of periodical shifts of the earth's crust provides both the link and the cause.

An interesting study of repeating geological cycles in a very remote period has been completed by Weller (440). He deals with the so-called "Pennsylvanian" period several hundred million years ago, which had a span of between 35 and 50 million years. He points out that in the study of this period numerous examples have been observed of the deposition of different kinds of sedimentary beds in the same order at irregular intervals of time. The changes in the composition of the beds imply changes both in climate and in the elevation of the areas above sea level. The cycles are not just local but can be traced over wide areas (440:110). Furthermore each complete cycle represents an advance, retreat, and readvance of the sea. Weller accounts for the cycles by diastrophism—that is, by some

* The effects of crust displacements in changing sea levels will be discussed in Chapter IX below.

sort of upheaval in the earth, some activity within the earth's body—but is not able to specify its nature. He recognizes about 42 cycles during the period, with each cycle having a duration of about 400,000 years.

These cycles would appear at first glance to be considerably longer than those that might result from crust displacements. However, there are a number of factors that tend to lessen the apparent difference between them. First, Weller points out that discontinuities in the deposits he is discussing are far more numerous than is generally supposed (440:99–101). This means that a part of the record is missing. Then, we must remember that a complete cycle, involving the retreat and advance of the sea (probably in a number of stages), would call for several movements of the crust. At any one point on the earth's surface, several movements might be required to bring the sea level to its lowest point, and several more to bring it to its highest point. Moreover, Weller points out that in each of his cycles deposition has been interrupted twice, thus reducing the average length of the subdivision of the cycle to periods of the order of 75,000 to 250,000 years. But it must be remembered that we have only averages; the cycles differ greatly and their subdivisions also differ greatly in length. When we consider the fact that the intervals and directions of crust displacements are necessarily irregular, there appears to be a very good agreement between our theory and the facts of the Pennsylvanian cycles. At least it will hardly be denied that the theory offers the first possibility of understanding the cycles. Moreover, if our recent experience of the shortening of our estimates of geological time in the Pleistocene is a valid basis for extrapolating to earlier periods, it may well be that Weller has attributed too great an average duration to his cycles. It appears therefore that crust displacements may have been occurring through the whole of one of the major subdivisions of the Paleozoic Era.

It is impossible in the present state of the evidence to say that displacements of the crust have been going on uninterruptedly all through geological history. It may be that there have been times of quiet. The important thing at the moment is that investigators should be willing to undertake further inquiry without preconceptions based on outmoded ideas of gradual change. We may note a serious warning against this bias uttered by no less an authority than Sir Charles Lyell, the greatest geologist of the first half of the nineteenth century, and the father of gradualism in geology. In the course of a discussion of some evidence of recent folding of rock strata on the Danish island of Moen, he remarked:

It is impossible to behold such effects of reiterated earth movements, all of post-Tertiary date, without reflecting that, but for the accidental presence of the stratified drift, all of which might easily, where there has been so much denudation, have been lacking, even if it had once existed, we might have referred the verticality and flexures and faults of the rocks to an ancient period, such as the era between the chalk with flints and the Maestricht chalk, or to the time of the latter formation, or to the Eocene, or Miocene or older Pliocene eras . . . Hence we may be permitted to suspect that in some other regions, where we have no such means at our command for testing the exact date of certain movements, the time of their occurrence may be far more modern than we usually suppose (281:393-94).

And let us also recall the following words of the greatest geologist of the second half of the nineteenth century, Eduard Suess:

The enthusiasm with which the little polyp building up the coral reef, and the raindrop hollowing out the stone, have been contemplated, has, I fear, introduced into the consideration of important questions concerning the history of the earth a certain element of geological quietism—derived from the peaceable commonplaceness of everyday life— an element which by no means contributes to a just conception of those phenomena which have been and still are of the first consequence in fashioning the face of the earth.
The convulsions which have affected certain parts of the earth's crust, with a frequency far greater than was until recently supposed, show clearly enough how one-sided this point of view is. The earthquakes of today are but faint reminiscences of those telluric movements to which the structure of almost every mountain range bears witness. Numerous examples of great mountain chains suggest by their structure the possibility, and even in some cases the probability, of the occasional intervention in the course of great geological eras of processes of episodal disturbances of such indescribable and overwhelming violence, that the imagination refuses to follow the understanding and to complete the picture of which the outlines are furnished by observations of fact. (398: I, 17-18).

The great work from which the foregoing statement was taken is entitled *The Face of the Earth*. The prospect that unfolds before us, as we contemplate the possibility that total displacements of the earth's crust have been a feature of geological history since the formation of the crust itself, is nothing less than the discovery of the formative force, of the shaping factor, that has been responsible not only for ice ages, not only for the mountain ranges, but possibly for the very history of the continents and for all the principal features of the face of the earth.

chapter **9**

THE SHAPING OF EARTH'S SURFACE FEATURES

Part I.

MOUNTAINS AND GREAT RIFTS

BY FAR the most magnificent features of the lithosphere are the lofty mountain ranges that are found on all the continents, exciting the wonder of man, and those other, equally tremendous mountain ranges that lie drowned in the silent depths of the sea. These mountain ranges carry in their intricate formations much of the history of the lithosphere. If we could know the forces that produced them, we could grasp the basic dynamic principles of the earth's development. Unfortunately, though the mountains have long been the subject of intensive scientific investigation, they have preserved their secrets well. The most important of these secrets is the secret of their birth. What forces within the earth were responsible for their formation? As of now we do not know.

Nothing could better betray the extent of our ignorance of the dynamic processes that have shaped the face of the earth than this confession of ignorance. Yet it is agreed by geologists that no theory has so far satisfactorily explained mountain building. Daly, for example, has referred to the process of the folding of the rock strata, a phase of mountain building, as "an utterly mysterious process" (70d:41). Gutenberg has concluded that none of the present theories will do. He remarks that "all the forces discussed so far seem to be insufficient to produce the formation of mountains" (194:171), and this includes, of course, the long-exploded (but still widely current) theory that ascribes mountains to the cooling and shrinking of the earth. As to this, Gutenberg remarks, ". . . other scientists have pointed out that the cooling of the earth is not sufficient to produce the major part of the crumpling, especially since investigations of the radioactive heat which is produced inside the earth have indicated that the cooling of the earth is less than it had been originally believed . . . " (194:192). Bullard, reviewing the third edition of Harold Jeffrey's basic work, *The Earth*, notes the absence of progress toward solving the problem of mountain building since the second edition thirty-seven years ago (59). Pirsson and Schuchert, the authors of a general text on geology, conclude a section on the cause of mountain building with the statement: "It must be admitted, therefore, that the cause of compressive deformation in the earth's crust is one of the great mysteries of science, and can be discussed only in a speculative way" (345:404).

More recently, P. Chadwick, after quoting a statement by F. D. Adams, written in 1938, on the failure of attempts to solve the problem of mountain building, added:

Since these words were written, the earth's surface features have been studied more intensively than ever before and much thought has been given to the processes by which they have evolved. But a deeper understanding has brought with it an increased awareness of the extreme difficulty and complexity of the subject, and the possibility of a final solution appears, if anything, to have receded (363a:195).

What is the nature of this problem that has so far baffled science?

1. THE PROBLEM OF LITHOSPHERIC FOLDING

It is important to take into account the fact that there are several kinds of mountains, and that their origins may be ascribed to somewhat different circumstances, even though (as we shall see) they may be related to one underlying cause. Some mountains are caused by volcanic eruptions. These consist of piles of volcanic matter. Some of the greatest mountains on the earth's surface are volcanic mountains. Many of them are found on ocean bottoms, and when they rise to the surface they form the island chains (such as the Hawaiians) that are especially numerous in the Pacific. Sometimes volcanic islands or mountains can be formed quickly, as was the case recently in Mexico, where a large cinder cone, Paricutín, was developed in a few years to a height of several thousand feet from an eruption that started in a cornfield on the level ground. Some mountains result from a vast flow of molten rock that gathers under the lithosphere at one spot and domes it up. The causes of these events are unknown.

Many mountains, and even whole ranges of mountains, are brought into existence in part by the cracking of the earth's lithosphere, accompanied by the tilting of the separated blocks. The Sierra Nevada Mountains of California appear to have been formed in this way. According to Daly they represent the tilting of a single block of the lithosphere some 600 miles long (98:90). Some folding of the lithosphere, however, had previously taken place. Many great chasms, extended cliff formations, and rift valleys appear to have been formed by the cracking and drawing apart of the lithosphere and by the elevation or subsidence of the different sides. The great African Rift Valley is perhaps the best-known example of this sort of formation; the rift of which it is a part, as we shall see below, has recently been connected with a worldwide system of great submarine rifts. The cause of all this cracking and tilting is still one of the mysteries of science.

The greatest mountain systems on the earth's surface have been formed as the result of the lateral compression and folding of the lithosphere. Since folding is the cause of most mountain building it must hold our particular attention. As already suggested, science is at a loss to explain the folding. A number of suggestions have been advanced, but they all, including the convection-current hypothesis we have already mentioned, are deficient for various reasons.

A part of the public is under the impression that mountains have been formed by the action of running water, wearing away the stone, eroding the tablelands, and depositing layers of sediment in the valleys and in the sea. Although it cannot be denied that erosion has been a powerful factor in shaping many mountains, and may have been the main factor in shaping some of them (for example, Mt. Monadnock, in New Hampshire, which I can see from my window as I write these words), it cannot have been the principal cause of the formation of our great folded mountain ranges.

Geologists who have argued in favor of this theory have pointed out that the deposition of sediment in narrow lithospheric depressions may have been a cause of the folding of the lithosphere. The folding could have resulted in part from the sinking of the valley bottoms under the weight of the sediments. The process will be found described in detail in almost any textbook of geology. There are serious objections to it, and no geologist today considers it a satisfactory explanation. One objection is that this process of folding is essentially local. It cannot explain the greatest mountain systems, some of which virtually span the globe. It cannot explain, for example, the almost continuous line of mountain ranges that includes the Rockies, the Andes, and the Antarctic Mountains, and which extends for a total distance of almost half the circumference of the earth. Neither can this theory explain the numerous submarine mountain ranges that have, in recent years, been discovered on the bottoms of the Atlantic, Pacific, and Arctic Oceans. Moreover it has been pointed out that in many cases folding of the lithosphere has taken place without any deposition of sediment and therefore must have been due to other causes. The geologist Henry Fielding Reid remarked:

. . . There are many deeps in the ocean, such as the Virgin Islands Deep, the Tonga Deep, and others, which appear to have sunk without any material deposit of sediments. . . . (354).

For these various reasons, then, geologists have come to the conclusion that erosion is only a secondary cause of mountain building (345:382–84). We shall consider this again.

Another common impression, as already mentioned, is that mountain formation has been due to the cooling and shrinking of the earth. It was reasonable, perhaps, as long as the theory of the cooling of the earth was unquestioned, to try to explain the origin of folded mountains in this way, for, of course, if the earth shrank in size, even only slightly, as a result of cooling, some wrinkling of the lithosphere must be the result. The fact that the pattern of wrinkles that would be produced in this way (and which could be deduced fairly clearly) bore no resemblance whatever to the patterns of the existing mountain ranges did not greatly diminish the currency of this theory, though it did bring about a devastating attack upon it by one competent geologist, Clarence Dutton.

We have seen that there is now an impressive body of evidence and opinion against the theory of a molten origin for the earth. The doubts that have gathered about this assumption are sufficiently serious to prevent us from basing any theory of mountain building upon it (for no theory can have greater probability than its own basic assumptions). But even if this were not the case, even if the molten origin of the earth were a demonstrated fact, still, it was pointed out years ago by Dutton that the shrinking of the globe would not explain the folded mountains. Dutton had two objections. First, he said that the calculated amount of the shrinkage that could have occurred since the lithosphere was formed, by the reduction of temperatures, would not account for the volume of the mountains known to have existed during geological history. Secondly, he pointed out that the kinds of pressures that would exist in the lithosphere as a result of the shrinking of the earth could not produce mountain ranges of the existing patterns. On this point, he said:

. . . As regards the second objection, which, if possible, is more cogent still, it may be remarked that the most striking features in the facts to be explained are the long narrow tracts occupied by the belts of plicated strata, and the approximate parallelism of their folds. These call for the action of some great horizontal force thrusting in one direction. Take, for example, the Appalachian system, stretching from Maine to Georgia. Here is a great belt of parallel synclinals and anticlinals with a persistent trend, and no rational inquirer can doubt that they have been puckered up by some vast force acting horizontally in a northwest and southeast direction. Doubtless it is the most wonderful example of systematic plication in the world. But there are many others that indicate the operation of the same forces with the same broad characteristics. The particular characteristic with which we are concerned is that in each of these folded belts the horizontal force has acted wholly or almost wholly in one direction. But the forces that would arise from a collapsing crust would act in every direction equally. There would be no determinate direction. In short, the process would not form long narrow belts of parallel folds. As I have not time to discuss the hypothesis further, I dismiss it with the remark that it is quantitatively insufficient and qualitatively inapplicable. It is an explanation that explains nothing that we want to explain. . . . (122:201–02).

It is indeed astonishing to note that though a third of a century has passed since this statement was made, and though leading geophysicists today sustain Dutton's views (194:192), the impression is still widespread, and not merely among laymen, that mountains are, more or less, understandable as the consequence of the cooling of the earth. The cause of this inertia is, very likely, the absence of any alternative, acceptable theory of mountain building.

In recent years many geologists have agreed with Dutton that the mountains were folded by some immense force operating horizontally on the lithosphere. Furthermore they have come to recognize that the force or forces involved in mountain folding acted on the lithosphere as a whole and at the same time. Thus one of our leading geophysicists, Dr. Walter Bucher, of Columbia, remarked:

Taken in their entirety, the orogenic [mountainous] belts are the result of world-wide stresses that have acted on the crust as a whole.

Certainly the pattern of these belts is not what one would expect from wholly independent, purely local changes in the crust (58:144).

The same thing was pointed out by Umbgrove:

. . . But the growing amount of stratigraphic studies make it increasingly evident that the terrestrial crust was subjected to a periodically alternating increase and decrease of compression. . . . I feel there is overwhelming evidence that the movements are the expression of a common, world-wide, active, and deep-seated cause. . . . (420:31).

Chadwick agrees with Umbgrove. He remarks:

. . . There is a strong presumption that orogenic movements in one part of the continent are nearly simultaneous with epeirogenic movements and growth of tensional fracture belts elsewhere, as though all were manifestations of the same ultimate force (363a:211).

Umbgrove was impressed by another characteristic of this worldwide force: It did not act continuously. It was not always acting to expand or squeeze sectors of the lithosphere to fold them into mountains. It acted only at certain times, and then, for other periods, it was inactive. There was a sort of periodicity to its operation. This periodicity extended also to other aspects of the earth's geological history:

> The geologist comes across periodicity in many of the pages which he is arduously deciphering—in the sequence of the strata, for instance, and their contents of former organisms. . . . He observes it elsewhere, in the deep-seated forces that bring subsidence first in one area and then in another . . . in the intrusion of liquid melts or "magma" rising from some deeper part of the earth's interior; in the rhythmical invasion of the continents by epicontinental seas and the subsequent retreat of the latter . . . [in] the pulsation of climates and in the rhythmical evolution of life (420:23).

Let us note that Umbgrove has here called attention to evidence suggesting a common causal factor acting upon (a) the formation of successive sedimentary beds, (b) changes in sea level, (c) the intrusion of molten material into the lithosphere, (d) the alternation of climates (including, of course, the occurrence of ice ages), (e) changes in the forms of life, and (f) mountain building. In Chapters II and III we have seen that lithosphere displacement appears to account for the phenomena of ice ages and climatic change. Later we shall see that it may also account for changes in the forms of life. In the present chapter we shall see that lithosphere displacement may explain mountain folding, magmatic instrusions, and changes in sea level.

The periodicity of these processes may be explained if it is assumed that the displacement of the lithosphere at comparatively short intervals is the underlying cause. This holds true even though it may appear that, in some cases, the periodic intervals were of very great length. I shall suggest reasons, later on, for holding that the concept of the occurrence of long-range cycles in earth history is an illusion produced by the meagerness of our information.

Periodicity, as accounted for by lithosphere displacements, should be considered under two separate headings. There is first the periodicity resulting directly from the successive displacements of the lithosphere and having a span determined by the average intervals between those displacements. Then there is the periodicity of much greater span resulting from the fact that many parts of the earth may escape serious geological or climatic changes during one or more successive displacements.

I have already pointed out that two areas at 90 degrees of longitude from the meridian on which the lithosphere turns will be essentially unaffected by a displacement, and that intermediate areas will be affected according to their distance from the meridian of maximum displacement. We shall return to this matter again. I need point out here simply that the effect of this is to cause perhaps the greater part of the earth's surface to be unaffected by the changes produced by one displacement. If this is true, then mathematical probability would favor the escape of any one part of the earth's surface from serious effects for a number of successive displacements. Here we have the basis for a variable periodicity of considerably longer range.

There have been numerous attempts to account for the observed geological periodicity, but they have come to nothing. Joly attempted to prove that the accumulation of radioactive heat in the earth resulted in mountain building at intervals of 30,000,000 years (244;235:153). Gutenberg, however, says that details of Joly's theory have been disproved (194:158) and, moreover, that the theory includes no mechanism to account for the 30,000,000-year intervals (194:188). It is impossible to see that the resulting upheaval of the surface could produce mountain ranges of the patterns that exist. Joly's theory does postulate a growing earth, but whether the lithosphere bursts occasionally or is continually collapsing because of shrinking, it all amounts to the same thing: Neither theory meets the requirements. Attempts have also been made to explain periodicity as the result of long-range astronomical cycles, but they have been unsuccessful (420:281–82). It is obviously difficult to explain mountain building by astronomical cycles.

For some years geologists have been looking for a mountain-folding force below the lithosphere. They have been investigating the possibility of the existence of currents in the semiliquid layers under the lithosphere, and speculating on the possible effects of such currents on the lithosphere itself.* It has been suggested that such currents, rising under the lithosphere, or sinking, might fold the lithosphere. A sinking current, for example, would have the effect of

* We have noted (Chapter I) the use of this concept to buttress the theory of continental drift, and what serious objections there are to it.

drawing the lithosphere together over it, and pulling it down, forming wrinkles, in long narrow patterns, like the mountain ranges. Calculations have been made of the forces that could be brought to bear upon the lithosphere in this way. Vening Meinesz prefers this way of accounting for mountain building.

If we examine the pattern of great geosynclines over the earth's surface, we cannot doubt that their cause must have a world-wide character. The geology in these belts points to horizontal compression in the crust, at least during the later stages of their development. The two main hypotheses suggested to explain these great phenomena are (1) the thermal-contraction hypothesis, and (2) the hypothesis of subcrustal current systems of such large horizontal dimensions that, vertically, they must involve at least a great part of the thickness of the mantle and probably the whole mantle (349:319).

Vening Meinesz summarizes the arguments against the thermal-contraction hypothesis (the cooling of the earth) and argues for the second theory. It is interesting, in passing, to note that one of his arguments against the contraction theory is that "In large parts of the earth's surface . . . tension seems to exist in the crust at the same time that folding takes place elsewhere, and this fact is difficult to reconcile with thermal contraction (giving compression) throughout the crust . . ." (349:320). He is here saying that the lithosphere was being stretched in some places and compressed in others at the same time, which is inconsistent with the cooling and contracting theory. It is, however, quite consistent with the lithosphere-displacement hypothesis.

Now, as to the sublithospheric current hypothesis, we may note that Meinesz is assuming currents traveling for great distances horizontally and moving in great depths of hundreds of miles below the lithosphere. Naturally the movement of such masses of rock could potentially create pressures to stagger the imagination. Gutenberg discusses the work of many men who are studying sublithospheric currents (194:186,191). The chief difficulty here is the absence of any real evidence for the existence of currents on the necessary scale.*

The problem that we are involved with here is that of the origin of the geosyncline. Geologists refer to a downward fold in the litho-

* Some reasons to doubt the existence of such currents of such magnitude have been discussed in Chapter I.

sphere of major proportions as a geosyncline. An upward fold (or arch) is a geanticline. They are sometimes associated, since a downward fold may be accompanied by an upward fold on either side. There is a developed body of geosynclinal theory. There have been two points of view concerning the origin of geosynclines. According to one view they have been the result of the deposition of heavy loads of sediment in basins or shallow seas, loads that have forced the lithosphere to subside and subsequently fold. A second opinion is that the lithosphere has originally been folded from other causes, and that the sediment has simply been deposited in the resulting depressions. Furthermore those who hold the second opinion maintain that the amount of the sedimentary accumulation and the rate of the accumulation are controlled by the magnitude and the speed of the lithospheric folding. It seems that the latter has come to be the dominant view. Krumbein and Sloss, referring to the main body of recent evidence on the development of geosynclines, remark:

These newer data [sample logs, electric logs from wells, detailed faunal studies] support Barrell's fundamental principle that sedimentation is controlled by subsidence, with the slight modification that the subsidence may range from discontinuous to continuous (258:319).

The successive stages in the development of geosynclinal theory, beginning with the earlier thinkers, are summarized by Krumbein and Sloss, as follows:

Hall . . . concluded that subsidence was caused by the weight of the accumulation of sediments, which automatically produced the folding. . . .

J. B. Dana of Yale University, on the other hand, argued that contemporaneous subsidence of the earth's crust was the reason for the accumulation of the thick sediments. . . .

In 1873, Dana published his classic paper on the origin of mountains and the nature of the earth's interior. He defined a geosynclinal as a "long continued subsidence," and went on to state:

"These examples exhibit the characteristics of a large class of mountain masses or ranges. A geosynclinal accompanied by sedimentary depositions, ending in a catastrophe of plication [folding] and solidification are the essential steps. . . ." (258:317).

Krumbein and Sloss accept Dana's statement of the case as essentially sound (258:318). It seems that the original folding of the lithosphere, forming the geosyncline, occurs for reasons unknown, and that it is followed, or rather accompanied, by the deposition of sediment in the geosynclines. Later the thickened masses of sediment in the geosynclines are folded into mountains. It is obvious that the fundamental point requiring explanation is the cause of the original folding. Whether, as the result of this original folding and the deposition of masses of sediment on the bottoms of the geosynclinal folds, there are also subsequent and secondary causes of folding, as may well be, is not important. It is sufficient that displacements of the lithosphere can, as we shall see, explain the original folding of the lithosphere, both on land and under the sea; it is not necessary to prove that all folding whatever is due to that cause.

2. CAMPBELL'S THEORY OF MOUNTAIN BUILDING

James H. Campbell is responsible for the elaboration of a theory of mountain building based on the premise that the original active factor in the process is lithosphere displacement. We shall now examine his thought in detail.

The reader will understand that in any displacement of the whole lithosphere, some areas must be shifted toward the equator and others simultaneously toward the poles. To be exact, two quarters of the surface, diametrically opposite each other on opposite sides of the earth, must move equatorward, while the other two quarters move poleward. This may best be visualized by looking at a globe.

Since the earth is oblate—a slightly flattened sphere—the parts of the surface that move equatorward will have to pass over the slight equatorial bulge, thereby being stretched, while those being displaced poleward will have to undergo an equal degree of compression, or squeezing together, forming both synclines and anticlines. These deformations of the lithosphere may lead eventually, through the deposition of sediments and possibly through a number of successive displacements of the lithosphere, to the formation of folded mountain ranges.

The systematic presentation of this theory requires us to consider the two different phases of displacement—equatorward and poleward—separately, for they have very different results. We will begin with the consideration of the effects of a displacement of a lithospheric sector toward the equator.

In a shift in that direction a lithospheric sector is submitted to tension (or stretching), and this tension is relieved by the fracturing that takes place when the bursting stress exerted on the lithosphere has come to exceed the strength of the lithosphere. (For Campbell's calculations of the quantity of the bursting stress, as compared with estimates of lithospheric strength, see Note 3.) Until fractures appear and multiply, the lithosphere cannot move over the bulge. After the fracturing permits the movement to begin, the lithospheric blocks tend to draw slightly apart. The spaces between them are immediately filled by molten material from below.

Let us form a clear picture of this lithospheric stretching, from the quantitative standpoint. It is important to estimate the stretch per mile if we are to visualize the results. Taking the globe as a whole, the difference between the polar and equatorial diameters is about 13 miles. The circumferences, therefore, differ by about 39 miles. If the lithosphere were displaced so far that a point at a pole was displaced to the equator, the polar circumference would have to stretch 39 miles to fit over the equator. This would amount to about 9 feet in the mile. Since the magnitude of displacements, however (according to evidence presented previously), seems to have been of the order of no more than about 30 degrees, or one third of the distance from pole to equator, the average stretch per mile may have amounted to 3 feet, or about one foot in two thousand.

It would be a mistake to visualize this stretching of the lithosphere in the equatorward-moving areas as evenly distributed around the whole circumference of the globe. Obviously the real events would not correspond to this. The lithosphere would be under bursting stress, and this would be relieved spasmodically, during the movement of the lithosphere, by fractures at the weakest points. A fracture through the crust at one point would relieve the stress for perhaps hundreds of miles. Since the elasticity of the lithosphere is slight, the stretching or extension of the lithosphere would consist of the drawing apart, to varying distances, of the fractured blocks. Generally speaking, the fewer the fractures, the farther their sides would draw apart. It would be possible that the total amount of the stretching of the earth's circumference would be concentrated in relatively few critical areas such as those represented by the world-wide midoceanic rift.

It must also be kept in mind that some parts of this area being

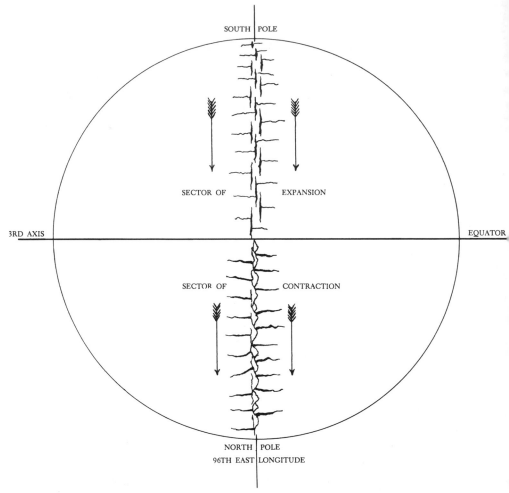

SOUTH POLE

SECTOR OF   EXPANSION

3RD AXIS   EQUATOR

SECTOR OF   CONTRACTION

NORTH POLE
96TH EAST LONGITUDE

displaced toward the equator will be displaced farther than others. The greatest displacement will occur along the line, or meridian, that represents the direction of the movement. As I have pointed out, at two pivot points on the equator 90 degrees away from this meridian there will be little or no movement, and the points in between will move proportionately to their distances from the meridian. The tension, or stretching, will be proportional to the amount of displacement. It therefore will be greatest along the central meridian of movement, and it is here that Campbell expects the first major fractures of the lithosphere to develop.

It is important to remember the nature of the lithosphere on which this tension is exerted. The lithosphere is comparatively rigid, having little elasticity, but it is not strong. It varies in thickness and strength from place to place. As we shall see, it is even now penetrated by great systems of deep fractures of unexplained origin.

Fig. 32. Mountain building by displacements of the lithosphere: patterns of fracture and folding.

The lithosphere is represented in a future movement resulting from the effect of the present ice cap in Antarctica.* Since the latter's center of mass is on (or near) the meridian of 96° E. Long., the lithosphere is represented as moving in that direction from the pole. The sector of expansion is moving equatorward and therefore being extended. The sector of contraction is moving toward the North Pole from the equator and therefore being compressed.

In the sector of expansion, parallel major faults can be observed, with minor faults at right angles. The wavy lines suggest the effects of local differences in lithospheric strength. The pattern of the fractures is indicated, but not their number; a very large number of meridional fractures might be formed, while the minor fractures would be even more numerous.

In the sector of contraction, lithospheric folding is shown only schematically. It is represented as if all the folding is taking place along one meridian, although in reality there would probably be many parallel zones of mountain folding at considerable distances from each other. Campbell indicates that this movement will be accompanied by fracturing of the lithosphere, with faults running at right angles to the main axes of the folds. The third axis, which runs through the equator, is considered to be the axis on which the lithosphere turns. The points directly at the two ends of this axis do not move.

These inequalities of strength will be very important in determining the reactions of the lithosphere from place to place to the tension exerted upon it; they will determine the precise locations, and to some extent the patterns, of the fractures that will result.

Without attempting to anticipate a more detailed discussion, to be introduced later, of the forces involved in this fracturing of the lithosphere, I would like to remark that the forces required for the fracturing are by no means so great as might be at first supposed. It is a question of relatively slight forces exerted over considerable periods of time.

If we disregard the factors that may locally influence the locations and sizes of fractures, a general pattern may be indicated to which they will tend to conform. Campbell has worked out this pattern schematically and has indicated it in Figures 32, 33, and 34. The reader will note that the fractures take two directions. There are the north-south, or meridional, fractures, which Campbell refers to as the major fractures, and then there are minor fractures at right angles to them.

---

* Since the abandonment of the ice cap as the cause of the displacement of the lithosphere, the ice cap here can be taken to represent a net imbalance of the distribution of mass in the lithosphere. There will be a center of such imbalance and a direction of motion determined by it, as with the ice cap if it were uncompensated.

Campbell anticipates that numerous major fractures will occur parallel to each other as the lithosphere moves. The formation of very numerous minor faults at right angles to the major faults will form a gridiron pattern of fractures. Campbell has suggested a method for visualizing the process. If the reader will cup his hands and place them together, with fingertips touching and the fingers of each hand close together (as if they lay on the surface of a sphere), and then imagine the sphere growing, and causing the fingertips of both hands to spread apart, and at the same time the fingers of each hand to spread apart, he may visualize the process. The gap between his hands will now represent a major fracture, and the gaps between the fingers of each hand will represent the minor fractures at right angles to it. The reader will see, a little later on, how closely this projection of fracture effects corresponds to the real phenomena in the lithosphere.

Another important aspect of these fractures is shown in Figure 33. Campbell has indicated that, owing to the changing arc of the surface as the lithosphere sector moves equatorward, the fractures will tend to open from the bottom. This would, of course, favor the intrusion into them of magma from below, and accordingly Campbell shows them filled up (in black). At the same time, as the reader may see, fractures in areas moving poleward would tend to open from the top. These might be less likely to reach sources of molten rock; accordingly they are not shown filled up. Whether these fractures would or would not fill up (and perhaps the probabilities are that they would), the configuration of the resulting solidified "dikes" in the rocks would be very different from that in fractures that had opened from the bottom. Campbell has suggested that this way of explaining existing fracture patterns in the lithosphere could be an aid in prospecting for ores, most of which occur in such "dikes." It would be a question of ascertaining, for the general region, whether the "dike" being investigated were part of either a poleward type or equatorward type of pattern, and from this it might be possible to deduce whether the "dike" was to peter out or not. Campbell believes that the hypothesis provides numerous possibilities for the exploration of the lithosphere, some of which may prove eventually to be of commercial value.

The time element is essential to visualizing the general process

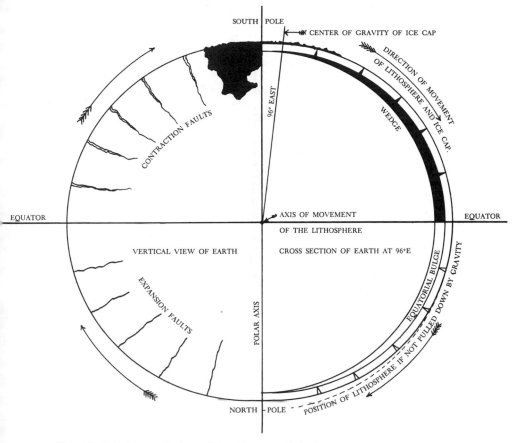

Fig. 33. Displacement of the lithosphere: vertical view.

This figure illustrates a number of simultaneous effects of displacement. The upper right-hand quadrant shows a sector of the lithosphere displaced toward the equator. Here the lessening arc of the surface will cause faults to open from the bottom. The lower right-hand quadrant shows a sector of the lithosphere displaced toward a pole. Here the increasing arc of the surface results in faults opening from the top. The lower left-hand quadrant, which is a vertical view of a sector moving equatorward, shows major meridional faults, which have opened from the bottom. The upper left-hand quadrant, which is a vertical view of a sector displaced poleward, shows meridional faults opening from the top.

The reader should visualize the left-hand quadrants as if looking straight down on the earth at the point where the central meridian of displacement (96° E. Long. in this case) crosses the equator.*

of a displacement. We have seen that the last one apparently took about 5,000 years. This means that this amount of time would be available for the creation of the system of fractures we are considering. It means, for example, that a single major fracture, which might

---

* See note, previous page.

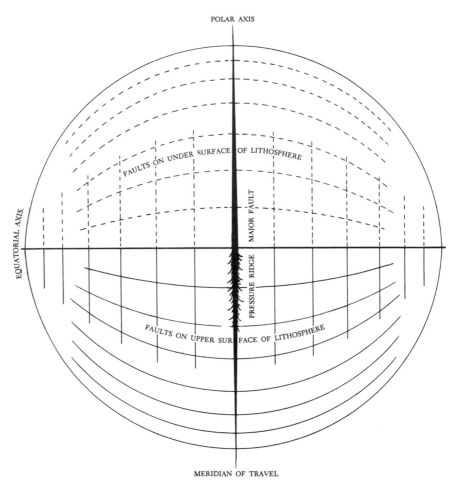

Fig. 34. *Displacement of the lithosphere: patterns of fracture.*

This figure indicates schematically the mechanics of faulting and folding in a displacement of the lithosphere. It is suggested, for purposes of illustration only, that all effects are concentrated on the meridian of maximum lithosphere displacement. Therefore, only one major meridional fault is shown in the upper hemisphere, which is moving toward the equator. Dotted lines indicate other faults opening from the bottom of the lithosphere as the arc of the surface diminishes.

Across the equator, where the surface is moving toward the pole, and compression is resulting, the continuation of the major expansion fault is shown as a pressure ridge, which may later become the main axis of a mountain range. Again, for purposes of illustration only, it is assumed that all folding will take place along the meridian of maximum displacement. If the major fault is filled with molten magma, and the magma solidifies, then this intruded matter, which has expanded the lithosphere, must add to the folding in the lower hemisphere, which is moving toward a pole.

In the lower hemisphere the unbroken lines indicate the fractures opening from the top, as the arc of the surface increases.

involve, let us say, the pulling apart of the lithosphere to a distance of several miles and the filling up of the crack with molten material from below, might be formed over a period of several thousand years, during which time there might be spasmodically renewed earthquake fracturing and volcanic effects, interrupted by periods of quiet. It is obvious that the amount of time available for the work of extension and fracturing of the lithosphere is sufficient to permit the process to complete itself without undue or incredible violence.

We must now consider a question that relates to mountain building and at the same time involves another of the major unsolved problems of geology. It is connected with our phase of equatorward lithosphere displacement. It has to do with the filling of the fractures by molten magma from below. Campbell considers that this filling of the fractures is the first step in mountain building, or at least in the formation of a geosyncline. Obviously it is possible to start the process at other points; this is therefore only a matter of convenience and for the purpose of drawing a clear picture of the process for the reader. But Campbell points out that the process of the filling of the cracks, and the later solidification of the intruded material, adds extension of the lithosphere; there is now more surface. When, in future shifts of the lithosphere, this area passes over the equator toward a pole, or moves poleward from where it is, the extended surface has to yield to the resulting compression by folding more than it would have had to do had there been no molten intrusions in the first place. It is therefore reasonable to call this the first step in mountain building, although there is as yet no folding, and no uplift of the rock strata.

But this question of molten intrusions into the lithosphere raises another vital point. It has been, until now, a very difficult thing to explain the rise of molten matter into the lithosphere. Geologists have speculated as to what force could have shot up the molten matter that formed the innumerable "dikes" and "sills," as the resulting veins are called. They have not been able to agree upon the question. No reasonable explanation of these millions of magmatic invasions of the lithosphere has been found.

Of course, it is realized that the lithosphere is, in a sense, a floating crust. The materials of which it is composed are lighter, it is assumed, than the materials below and are solid as compared with the plastic or viscous state of the underlying layers. The lithosphere can be thought of as floating in hydrostatic balance in the semiliquid lower layer. This is generally understood among geologists. It follows logically from this that, if two or more blocks of the lithosphere got separated with cracks between them, the molten material would rise in the crack and the blocks would sink, until the cracks were filled up far enough to establish hydrostatic balance. But this did not solve the problem; it did not help because nobody could imagine what could produce the necessary pulling apart of the blocks.

For those who like to see complicated problems made simple, Campbell's presentation of this matter is worth considering. He suggests that the concept of a great sector of the lithosphere being stretched, and thereby fractured in innumerable places at one time, permits a comparison to be made with an ice sheet which is floating on water and which undergoes fracturing. Just as the individual pieces of the ice floe sink until they have displaced their weight in water, and the water rises in the cracks between the pieces, so he visualizes the behavior of the lithosphere during its displacement equatorward. He sees this as the explanation of the fact that although the lithosphere is shot through with igneous invasions of all sorts, these are hardly ever known to reach the surface of the earth. He compares the behavior of the lithosphere during displacement with the behavior of ice as follows:

. . . As a matter of fact the lithosphere (or crust) can be likened to ice floating on water, a solid and lighter form of a substance floating in a liquid and heavier form of a similar substance. The solid and lighter substance sinks in the heavier and liquid substance until it displaces its own weight in the heavier and liquid substance and then floats with its surplus bulk above the surface of the heavier liquid, which in the case of ice would be one tenth of its bulk. To put it another way, if you were out on a lake where the ice was ten inches thick, and you were to bore a hole through the ice to the water, the water would rise in the hole to within one inch of the surface of the ice and remain there. Now, that is exactly what happens to the lithosphere. It sinks into the asthenosphere (or subcrustal layer) until it displaces its own weight of the substance of the asthenosphere and a state of equilibrium is reached. That will bring the substance of the asthenosphere far up into the lithosphere, wherever it finds an opening or a fault that reaches all the way to the bottom of the lithosphere (66).

Purely for purposes of illustration, and not as an accurate picture of the facts, Campbell has made a very rough calculation, as follows:

Assuming that the lithosphere is composed of granite that has a weight of 166 pounds to the cubic foot, and the asthenosphere consists of soapstone with a weight of 169 pounds per cubic foot, the lithosphere being three pounds lighter per cubic foot than the asthenosphere, it would float in the heavier asthenosphere leaving 1.775% of its volume above the surface of the asthenosphere, and as the lithosphere is assumed to be forty miles deep in this case, then 1.775% of forty miles would be 71% of a mile above the top of the asthenosphere. That is, the soapstone molten asthenopshere would rise up into the fault to within three quarters of a mile of the surface of the earth. . . . (66).

Campbell's suggestions appear to provide us with an explanation of the rise of molten magma into the worldwide midoceanic rift which has no need of the assumption of convection currents.

Summarizing his general thoughts regarding the effects of an equatorward displacement of a lithospheric sector, and the hydrostatic balance of the lithosphere itself, Campbell has remarked:

I think you should stress this point, for while the geophysicists have seen faults in the earth's crust, and have seen many of these faults that they knew had been filled up from below, they didn't have any logical solution of what caused the faults, nor did they connect the faults with the formation of our mountains (66).

3. THE EFFECTS OF POLEWARD DISPLACEMENT

In the poleward displacement of sectors of the lithosphere, compression, instead of extension, would be the rule. The magnitudes and the distribution of forces, and the time element, would, of course, be the same. Otherwise, the effects would be very different.

We should have, in the first place, some folding of the rock strata. As with the fractures, the precise locations of the rock foldings, their number, and their magnitudes would be controlled by the amount of the displacement locally, the local variations of lithospheric strength (which would be less where geosynclines already existed), and the distances of the areas concerned from the central meridian of movement. The amount of the folding would be increased as the result of any previous process of extension of that area of the lithosphere in any previous displacement.

The elastic properties of the lithosphere would probably be of much greater importance in this compressive phase than in the extensive phase of a displacement. This is because compression could lead to flexing or bending of the lithosphere, to a slight degree, without a permanent change of shape. It might be possible to bend or flex the lithosphere slightly, and hold it so for thousands of years, without fracture or folding of the rock strata, or even without much plastic flow of the materials. This would mean no permanent change in the conformation of the surface. A compressive tension might be exerted for thousands of years, causing a flexing, and then be relaxed, permitting the lithosphere to return to its original shape. It may be supposed that in areas sufficiently removed from the meridian of maximum displacement, the compressive tensions on the lithosphere might be contained by its tensile strength, and the lithosphere might yield elastically without deformation. If this occurred, however, the total amount of the compression for the whole circumference of the globe would probably be concentrated at comparatively few points, where the compressive stresses happened to be in excess of the strength of the lithosphere; here there would be a considerable amount of folding of the rock strata. It is obvious, also, that these points would tend to coincide with existing geosynclines, which would naturally represent comparatively weak zones, where the lithosphere would be less able to withstand the horizontal stress.

Campbell suggests that in an area displaced poleward, no fewer than four pressures will be operating simultaneously on the lithosphere. There will be, in the first place, two pressures developing from opposite directions toward the meridian of displacement. These will arise because of the diminishing circumference. Two other pressures will simultaneously develop at right angles to these as the result of the reduced radius. Since the radius is only one sixth of the circumference, the forces will be in proportion; the folds due to the first compression will tend to be six times as long (and accentuated) as those due to the second compression. The former may ultimately correspond with the long axes of the mountain ranges, and the latter to their radial axes. The long, narrow, folded tracts referred to by Dutton may be thus explained.

In Figure 32 Campbell has suggested an idealized representation of the formation of a mountain chain by a displacement of the lithosphere. The reader will note the long major axis and the shorter radial axes. That this is a fairly close approximation to the patterns of existing mountain ranges is obvious; however, a number of modifying factors must be recognized. In the first place, we do not contemplate that a mountain range can be completed in the course of one movement of the lithosphere. It is quite obvious, from the quantitative considerations already mentioned, that a single displacement could cause comparatively little folding even if, as the result of elastic yielding, most of the folding was concentrated in a few areas. It is certain that many displacements would be required to make a large mountain range, and since successive displacements will not necessarily occur in the same directions on the earth's surface, the resulting patterns might rarely conform to the idealized pattern. And yet, if most of the folding in one displacement happened to be concentrated in one area, and if one or more successive displacements happened to concentrate folding in the same area, a mountain range might come into existence in a comparatively short period.

It should not be thought that Campbell is in disagreement with Dutton's statement, quoted above, that the compressive mountain-folding forces have acted in one direction only on the earth's surface. The laws of physics require the operation of equal and opposite forces for the production of effects. A compression is the result of two equal and opposite pressures. There is still a definite orientation, such as northeast-southwest, along which the compression operates on the lithosphere.

4. THE MOUNTAIN-BUILDING FORCE

Campbell has a most interesting suggestion as to the identity of the force that actually accomplishes mountain folding. He suggests that this is none other than the force of gravity itself. It is his opinion that the shift of the lithosphere merely performs the function of sliding an area to a place where the force of gravity can act upon it. When an area is moved toward a pole, the radii of the earth are shorter, circumference is shorter, and the surface required is less. There is a surplus of surface, and this, being pulled down closer to the earth's center by gravity, must fold. From this point of view, it appears that the mountains *are not pushed up* at all, and therefore, no lifting force is required; instead, it is the surface of the earth *that is pulled down,* by gravity, nearer to the earth's center, as a sector of the lithosphere approaches a pole. Where this happens the surplus surface must fold. Thus it is the force of gravity, over a large area, that folds the lithosphere in a small area.

It may help the reader to grasp this idea if he will visualize a flat area on the equator which, in process of being displaced with the lithosphere as far as a pole, has been folded enough to produce mountains six miles high. Now, actually, the peaks of those mountains are no farther from the center of the earth than the flat area was at the equator. Their altitude, with reference to the earth's center, is unchanged. What has changed altitude, however, is the rest of the surface, outside the mountain chain. That has been pulled down six miles. What pulled it down, obviously, was the force of gravity, and the reason it was pulled down was that it was first shifted horizontally to a place where gravity could act upon it.

Here, then, is the answer to the long-standing enigma of the source of the energy for mountain folding. The mountains are not lifted up at all; the surface is pulled down, the force of gravity does the pulling, and folding results where there happens to be excess of surface.

5. EXISTING FRACTURE SYSTEMS AS EVIDENCE FOR THE THEORY

It is a strong argument in favor of a hypothesis if it enables one to anticipate discoveries. Campbell has shown that the theory of displacements of the earth's lithosphere calls for the existence of great systems of parallel fractures, intersected by other fractures at right angles to them. It was some time after Campbell began to consider this matter, and quite independently of him, that I became aware of the fact that such fracture patterns do, in fact, extend over the whole face of the globe, and that geologists are in agreement that their origin is unexplained. Many years ago Hobbs pointed out that they must have been the result of the operation of some worldwide force:

> The recognition within the fracture complex of the earth's outer shell of a unique and relatively simple pattern, common to at least a large portion of the surface, obscured though it may be in local districts through the superimposition of more or less disorderly fracture complexes, must be regarded as of the most fundamental importance. It points inevitably to the conclusion that more or less uniform conditions

of stress and strain have been common to probably the earth's entire outer shell (217:163).

As I have pointed out, Campbell's projected pattern of fractures is a sort of gridiron, with major fractures paralleling the meridians, and minor fractures at right angles to them. In the actual earth's surface, however, there are two such patterns. One of them consists of north-south fractures paralleling the meridians, intersected by east-west fractures paralleling the equator. The second gridiron is diagonal to the first; the lines run northeast-southwest, and north-west-southeast. Hobbs insists that the existence of these worldwide patterns points to a cause acting globally; they could not have been the result of local causes; the force causing the fracturing must have acted simultaneously, so to speak, over a great part of the whole surface of the earth:

. . . The results of this correlation possess considerable significance inasmuch as it is clear that over quite an appreciable fraction of the earth's surface, the main lines of fracture betray evidence of common origin. . . . (218:15).

The fracturing of the lithosphere under the operation of some global force has been accompanied by much tilting and relative movement of blocks of considerable size, resulting in the alteration of topographic features. One of the earlier geologists, Lapworth, remarked with considerable truth, though with some exaggeration, that

On the surface of the globe this double set of longitudinal and transverse waves is everywhere apparent. They account for the detailed disposition of our lands, and our waters, for our present coastal forms, for the direction, length and disposition of our mountain ranges and plains and lakes (420:296).

It is clear, I think, from what has already been said, that Lapworth was in error in ascribing the folded mountains to the effects of fracturing alone. However, it may well be that formation of block mountains, such as the Sierra Nevadas, can be accounted for in this way. Innumerable other features of the lithosphere have been formed or obviously much affected by the fracture patterns. Hobbs, for example, has maps of river systems in Connecticut and Ontario showing how closely the rivers and their tributaries follow the lines of the fracture systems (216:226). Many riverbeds, many submarine canyons, were never created by subaerial erosion; they were, instead, the results of deep fractures in the lithosphere, later occupied by rivers or by the sea.

A succession of theories to account for the worldwide or "planetary" fracture patterns were developed and rejected. As soon as it became clear that these patterns could not be explained as the result of local forces, the problem was recognized as very formidable. Sonder, a Swiss geologist, attempted to explain them as the result of a difference in the compressibility (or elasticity) of the rocks of the continents as compared with those under the oceans. But Umbgrove pointed out that this would call for independent fracture systems for each of the continents, whereas existing fracture patterns extend to several continents (420:300–01).

The Dutch geologist Vening Meinesz suggested that the fracture patterns could be explained mathematically by a displacement of the lithosphere. He postulated one displacement about 300,000,000 years ago through about 70° of latitude (194:204ff). Umbgrove rejected this theory because he saw that there were many features of the earth's surface that could not be explained by the particular displacement suggested by Vening Meinesz. This is not at all remarkable, since it is quite impossible to see how any one displacement, and particularly one 300,000,000 years ago, can be made to explain most of the earth's present topographic features. Umbgrove was justified in rejecting the Vening Meinesz theory, but he admitted that this left him with no explanation at all. ". . . On the other hand, it means that the origin of both lineament systems remains an unsolved problem" (420:307).

Some writers have suggested that the two fracture systems originated at different times, and this is a very important point. Umbgrove says:

It is a rather widespread belief that the origin of faults with a certain strike dates from a special period, whereas faults with a markedly different strike would date from another well-defined period. In certain areas this conviction is founded upon sound arguments. . . . (420:298).

He continues:

Some authors, however, have doubtless overrated the relation between the direction and the time of origin of a fault system. As a typical example, I may mention Philipp, who once advanced the opinion that the

direction of the principal fault lines of northwestern Europe changed
from W. and W.N.W. in the Upper Jurassic toward N. or N.N.E. in the
Oligocene, and thence E.N.E. in the upper Tertiary and Pleistocene.
He added the hypothesis that their rotation could have been caused by
a large displacement of the poles. In the meantime it has been shown
that some faults with a meridional strike date from much older periods.
Moreover, large and well-defined faults with a N.N.W. direction dating
at least from the Upper Paleozoic appear to have been of paramount
influence in the structural history of the Netherlands. Therefore Philipp's
hypothesis has to be abandoned because it is inconsistent with well-
established facts (420:298).

"Abandoned" much too easily! The reader can easily see that the
objections Umbgrove raises to Philipp's theory are removed by the
present theory of lithosphere displacements. With this assumption,
it would be inevitable that, in the long history of the globe, the poles
would often be found in about the same situations. If the strikes of
the fault systems are related to the positions of the poles, those of
later periods would often coincide approximately with those of
earlier periods.

We find in this very fact the answer to another of the mysteries
of geology, the so-called "rejuvenation" of similar features in the
same geographical situations at various times. The term "rejuvenation" is a commonplace of geological literature and is especially
emphasized by Umbgrove. He is puzzled by the fact that old
geological features have repeatedly been called back to life. It seems
that this renewal of old topographies may be explained by the accidental return of the poles to approximately the same places.

There is nothing remarkable about the fact that only two world-
wide fracture systems can now be recognized in the lithosphere. If
each successive displacement produced a new gridiron pattern of
fractures and resulting surface features, it must, in addition, have
disrupted the evidence of previous patterns. In a long series of displacements, the older fracture patterns must soon be reduced to an
indistinguishable jumble. It is probable that the two systems now
recognizable date only from the last two displacements of the lithosphere, even though many of the fractures and individual topographic
features now coinciding with these systems may date from remote
periods.

It is not true, of course, that in one displacement of the lithosphere all fractures all over the earth will form a single rectilinear

pattern. This can be made clear from an example. Let us suppose that North America was moved directly southward at the end of the last ice age. Campbell has suggested that major fractures would run north and south (meridionally) and minor fractures east and west, and this would be true of the whole western hemisphere, which was, presumably, moved southward, and of the opposite side of the earth, which was equally displaced northward. But what about Europe? If, before the last displacement, the pole was situated in or near Hudson Bay, it seems that the last displacement must have created diagonal and not meridional fractures in Europe, for the reason that Europe was nowhere near the meridian of displacement. Thus, in one given displacement, a meridional fracture pattern will be created near the meridian of displacement, a diagonal fracture pattern in very large areas approximately 45 degrees from this meridian, and, of course, no fracture pattern in the "pivot" areas, 90 degrees from the meridian of displacement, where no displacement will occur.

We have mentioned the oceanographic research work which has recently resulted in tracing a globe-encircling crack on the bottoms of the Atlantic, Indian, and Pacific Oceans, and connecting it with the Great Rift Valley in Africa. The pattern that has been traced out is about 45,000 miles long; it is reported that there is seismic activity at present along the whole length of the crack, suggesting recent disturbance of the area and a still-continuing process. The rift valley, associated with the midoceanic ridge, appears to average two miles in depth and twenty miles in width. The fact that it is connected with the Rift Valley in Africa, that it bisects Iceland, and apparently invades Siberia indicates that it is not a phenomenon of ocean basins only. *The Columbia Research News*, published by Columbia University, in its issue of March, 1957, described the discovery thus:

In January, Columbia University geologists announced the discovery of a world-wide rift believed to have been caused by the pulling apart of the earth's crust. The big rift traverses the floors of all the oceans and comes briefly to shore on three continents in a system of apparently continuous lines . . . 45,000 miles long.

Throughout its vast length the world-wide rift seems to be remarkably uniform in shape, consisting of a central valley or trench averaging 20 to 25 miles in width and flanked on either side by 75 milewide belts of jagged mountains rising a mile or two above the valley. The peaks of the highest mountains in the system are from 3,600 to 7,200 feet below the ocean's surface while the long undersea stretches of the rift valley itself lie from two to four miles down. In addition to being marked by its topography, the globe-circling formation is the source of shallow earthquakes that are still going on along its entire length—an indication that, if the rift is due to a pulling apart of the earth's crust, the geological feature is a young and growing one.

We have already suggested (Chapter I) that a series of displacements of the lithosphere can explain these fractures and fissures in the oceanic crust. It seems that in this system of fractures extending, without doubt, to the very bottom of the lithosphere, we may have evidence of the existence of the zones of weakness along which, again and again, the fracturing has taken place that has permitted the displacements of the lithosphere.

We have already considered (Figs. 9-10, pp. 32, 37) the evidence of the fracture patterns that have been discovered on the ocean floors, patterns that agree with those postulated by Campbell and with those found on the continents by Umbgrove, Vening Meinesz, and others. These fracture systems of the ocean floor are obviously related genetically to the midoceanic ridge but appear to reflect a number of shifts of the lithosphere in different directions.

To return briefly to the question of block mountains, Campbell has a further suggestion as to the way in which compression in a poleward displacement may combine with subsequent fracturing to cause them. One of the problems that await solution in geology is the cause of the widespread doming and basining of the lithosphere that occurs from place to place. The domes are sometimes of considerable extent. Examples of basins include the Gulf of Mexico and the Caspian and Black Seas. Campbell points out that if an area is displaced poleward and is thereby subjected to four compressions, as already mentioned, limited areas will be entrapped by these compressions, and doming must result; conversely in areas moved equatorward the reverse must occur, and larger or smaller basins will tend to be produced.

A block mountain might tend to be produced, Campbell thinks, if a major fault should bisect a domed-up area. This would create the possibility that the abutting rock sections of one half of the dome might give way, allowing half the dome to collapse and pushing sublithospheric viscous or plastic rock under the other half of the dome, thus rendering the latter permanent. This effect, however, would depend upon many local circumstances.

Part II.
VOLCANISM AND
OTHER QUESTIONS

In the preceding part of this chapter I have sketched the principal problems that are basic to the formation of the folded mountains and block mountains, and have examined the planetary fracture systems in the light of Campbell's mechanism for lithosphere displacement. There are, however, a number of other aspects of this general problem that must now engage our attention. We must consider the remarkable phenomena of volcanism in their relationship to lithosphere displacement. In connection with the creation of volcanic mountains we must consider briefly the question of the origin of the heat of the earth, an unsolved problem of great interest. We must then examine the relationship of lithosphere displacement and mountain building to the question of changes in the sea level. Finally we must consider the problem of the chronology of mountain building.

1. VOLCANISM

We have seen that one kind of mountain is the volcanic mountain. Volcanic phenomena cover a wide range; all of these must be considered in order to see how closely they can be related to a general cause. The phenomena that need explaining include volcanic eruptions, the creation (sometimes rapid) of volcanic mountains on land or in the sea, the genesis of volcanic island arcs, and last but not least the vast lava flows or lava floods that have at times in the past inundated great areas of the earth's surface.

Since volcanoes occur frequently and are the most dramatic manifestations of volcanism, they have been thoroughly studied, and a whole literature has been devoted to them. It is astonishing, therefore, that neither the causes of volcanoes nor the present distribution of volcanic zones on the earth's surface has as yet received an acceptable explanation. As in the case of other unsolved problems, the absence of certainty has led to a multiplicity of theories. Jaggar, one of the best field observers of volcanoes, refers to the two leading theories thus:

It would be hard to imagine any more completely different explanations for the same phenomenon than is R. A. Daly's doctrine of the causes of volcanic action, as compared with the crystallization theory of A. L. Day (235:150).

Dr. A. L. Day was formerly director of a geophysical laboratory in Washington; his theory is based upon geophysical experiments conducted in the laboratory. He observed that the crystallization of rock from the molten state resulted in some increase in volume. He assumed that the whole lithosphere was once molten, and that as it cooled it continued to contain, here and there, comparatively small pockets of molten rock. When such pockets of molten rock finally were cooled to the crystallization point, then expansion would occur, and great pressures would be set up, which might lead to eruption at the surface. This theory is based upon the assumption of the molten origin of the earth and carries with it the corollary that volcanic eruptions are essentially local phenomena. Day insisted that volcanoes were not connected with a molten layer under the lithosphere and were not related to events occurring over large areas.

Professor R. A Daly based his opposed theory on his observations of the field evidence of geology. He insisted that only the assumption of a molten layer under the lithosphere could account for the countless facts of igneous geology. His theory is reconcilable either with the assumption of the molten origin of the globe or with the theory of a growing and heating earth.

Jaggar objects to Day's view that volcanoes are purely local. He says:

There is some reason to think that a very long crack in the bottom of the Pacific Ocean, with interruptions by very deep water, extends all the way from New Zealand to Hawaii, because there are striking sympathies of eruptive data between the volcanoes of New Zealand, Tonga, Samoa and Hawaii (235:23).

He lists a number of eruptions with their dates to show their intimate connection. In particular he mentions the erupton of August 31, 1886, on the island of Niuafoo, Polynesia:

. . . Only two months before, Tarawere Volcano was erupted disastrously in New Zealand, indicating volcanic sympathy between two craters hundreds of miles apart on the same general rift in the earth's crust (235:95).

These observations imply that a connection may exist, at least in some cases, between volcanoes at great distances from each other, because of their being located along the same crack in the lithosphere. This implies a connection between the deep fracturing of the lithosphere and volcanism. We have seen that Columbia scientists discovered a vast connected system of rift valleys, or cracks in the lithosphere, extending over the surface of the whole planet and associated with constant seismic disturbances. Jaggar makes it clear that volcanic eruptions, as well as earthquakes, may be associated with such rifts. Since the lithosphere is relatively thin, it is reasonable to suppose that the molten rock erupting in volcanoes at great distances from each other must come from below the lithosphere and that it is not created by any processes occurring within the lithosphere itself. All this is confirmation of Daly's position.

Another theory of volcanic action that should be mentioned briefly is that associated with the name of W. H. Hobbs. It was his view that volcanic action could result from horizontal pressure arching up a sector of the lithosphere. This is based on the fact that if a rock that is too hot to crystallize at normal pressures is subjected to great pressure, it may take the solid state. Subsequently the release of the pressure is all that is required to restore the rock to its liquid condition. In the lithosphere considerable amounts of rock may be held in the solid state by the pressure of overlying strata. Then, if horizontal pressure arches the lithosphere, the pressure on the rock below will be relieved, and the rock will resume a liquid state. If the arching results in cracking at the surface, or in sufficient lateral squeezing of the liquid pockets, eruption may take place. This effect may account for the vast masses of igneous rock that are found associated with the folded mountain ranges (215:58), but it is necessary to ask the question, What causes the arching of the lithosphere? Obviously volcanism, according to this theory, must be traced to the cause of the arching. Hobbs's theory is not very satisfactory because he cannot explain the arching.

It is clear that volcanism might occur as the result either of the process imagined by Day or of that imagined by Hobbs, for several different causes might produce liquid pockets in the lithosphere. But it is equally clear that neither they nor Daly has advanced a theory to account for volcanoes, volcanic zones, plateau basalts, and volcanic mountains. Einstein, when he first received some material outlining the theory proposed in this book, wrote me that it was the only theory he had ever seen that could explain the volcanic zones (128). These, of course, can be explained as zones of fractures (such as the rift valleys just mentioned) resulting from lithosphere displacements.

2. THE VOLCANIC ISLAND ARCS

Campbell has suggested an explanation for the formation of the volcanic island arcs, so many of which are found in the Pacific, and which consist of chains of volcanic mountains in the sea. He shows not only how our theory of lithosphere displacement may account for the formation of these volcanic mountains but also how it may account for their occurrence in graceful curves:

As a sector of the lithosphere, or crust, moves toward the equator, the motion is fastest and the tension is greatest on the meridian of movement, and great north-south faults will open up, beginning there and spreading east and west. At the same time transverse faults of lesser extent will occur, but here again, owing to the different rates at which the lithosphere is moving in different longitudes, the central sector (abutting the meridian of movement) will approach the equator first and, suffering greatest extension, will fault. On either side, other "bands," moving more slowly, will fault farther back, or at a higher latitude, so that a sort of step effect will be created. A line drawn to connect the intersections of these stepped transverse faults with the main meridional fault will form an arc, and the intersections can be expected to be the loci of volcanic islands or similar features on the continents (66).

Another way of expressing the geophysics of the matter, which is somewhat more inclusive and perhaps more easily grasped, may be put thus (again, the formulation is Campbell's):

In any general movement of the lithosphere, one area moves toward the equator and must cover a greater area; there is insufficient surface, while on the farther side of the equator an area of equal size is being subjected to contractions: there is excess of surface. As a result there is an effect whereby an area of deficient surface tends to borrow surface from the area of excess surface. This takes the form of a lag, which reaches its maximum at the meridian of travel. As it diminishes on either side, an arc is formed. This arc determines major parallel fault lines in the earth's crust. It is bisected at intervals by meridional faults, running north and south. The intersections of the two sets of faults will create points of special crustal weakness, which, coinciding with the general downward pressure of the crust in the extension area, will be apt to lead to large-scale eruptions, to the formation of volcanic islands, and to similar features on the continents (66).

It would seem that volcanic island arcs, so formed, may be related to the origin of the geosynclines already discussed. Some recent research appears to have indicated that the formation of a volcanic island arc may be followed by erosion and deposition of sediments on the adjacent sea floors, with subsidence of the floors and ultimate folding. Such a process might be a part not only of mountain building but also of continent building. Krumbein and Sloss point out:

In 1947, Eardley re-examined the structural and stratigraphic implications of the Paleozoic Cordilleran geosyncline. He showed that the associated sediments and volcanics in the geosynclinal deposits can be logically explained by postulating a volcanic island arc system along, or slightly west of, the present Pacific coast (258:330).

If the foregoing considerations are sufficient to suggest the relationship between lithosphere displacement and some kinds of volcanic phenomena, they do not yet provide an adequate explanation of the plateau basalts. Before we can see the bearing of lithosphere displacement on the latter question, we must consider briefly one more of the great unsolved problems of the earth.

3. THE HEAT OF THE EARTH

The origin of the earth's heat is one of the most important of the unsolved problems of geology. Gutenberg says:

Several hypotheses have been proposed to explain the origin of the earth's internal heat, but at present only two fundamental heat sources are postulated—radioactivity, and gravitational contraction. . . . A vast amount of research has been devoted to this subject, but the fact remains that the origin and maintenance of the earth's internal heat continue to be one of the outstanding unsolved problems of science (194:107).

It may be noted that, according to Gutenberg, the assumption of an original molten condition of the earth plays no part in the present attempts to explain the earth's heat. This is a measure of how very far geological science has moved from that conception, and serves to underline still further the danger of falling back upon it for the solution of problems in geology. Gravitational contraction has been deemed insufficient to account for the earth's heat, even if augmented by heat produced by radioactive elements in the rocks. There is serious doubt that radioactivity adds much to the heat of the earth. Smart, for example, is of the opinion that radioactivity cannot produce heat in the earth as fast as it can be radiated through the lithosphere into outer space (386:62).

We have already considered the problem of the earth's heat in a general way (Chapter I). We now have to consider it in relationship to our assumed displacements of the lithosphere.

Daly thought he saw evidence that the heat gradient within the earth in America differs from that in Europe, being somewhat steeper in North America (94:139). This would imply that there is more heat in North America than in Europe. Benfield produced much more evidence of variations in heat from place to place (28); which are difficult to reconcile with a uniform heat gradient in the earth.

A matter of great importance for the general problem is the rate at which heat migrates through the lithosphere and is dissipated into outer space. Geophysicists have determined that the rate of heat migration through the lithosphere is extremely slow. Jeffreys calculated that it would take 130,000,000 years to cool a column of sedimentary rock 7 miles below the earth's surface by 250° C. (241:136). As a result of this, the climate of the earth's surface is determined entirely by the radiant heat of the sun and is uninfluenced by heat from within the earth. We shall have to consider the bearing of this on another well-known fact, which is that earthquakes, and other movements within the lithosphere, are known to produce heat as a consequence of friction between the moving lithospheric blocks (194:158). Then, earthquakes are most frequent in areas where there are distortions of the gravitational balance of the lithosphere, while heat gradients are steeper in such areas (194:141). This indicates that any factor causing such distortions may be a factor in the production of the earth's heat.

Considering these facts, what are the implications, so far as the earth's heat is concerned, of a displacement of the lithosphere? Can there be any doubt that a movement continuing over a period of several thousand years must generate an immense quantity of heat within itself? There can be no doubt of this. The widespread fracturing, the friction between lithospheric blocks resulting from the increased number of earthquakes could have no other result. Moreover, Frankland has pointed out that friction between the lithosphere and the layer over which it moves must produce heat, which may itself facilitate the displacement (168).

The heat thus produced would migrate both inward into the body of the earth and outward into space. But since the rate of dissipation of this heat is so extremely slow, it follows that displacements at relatively short intervals might produce heat more rapidly than it could be dissipated. Over hundreds of millions of years slight increments of heat from this source may have accumulated to produce the earth's present temperature. The assumption of frequent lithosphere displacements thus suggests a third possible source of the earth's heat, in addition to those mentioned by Gutenberg.*

If it is true, as Daly thought, that the heat gradient is steeper in North America than in Europe, this fact serves as additional confirmation of a displacement of the lithosphere at the end of the Pleistocene. I have already presented evidence to suggest that the lithosphere moved at that time in such a direction as to bring North America down from the pole to its present latitude. If this occurred, it meant a displacement of about 2,000 miles for eastern North America but of only about 500 miles for western Europe. Quite obviously the friction must be proportional to the amount of the displacement, and therefore friction and resulting heat could be expected to be somewhat greater in America.

To return, now, to our plateau basalts, we may observe that, in a situation where the lithosphere was continuously in motion over an extended period, a build-up of heat in the lithosphere might cause considerable melting in its lower parts where the temperature was already very close to the melting points of the rocks. This increase of heat would link itself quite naturally, therefore, to an increase in the number and intensity of volcanic eruptions and to lava flows of all kinds. By means of these eruptions and flows some of the heat would be dissipated into the air; much of it, however, imprisoned in the lower part of the lithosphere, would simply increase the volume of the molten magmas.

* The geomagnetic evidence discussed in Chapter I suggests the possibility of there having been several hundred displacements since the Precambrian.

While the increase of heat in the lithosphere would naturally favor larger lava flows, another factor would create the possibility of massive flows, or lava floods.* A massive displacement of the lithosphere, because of the oblateness of the earth, must produce temporary distortions of its shape and gravitational imbalance of the lithosphere. The force of gravity subsequently must gradually force the lithosphere to resume its normal position. This, of course, involves great pressure upon the lithosphere and upon the molten or semimolten material under or within the lithosphere. Pressures of this kind might occasionally lead to the eruption of plateau basalts. It must not be supposed, however, that every displacement of the lithosphere must inevitably produce lava floods. The latter would perhaps be the result of an unusual combination of pressures and fractures. The same combination of forces which might, in one situation, produce volcanic mountains and island arcs might, under other circumstances, produce a doming up of the lithosphere in a local area or a lava flood.

4. CHANGING SEA LEVELS

An important problem closely related to that of mountain building is that of the cause of very numerous, and in some cases radical, changes in the elevations of land areas relative to the sea level. Umbgrove finds that mountain folding has been related, in geological time, to uplift of land areas or to withdrawal or regression of the sea (420:93). However, it is clear that the uplifts were not confined merely to the folded areas—that is, to the mountains themselves—but affected large regions. Such uplifts, where whole sections of the lithosphere were elevated without being folded, are referred to as epeirogenic uplifts, to distinguish them from the uplifts of the folded mountain belts which may have resulted from the folding itself and which are referred to as orogenic uplifts. As to the extent of the resulting changes in sea level, Umbgrove says:

. . . The most important question concerns the depth to which the sea-level was depressed in distinct periods of intense regression, in other words, the extent of the change to which the distance between the surface of the continents and the ocean floors was subjected during the

* Covering hundreds of thousands of square miles with molten lava as much as a mile deep!

pulsating rhythm of subcrustal processes. Joly was the only one who approached this question from the geophysical side, and he arrived at an order of 1000 meters. . . . (420:95).

It becomes necessary, therefore, to find a connection between the cause of the folding of the lithosphere and the cause of general, or epeirogenic, changes of elevation of continents and sea floors. Fortunately this problem is not really so difficult as it may seem at first glance. That it can be solved in terms of the assumption of displacements of the lithosphere is, I think, clear from the following considerations.

Gutenberg has pointed out that if a sector of the lithosphere, in gravitational equilibrium at the equator, is displaced poleward by a shift of the whole lithosphere, it will be moved to a latitude where gravity is greater, because gravity increases slightly toward the poles. Its weight will be thereby increased, and to remain in gravitational equilibrium it must seek a lower level: It must subside. The water level in the higher latitude adjusts easily, of course. Gutenberg points out, however, that if the movement of the lithosphere occurs at a rate greater than the rate at which the sector may sink, by displacing viscous material from below itself, the result will be that the sector will stand (for a time) higher relative to sea level than it did before. I give Gutenberg's own words:

> Movements of the earth's crust relative to its axis must be accompanied by vertical displacements. A block with a thickness of 50 kilometers in equilibrium near the equator should have a thickness of 49.8 near the poles to be bounded by the same equipotential surfaces there. If it moves toward a pole, it must sink deeper to keep in equilibrium. If the process is too fast for maintenance of isostatic equilibrium, positive gravity anomalies and regressions are to be expected. Thus regression may be an indication that an area was moving toward a pole, and transgressions that it was moving toward the equator (194:204–05).

According to Gutenberg, an area moved about 6,000 miles from the equator to a pole would stand about 1,200 or 1,400 feet higher above sea level, if the speed of the displacement was too rapid for maintenance of gravitational equilibrium. The speed of displacement that is suggested by the evidence I have presented is such as to eliminate the possibility that the lithospheric sector could sink and remain in gravitational equilibrium. Consequently, by our theory, a poleward movement of any sector of the lithosphere will result in uplift, and in regression of the sea. In addition, it appears to me that since any sector displaced poleward would also be compressed laterally, this would offer another obstacle to its subsidence. It would have to overcome the lateral pressures as well as displace underlying material.

The amount of the uplift of an area displaced poleward would depend, of course, on the amount of the displacement. As has already been made clear, much geological evidence appears to suggest that displacements may have amounted, on the average, to no more than a third of the distance from a pole to the equator. If this is true, then the resulting uplift to be expected should be of the order of about one third of the uplift Gutenberg suggested, or from 400 to 500 feet.

There is another factor that would operate in the same direction as the effect mentioned by Gutenberg, to alter the elevation of land areas and sea bottoms. Unlike the gravitational effect, however, this second factor would tend to a permanent change in sea levels and might therefore, cumulatively, result in important changes in the distribution of land and sea. It is a question of the permanent consequences of the stretching or compression of the lithosphere. As we have seen, an area displaced poleward must undergo compression because of the shortened radius and circumference of the earth in the higher latitudes. This compression must result in the folding of rock strata, which will be likely to occur mainly in areas where the lithosphere has already been weakened by the formation of geosynclines. The effect of the folding will be to pile up the sedimentary rocks that have been formed from sediments deposited in the geosynclines, causing them to form thicker layers. These thicker layers of lighter rock will tend, even after gravitational adjustments have taken place, to stand higher above sea level. The effect of one displacement in this respect would be slight, but the accumulation of the effects of many displacements through millions of years could lead to extremely important changes in the distribution of land and sea areas. Numerous displacements of the lithosphere could, in fact, constitute an essential, and perhaps even the basic, mechanism for the growth of continents.

Equally important for the general question of sea levels are the effects to be expected from a displacement of a sector of the lithosphere toward the equator. Here the lithosphere will be subjected to tension, or stretching. We have already noted that in this process innumerable fractures will be created in the lithosphere, and these will tend to be filled up with magma from below. Since this magma, invading the lithosphere, will average higher specific density than the rocks of the lithosphere, it may increase the general weight of the lithosphere and thus depress it, causing a deepening of the sea. It is also true, as we have noted, that massive lava flows may occur on the sea bottoms as a result of displacement of the lithosphere. These could have the effect of weighting the lithosphere. Moreover an equatorward displacement of an area must result in a gravitational effect opposite to that of the poleward displacement mentioned by Gutenberg. In this case, the lithosphere must rise to achieve gravitational balance. In so doing it may have to absorb a considerable amount of the heavier rock underlying the lithosphere. This obviously would tend to weight the lithosphere.

The foregoing factors, added together, may account for the observed deepening of the oceans, and the increase of their total surface area, from the poles to the equator. A survey indicates that this deepening is on the order of one kilometer or, perhaps, 4,000 feet (233).

There is still another factor that may affect sea levels but in an unpredictable way. It seems clear, for several reasons, that a displacement of the whole lithosphere must result in considerable readjustments and redistribution of materials of different densities on the underside of the lithosphere. While these can hardly be predicted, they must affect the elevation of points at the earth's surface.

Geologists believe that the underside of the lithosphere has unevennesses corresponding to those at the surface, and that the lithosphere varies considerably in thickness from place to place. They think, for example, that the lithosphere is thicker under the continental surfaces and thinner under the oceans, and that it is thickest of all under mountain ranges and high plateaus. Continents and mountain ranges not only stick up higher but they also stick down deeper. That is because they are composed, on an average, of lighter rock. The analogy is to an iceberg. An iceberg floats with one tenth of its mass above sea level and nine tenths of it submerged. It is lighter than water per unit volume and floats in the water, displacing its own weight and leaving its own excess volume above the surface. Continents and mountain chains stand in the same sort of hydrostatic balance. Their downward projections are thought to be much greater than their upward, visible projections. The downward projections of mountain chains are called "mountain roots."

The underside of the lithosphere, then, has a sort of negative geography. The features of the upper surface are repeated in reverse on the undersurface, although, naturally, the details are missing. The effects are rather smoothed out. We should expect that the Rocky Mountains would make a sizable bump on the underside of the lithosphere, but we couldn't expect to find any small, sharp bump just under Pikes Peak. The tensile strength of the lithosphere, though limited, is sufficient to smooth out the minor features.

As we attempt to envisage the situation at the bottom of the lithosphere, we must remember that the rocks are subjected to increasing pressure with depth, and probably to increasing heat, and as a consequence they must tend to lose their rigidity and strength. We do not just come suddenly to the boundary of the lithosphere at a given depth. On the contrary, the lithosphere just fades away. The rocks of the lowest part of the lithosphere must be very weak indeed, so that a very slight lateral pressure may suffice to displace them.

It follows that when lateral pressures develop during a displacement of the lithosphere, as the downward projections of continents and mountains are brought to bear against the upward extension of the viscous layer below the ocean basements, large blobs of this soft rock of lesser density will be detached from the undersides of the continents, or mountain ranges, and will get shifted to other places. If, as a result of this shifting around, the average densities of vertical columns extending from the bottom to the top of the lithosphere get changed, then there will eventually be corresponding changes of elevation at the surface. Some areas might, as a result, tend to rise and others to sink. This could account, naturally, for changes of sea level and for many topographical features such as basins and plateaus.

To sum up the question of sea levels, it appears that the assumption of displacements of the lithosphere (especially if they are considered to have been numerous) may help to explain them. It seems able to explain why glaciated areas (which we consider to have been areas displaced poleward) appear to have stood higher relative to sea level, and why periods of warm climate in particular regions appear to have been associated with reduced elevation of the land and transgressions of the sea. The theory seems to satisfy Umbgrove's conclusion that sea-level changes have resulted from some "world-embracing cause" (420:93). It accounts, too, for Bucher's suggestion that regressions of the sea have resulted from sublithospheric expansion, and transgressions from sublithospheric contraction, for this, obviously, is only another way of looking at a displacement of the lithosphere (58:479). (If an area is displaced poleward, the effect of sublithospheric contraction is created; if it is displaced equatorward, the effect of sublithospheric expansion occurs.) At the same time it provides an explanation for the rhythmic changes of sea levels through geological history that so mystified Grabau:

This rhythmic succession and essential simultaneousness of the transgressions as well as the regressions in all the continents, indicates a periodic rise and fall of the sea-level, a slow pulsatory movement, due apparently to alternate swelling and contraction of the sea-bottom (183).

5. UNDISTURBED SECTIONS OF THE LITHOSPHERE

It has been objected that there exist extensive areas where rock formations appear to have been little disturbed over very great periods of time. If the lithosphere has been displaced as often as is required by this theory of displacement, why would not the lithosphere be universally folded to a far greater extent than it is?

I think this objection has been partly answered where I pointed out that in a single displacement of the lithosphere the folding would be comparatively slight, and that it would be confined to a small part of the earth's entire surface. I have suggested that it would be greatest along the meridian of the lithosphere's maximum displacement, but that at some point between this meridian and the two areas suffering no displacement, the compressions would tend to fall below the elastic limit of the lithospheric rocks, so that the lithosphere would simply bend elastically and then return to its original, apparently undisturbed, position in some subequent movement. It may be added that most of the changes of elevation resulting from a displacement of the lithosphere would tend to be epeirogenic—that is, they would be broad uplifts or subsidences of large regions resulting from the tilting of great segments of the lithosphere, rather than merely local deformations of the rock structures.

Another point that may be urged in answer to this objection is that, apparently, over considerable periods the poles have tended to remain in approximately the same areas, resulting in leaving some areas far removed for long periods from the meridian of maximum displacement of the lithosphere.

6. THE CHRONOLOGY OF MOUNTAIN BUILDING

Another objection that may be raised to this theory of mountain building is that there are supposed to have been only a few great mountain-building epochs in the world's history of three or four billion years, and that these epochs have been separated by very long periods when mountains were eroded away, and no new ones formed. I shall indicate two reasons for holding that this concept is an illusion.

The first reason is that the record of the rocks is incomplete. It has been estimated that if all the original sedimentary beds of all geological periods were added together (that is, the entire amount of sediment that has been weathered out of the mountains and continents and accumulated to make sedimentary rocks since the beginning of geological time), the total thickness of sediment would be about eighty miles. At the present time, however, the average thickness of the existing sedimentary rocks at any one point on the lithosphere is estimated to be no more than a mile and a half. What has happened to all the missing sediment? The answer is that it has been used over again. At the present time, all over the earth, the forces of the weather and the sea are busy wearing away or grinding up rock, and most of the rock they are destroying is sedimentary rock. Thus more than 95 percent of all the sedimentary rocks formed since the beginning of the planet have been destroyed. As a result of this, geologists have been forced to piece together this geological record from widely separated beds. They find a part of the Silurian sediment in the United States and another part in Africa, and so on.

The enormous difficulty of piecing together the geological record from these discontinuous and scattered beds is rendered even greater by the fact that vast areas of what were once lands are now under the shallow epicontinental seas, and even under the deep sea (as we shall see in the next section). Let us remember, too, that even among the still-existing beds now to be found on the lands, only a tiny percentage are at or near the surface and thus available for study. And of these a large proportion are in such remote and geologically unexplored areas as Mexico, the Amazon, and Central Asia. And still, despite these enormous handicaps, new periods of mountain formation are constantly being discovered. Umbgrove remarks that a long list of them has been "gradually disclosed to us" (420:27). It seems to me that there is unjustifiable complacency in the assumption that the list of mountain-forming epochs is now complete. How can we reach a reasonable guess as to the number that remain undiscovered?

The second reason for holding that the idea of rare mountain-building periods is quite illusory is perhaps even more persuasive. It seems that a remarkable error has vitiated the interpretation of the evidence regarding these alleged periods. The error has been exposed by the development of nuclear methods of dating recent geological events, already referred to, which have revealed an unexpectedly rapid rate of geological change. The error, I think, consists in interpreting the geological evidence on the assumption that conditions as revealed in a particular deposit in one area necessarily determine worldwide conditions. Thus evidence of an ice age in a particular deposit in one place has been interpreted as meaning a period of lowered temperature for the whole world at that time. In the same way, mountain-building revolutions were assumed to affect all parts of the world at once. The idea that mountain building might go on on one continent while another went scot-free was not entertained.

The contemporaneousness of these events in different parts of the world rested, as we shall see, upon a very vague idea of geological time. The techniques for dating the older geological formations never did, and do not now, allow reliable conclusions regarding the contemporaneousness of mountain building on different continents any more than they permit such conclusions regarding climatic changes. Margins of error amounting to millions of years must always be allowed. Triassic folding in India need not be contemporary with Triassic folding in North America, because the Triassic Period is estimated to have lasted about 35,000,000 years! Calculations of the rates at which the weather wears away mountains have shown that mountain ranges may be worn away in much less time than that.

Thus we cannot place reliance on the accepted notions of the occurrence of mountain-building revolutions in time and space, but must hold that the process was, in all probability, much more continuous than has been supposed, but confined to smaller parts of the earth's surface at any one time. Further support for this view is provided by the geologist Stokes, who remarks, in connection with the history of the Rocky Mountains:

> Although the Rocky Mountain or Laramide Revolution is popularly supposed to have occurred at the transition from the Cretaceous to the Tertiary, it has become increasingly evident that mountain building was continuous from place to place from the late Jurassic or early Cretaceous and that deformation continued through the early Tertiary and Quaternary (395:819).

In other words, mountain building went on continuously in North America from the Jurassic Period, about 100,000,000 years ago, into the Pleistocene Epoch, which is considered to have come to an end 10,000 years ago! This is excellent evidence in support of the conclusion that, in all probability, none of the alleged mountain-building revolutions occurred at all.

Krumbein and Sloss point out that this view is, in fact, becoming widely accepted by geologists. They remark that "Gilluly . . . recently examined the evidence for and against periodic diastrophic disturbances, and he showed that such disturbances are much more nearly continuous through time than is generally supposed," and they conclude:

> Added complexity arises as additional stratigraphic studies afford data which imply that tectonic activity is continuous through time. The classical concept that a geological period represents a long interval of quiescence closed by diastrophic disturbances is not fully supported by these newer data (258:343).

7. SUNKEN CONTINENTS VERSUS DRIFTING CONTINENTS

There is an extraordinary contradiction in the very fact that, while continents are supposed to have been permanent, nearly all the sedimentary beds that compose them were laid down under the sea. There is no denying this fact. According to Schuchert, North America has been submerged no less than seventeen times (369a:601). According to Humphreys, the sea has covered as much as 4,000,000 square miles of North America at one time (231:613). Termier argued that the sedimentary beds composing the mountain ranges extending eastward from the Alps to Central Asia, which were laid down under the sea, would have required that the ancient Tethys Sea, in which they were laid down, should have been about 6,000 kilometers (or perhaps 4,000 miles) across (419:221–22).

Geophysicists tend to argue that such seas, which clearly did exist, were merely shallow affairs, invasions of the continents by the ocean owing to some unknown cause. The positive evidence for this, based on the apparent absence from the sedimentary rocks of sediments formed in the very deep sea, has a fallacy in it, as will be made plain. The positive evidence against the assumption that all these seas were shallow seas is, on the other hand, enormously strong. Umbgrove, for example, remarks:

. . . Not only have parts of the continents foundered below sea-level since pre-Cambrian times but they have even done so until quite recently, and their subsidence occasionally attained great depths! The present continents are but fragments of one-time larger blocks. . . . (430:30).

The Soviet geophysicist V. V. Beloussov is in agreement with this (25a).

A particularly important example of such foundering seems to have occurred in the North Atlantic, off the northeastern coast of the United States. It has been found that the sediments that compose the northeastern states were derived in ages past from a land mass to the eastward in the present North Atlantic.

Some geologists have attempted to argue that these sediments might have been derived from a land mass situated on the present continental shelf, but the argument fails from every point of view. Brewster, for example, comments:

It must have been a large continent, for the sand and gravel and mud which the rivers washed out to sea and the waves ground up on the shore have built up most of half a dozen big states, while in some places the deposits are a mile thick (45:134–35).

Umbgrove says that while it is impossible to estimate the size of the land mass (called "Appalachia" by the geologists), it was clearly large, to judge from the fact that it has been possible to trace out in the sedimentary beds of the Appalachian Mountains the outline of an enormous delta formed by a giant river flowing out of the land mass to the east (430:35–38).

Now the continental shelf of North America ends abruptly a very short distance from the coast. It is an extremely narrow strip between the coast and the so-called "continental slope," where the rock formations dip down suddenly and steeply into the deep sea. Its average width is only 42 miles, and its maximum width does not exceed 100 miles (46). If the sediments had been derived from a land mass on this continental shelf, this very narrow land mass would have had to carry huge and repeatedly uplifted mountain ranges. Furthermore, since drainage would naturally have carried sediments down both slopes of these mountain ranges, a large proportion of the material would have been carried eastward and deposited in what is now the deep ocean; but there is no evidence of this.

The suggestion that the enormous volume of sediments forming the northeastern states of the United States came from the continental shelf must be considered improbable. On the other hand, it is plain that the former continent in the North Atlantic could not have been eroded away by rivers any farther down than approximately sea level. Erosion did not dispose of the continent, nor create the deep-sea basin. After erosion had finished its work, the continent itself sank to a great depth. Umbgrove has cited recent oceanographic research by Professor Ewing of Columbia, showing that this ancient land mass of Appalachia now lies subsided about two miles below the continental shelf (430:35–38).

This extraordinary case is by no means unique, for Umbgrove has pointed out that the sediments composing much of Spitzbergen and Scotland come from the ocean west of them, while those composing the west coast of Africa come from a former land mass in the present South Atlantic. Most interesting of all, he indicates that the deepest of the world's deep-sea troughs (east of the Philippines), about seven miles deep, gives evidence that it was once part of a very large continent (420:38).

The origin of the sediments on either side of the Atlantic has a very important bearing on the theory of continental drift. If America and Europe split apart, just where did these sediments come from? The problem is insoluble without assuming a continent in between.

According to Umbgrove, there is ample evidence of repeated upward and downward oscillations of the floor of the entire Pacific (420:236). In a kind of rhythm, the great ocean has become alternately shallower and deeper. In the absence of any explanation of this phenomenon, Umbgrove becomes "geopoetic." There seems to him to be something almost mystical in this slow pulsation of the living planet. He finds that the unexplained upward and downward movements are not limited to sea areas:

> . . . It should be noted that blocks that were first submerged, then elevated, and then once more submerged and elevated, are also met with on the continents. The sub-Oceanic features and the similar continental characteristics cannot be explained at present, for our knowledge of pre-Cambrian history and terrestrial dynamics is not yet extensive enough. . . . (420:241).

Comparatively radical vertical changes in the positions of land masses are evidenced by a considerable number of ancient beaches (some of them, however, not very old) which are now found at great elevations above sea level, and sometimes far inland from the present coasts. Thus the geologist P. Negris claimed to have found evidences of beaches on three mountains of Greece: Mt. Hymettus, Mt. Parnassus, and Mt. Geraneia at, respectively, 1,400 feet, 1,500 feet, and 1,700 feet above sea level. He found a beach on Mt. Delos at 500 feet (324a:616–17). William H. Hobbs cited a particularly interesting case of a beach of recent date now 1,500 feet above sea level, in California:

> Upon the coast of Southern California may be found all the features of wave-cut shores now in perfect preservation, and in some cases as much as fifteen hundred feet above the level of the sea. These features are monuments to the grandest of earthquake disturbances which in recent times have visited the region (216:249).

It would be possible to multiply endlessly the evidence of the raised beaches, which are found in every part of the world. Many of them may imply changes in the elevations of the sea bottoms, such as are suggested by Umbgrove.

One of the most remarkable features of the earth's surface is the Great Rift Valley of Africa. The late Dr. Hans Cloos pointed out that the high escarpment along one side of this valley was once, quite evidently, the very edge of the African continent: not just the beginning of the continental shelf but the very edge of the continental mass. In some vast movement that side of the continent was tremendously uplifted, and the sea bottom was uplifted with it as much as a mile and became dry land. This is so interesting a matter, and of such special importance for our theory, that I quote Cloos at length:

> There are two rims to the African continent. Twice the fundamental problem arises: why do the continents of the earth end so abruptly and plunge so steeply into the deep sea? . . . Even more astounding, what is the meaning of the high, raised and thickened mountain margins that most continents have? (85:68).
>
> . . . The short cross-section through the long Lebombo Chain looks unpretentious, but it illuminates events far from this remote plot of the earth. For here the old margin of the continent is exposed. Not so long ago, during the Cretaceous Period, the sea extended to here from the east. The flatland between the Lebombo hills and the present coast is uplifted sea-bottom. . . . What we see are the flanks of a downward bend of High Africa toward the Indian Ocean. . . .
>
> But we see much more: the sedimentary strata are followed by volcanic rocks to the east of the hills. Some parallel the strata like flows or sheets, poured over them and tilted with them. Others break across the sandstone layers and rise steeply from below. This means that as the continent's rim was bent downward at the Lebombo hills, the crust burst, and cracks opened through which hot melt shot upward and boiled over.
>
> So the eastern margin of Africa at the turn of the Paleozoic Period was a giant hinge on which the crust bent down, to be covered by the ocean. What we see here is merely a cross-section . . . one can go further north or south, and even to the other side of the continent and discover that great stretches of this unique land have suffered the same fate. The oceans sank adjacent to the continents, and the continent rose out of the ocean (85:73–74).

Cloos makes it clear that in one geological period the continent was bent down so that a part of it became sea bottom (not merely continental shelf) and that at a later period it was uplifted some 6,000 feet, the sea bottom became land, and the continental margin was shoved far to the east. When we contemplate gigantic movements of this sort, it seems reasonable to take the geophysical objections to changes in the positions of the continents with a grain of salt. If a large part of a continent can be shown not to have been permanent, it is unnecessary to concede the permanence of any of it. On the other hand, such changes need to be explained, and they need to be reconciled with basic principles of physics. The fact that theories of continent formation and history hitherto proposed have failed to solve the problem recently reduced L. Sprague de Camp to the following confession of ignorance:

> Since somebody can bring good, solid objections on one ground or another against all these hypotheses, however, we had better agree that nobody knows why continents or parts of continents sink, and let it go at that. No doubt a sound explanation, perhaps combining features of the older theories, will be forthcoming some day (64:161).

It may be useful to consider, in juxtaposition, the African Rift with the question of the North Atlantic land mass already discussed. In a sense, the two are complementary. In one case a continent apparently subsided; in the other it first subsided and then was raised up. Quite obviously the movements in both directions must have been related to one fundamental dynamic process. The physical geology of the Rift, which can be directly examined, shows that the indirect evidence of the sedimentary rocks of the northeastern states of the United States, and of Scotland and Spitzbergen (and the paleontological evidence), must be taken seriously. The evidence in favor of an important land mass in the present North Atlantic cannot be dismissed. The theory of drifting continents does not offer a solution to these problems.

8. DEEP-SEA SEDIMENTS

One of the most impressive arguments in favor of the permanence of the continents is that almost all the sedimentary rocks that compose the continents appear to be made of sediments that were laid down in comparatively shallow water, on or near the continental shelves. We have already seen, however, that parts of continents (at least) have been submerged to great depths, and that parts of the deep-sea bottom have been uplifted to form land. Why, then, have rocks composed of typical deep-sea sediments not been found?

A number of factors may account for this. The primary factor may be the rate of sedimentation. In the deep sea this is extraordinarily slow—as low as one inch in 2,500 years. Near the coasts it can be hundreds of times more rapid.

The theory presented in this book provides a mechanism that would tend to operate against the consolidation of this deep-sea sediment into rock. Frequent displacements of the lithosphere would naturally be accompanied by increased turbulence on the ocean bottom, by which sediments would be dispersed and mixed with other sediments. There has been in recent years a great extension of our knowledge regarding the operation of turbidity currents (137, 139, 141) caused by the slumping of sediments from the continental slopes and by other forces. It seems that such currents, even now, are powerful enough to bring about considerable rearrangement of the unconsolidated sediments of the ocean bottom. A displacement of the lithosphere would greatly magnify their force, for it would cause extensive changes in the directions of major ocean currents, changes in sea and land levels, extensive volcanism in the sea as well as on land, and an increased number and a greater intensity of earthquakes, which would occasion extensive slumpings of sediments along the continental slopes. If we consider that one such displacement would, in all probability, keep the turbulence at a high point for several thousand years (see Chapters V, VI), we can conclude that the resulting dispersal of deep-sea sediments would probably be on a considerable scale.

Finally, since we cannot suppose that any area would be uplifted rapidly from the deep sea to the surface (that is, all the way in the course of a single displacement), it follows that in most cases deep-sea sediments would be raised into shallow water, where they would be exposed for a long time in an unconsolidated state to the erosive action of the much more rapid currents near the surface before they would be likely to be raised above sea level. A very small proportion of deep-sea sediment would then be mixed by the currents with a large proportion of sediment typical of shallow seas, and would, in most cases, entirely disappear. These factors together dispose of this argument for the permanence of continents, whether they are fixed or drifting.

Another interesting line of evidence with respect to this problem is provided by the discovery in recent years, on the bottoms of the oceans, of several hundred mountains of varying heights, which have been given the name of "seamounts." These have the common characteristic of being flat-topped. Apparently their tops were made flat by the action of the sea at the time they were at the sea level. Now the flat tops are submerged anywhere from a few hundred feet to three miles below sea level.

When these seamounts were first discovered, they were explained in accordance with the theory of the permanence of ocean basins (210). It was proposed that as the sediments gathered in enormous thickness on the ocean floor through hundreds of millions of years, the floor actually gave way and sank, taking the seamounts down below sea level. This theory was undermined, of course, by the recent discovery that no such thick layer of sediments exists on the ocean floors, but that, on the contrary, the layer of sediments is in some places extremely thin or even virtually nonexistent.

Another line of evidence helps to dispose completely of this explanation of seamounts. Foraminifera are minute protozoa that live in the sea. Their species vary with differences in the depth and temperature of the water in which they live, and those of past geological periods, found in fossil state, differ from living species. Studies of fossilized foraminifera from the tops of some of the seamounts have revealed that they are much younger than the seamounts themselves (197). Comparatively recent species have been gathered from the tops of seamounts that have subsided to great depths. Unless turbidity currents could suffice to carry such deposits long distances across the ocean floor and then upward to the tops of the seamounts, another cause for the subsidence of the seamounts must be found. When we remember that Umbgrove referred to frequent upward and downward oscillations of the floor of the Pacific, resulting from an unknown cause, we can see that the idea of a gradual and continuous subsidence of the seamounts is a singularly weak one, for even if the supposed layer of sediments existed, the theory still unaccountably ignores the recurring uplifts of the sea bottom.

The foregoing considerations reveal the essential weakness of the conclusion that the seas that periodically invaded the continents were always shallow seas—"epicontinental" seas, flooding the permanent continents. First, there was land; then, no doubt, shallow seas; after that, in some cases, very deep sea, then again shallow sea, and finally again land, all in the same place. But the interludes of deep sea may have been, in many cases, very short, and the sedimentation resulting may have never been consolidated. Thus the deep sea could come and go, with nobody the wiser. New evidence bearing on this problem is now available as the result of recent Soviet oceanographic work in the Arctic. Soviet scientists have found evidence that the Arctic Ocean itself has existed only since the comparatively recent Mesozoic Era (364:18).

It seems reasonable to conclude that at least some of the problems presented by the continents and ocean basins are soluble in terms of the principles described in this and previous chapters. Land links may be explained as the consequences of mountain formation on the sea bottom; temporary and limited uplift or subsidence of large areas may result directly from their poleward or equatorward displacement. The major changes, however—the enormous elevations and subsidences, the destruction and creation of continents—require us to examine the deepest possible consequences and implications of lithosphere displacement. We must now undertake this deeper examination. This requires us to take another glance at the nature and structure of the lithosphere, to its full depth, as far as our present geophysical knowledge permits.

9. DENSITY CHANGES IN THE LITHOSPHERE

It has, until lately, been the impression that the lithosphere, considered to be a crystalline layer between 20 and 40 miles thick, was itself composed of various layers, with rocks of increasing density at increasing depths. This would have been a natural development with a cooling earth, for it might be supposed that the lighter materials in a liquid would tend to float on the heavier ones and would solidify in the order of their density, with the lightest on top. Daly, in 1940, proposed the following layering of the lithosphere, at least under the continents:

a. The sedimentary rocks, at the surface.
b. Below these a layer of basement rocks of granitic type, ending about 9 miles down.
c. A third layer of rock, of somewhat greater weight, about 15 miles thick.
d. A fourth layer, about 6 miles thick, of still heavier rock.

Having proposed these layers, however, he added that "the exact depths of the discontinuities are not easily demonstrated" and "it seems clear that each of the breaks varies in its depth below the rocky surface" (97:17). In another place he gave evidence of light matter at the very bottom of the crust (97:223). These layers, then, according to Daly, are very peculiar. There is nothing regular about them. On the one hand, he gives rough estimates of their thicknesses. On the other hand, he indicates that these estimates are of little value. They are, in fact, mere rough averages; they indicate a trend toward increasing density with depth, together with enormous confusion in the distribution of materials.

In view of the extreme uncertainties of Daly's view of the structure of the lithosphere, it can hardly come as a complete surprise that the most recent geophysical investigation of its structure by the method of sound-wave surveying has failed to show any distinct layering at all.

The geophysicists Tatel and Tuve have reported that the results of the most recent studies, using the most recent techniques, indicate that rocks of greater or less density are intermixed in utter confusion, that the essential structure of the lithosphere is really that of a rubble on a large scale (406:107).

The main argument of the geophysicists who speak in favor of the permanence of the continents, whether or not they drift, is based on the observed difference in composition of continents and of the lithosphere under the oceans, a difference that has been verified for the uppermost few miles of the continental and oceanic sectors of the lithosphere but not for the greater depths.

Now, if everything below, say, a depth of ten miles were layered everywhere at equal depths with rock of equal densities, no quarrel whatever could be had with the geophysicists who argue for the permanence of the continental shields. For in that case granitic or sedimentary rocks at the surface would have to be destroyed or created in enormous quantity to destroy or create a continent.

It is entirely otherwise with the structure as suggested by Daly and as revealed in the recent geophysical surveys. To understand this it is necessary only to visualize that the relative elevation of the surface at any point is determined by the *average* density of the entire column of matter between the surface at that point and the bottom of the lithosphere, where, presumably, the inequalities at different points are pretty well averaged out.

If the lithosphere is not definitely layered, if, as both Daly and recent geophysicists agree, there are radical variations in the structure, then the vital changes may occur at any depth, deep down as well as at the surface. There is reason to believe that massive changes may occur more easily deep down than at the surface. Thus the lithosphere might be weighted in its lower parts by an intrusion of a great mass of molten rock of high density from below, a very likely result of a displacement. In either case, whether the addition of the heavy matter occurred near the surface or far below it, the result would have been a depression of the surface, with a consequent encroachment of the sea. Obviously a repetition of such movements could cause a continent to subside to a great depth without altering the composition of the superficial formations.

On the other hand, a shifting of the masses of lighter rock, which might have formed the downward projections of continents and mountain chains, as the result of a displacement, could lighten certain sectors of the lithosphere and result in their uplift.

That rocks of light weight are to be found at the very bottom of the lithosphere (and even in the downward projections under continents and mountain chains that extend to greater depth) might, as a matter of fact, have been deduced from Daly's observation that the process of mountain folding has involved the whole depth of the lithosphere and not just its surface layers (97:399). His suggestion is that since the folding of the lithosphere to form mountain ranges involves its horizontal shortening, the horizontal shearing movement has to take place at the level where displacement will be easiest, which will be at the bottom of the lithosphere where the rock has minimum, or zero, strength. For it is plain that at a level at which the rock possessed any considerable tensile strength, the shearing of one layer over another horizontally would be practically out of the question.

It would seem, from these considerations, that the commonly used terms "sial" and "sima" to differentiate lighter from heavier material in the lithosphere, especially when they are presumed to indicate different layers, have very little meaning. They amount to trends merely and take no account of the detailed distributions of the materials of different density either vertically or horizontally. It would be wrong, therefore, to assume that just because we find a layer of basalt on the floor of the Atlantic this layer necessarily extends to the bottom of the lithosphere and is not underlain, at greater depth, by sedimentary and granitic rocks of less density. It is even true that the layer of basalt may have been extruded during the subsidence of the sea bottom in the last movement of the lithosphere, and have been, in itself, a factor in increasing the amount of the subsidence.

Jaggar, for one, considered that it was far more reasonable to account for subsidence or elevation at the surface by changes of weighting deep in the lithosphere than by erosion and sedimentation of the surface. He remarked:

> . . . It would seem possible that intrusive and extrusive processes may lighten or weight the crust much more profoundly than the movement of sediments (235:153).

How, precisely, would these processes be apt to be set in motion by a displacement of the lithosphere? We saw that a displacement would cause a general fracturing of the lithosphere, the creation of a new worldwide fracture pattern. In areas moved toward the equator, the extension of the surface area would involve some pulling apart of the lithosphere, the separation of the fragments, their subsidence into the semiliquid melt below, and the rise of the magma into the fractures, with, at some points no doubt, massive eruptions on the surface. Differential movements of blocks of the lithosphere would occur as each sought its gravitational equilibrium, some rising and others subsiding. In areas moved poleward compression would be the rule, with some folding of the lithosphere with block faulting accompanied by tilting of larger or smaller blocks. In these areas the strata of lighter rock would grow thicker, from being folded upon themselves; in equatorward-moving areas, on the other hand, they would tend to grow thinner, because much of the lighter rock might be engulfed in the rising heavy magma.

However, the destruction or creation of continents requires more massive changes. These might be of two sorts, though both of them must, in the nature of things, remain speculative for the present. One of these would be the massive intrusion of immense quantities of heavy magma into the lithosphere (resulting sometimes in plateau basalts). Such an effect could be produced by sublithospheric currents set in motion by a displacement and would have the effect of causing a major subsidence at the surface. The other cause of massive change in the average density of a given column under the lithosphere (that is, a section extending from top to bottom of the lithosphere) would be a shifting of light matter from one point to another under the bottom of the lithosphere.

What might be the upshot of all these changes during a displacement? The result might well be that while the distribution of light and heavy matter near the surface would be unchanged, its distribution between the surface and the bottom of the lithosphere would be materially changed. This, rather than the theory of continental drift, is probably the direction in which we must look for a solution to the problem of continents and ocean basins.

I cannot close without reference to a singular confirmation of the line of reasoning adopted in this chapter, which I find in Umbgrove's discussion of the work of the geologist J. Barrell, with whom he disagreed.

Umbgrove is considering the question of the submergence of continents. It is clear from his discussion that Barrell's conception of the process requires a theory of lithosphere displacement. Umbgrove states the problem thus: If we are to suppose the submergence of continents, we must either suppose a change in the amount of the ocean water, which, if it increased, could flood a continent (or several at once), or the submergence of one continent balanced by the elevation of another. He presents the findings of specialists to show that the quantity of water on the earth's surface has remained about the same from the earliest times, and adds:

> Should one, nevertheless, cling to the theory of submerged continents, the only alternative would be to assume that while vast blocks were being submerged in one area, parts of the ocean floor of almost identical size were being elevated in others. . . .
> It is not quite clear, however, why such opposed movements should have occurred in areas of almost equal extent. Nor is it clear why these movements should have occurred in such a way that the sea-level remained comparatively stable. . . . (420:235–36).

It is clear that in a displacement two quarters of the surface, opposite each other, must move toward the poles, while the other two quarters must move toward the equator. Whatever forces tend to produce uplift in the poleward-moving areas will be balanced by equal forces producing subsidence in the quarters moving equatorward. And the sea level would be stable, except for very minor fluctuations.

Barrell himself suggests that subsidence of continental areas would be aided by liquid intrusions, "the weight of magmas of high specific gravity rising widely and in enormous volume from a deep core of greater density into these portions of an originally lighter crust. . . ." (420:235–36).

Barrell's suggestion points to the chief weakness of the geophysical argument in favor of the permanence of the continents, whether or not they drift. As I have already pointed out, geophysicists seem, too often, to take as the frame of reference only the outermost ten miles or so of the lithosphere. Theoretically they base their calculations on the full depth of the lithosphere, but practically this assumption is canceled out by the assumption that the lithosphere is arranged in layers of equal density, so that significant changes of density in depth are excluded. But if the real possibilities of changes of average density in the full depth of the lithosphere are taken into account, the difficulties in the face of the subsidence and elevation of continents vanish, and the theory of continental drift becomes superfluous.

chapter **10**

THE EXTINCTION OF
THE MAMMOTHS AND
THE MASTODONS

WHEN this theory of crustal displacements was first presented to a group of scientists at the American Museum of Natural History, on January 27, 1955, Professor Walter H. Bucher, former President of the Geological Society of America, made an interesting observation. I had presented evidence to support the contention that North America had been displaced southward and Antarctica had been moved farther into the Antarctic Circle by the movement of the crust at the end of the ice age. Professor Bucher pointed out that, if this were so, there must have been an equal movement of the crust northward on the opposite side of the earth. He asked me whether there was evidence of this. I said I thought there was. I am presenting the evidence here.

1. THE EXTINCTION OF THE MAMMOTHS

The closing millennia of the ice age saw an enormous mortality of animals in many parts of the world. Hibben estimated that as many as 40,000,000 animals died in North America alone (212:168). Many species of animals became extinct, including mammoths, mastodons, giant beaver, sabertooth cats, giant sloths, woolly rhinoceroses. Camels and horses apparently became extinct in North America then or shortly afterward, although one authority believes a variety of Pleistocene horse has survived in Haiti (365). The paleontologist Scott is enormously puzzled both by the great climatic revolution and by its effects:

> The extraordinary and inexplicable climatic revolution had a profound effect upon animal life, and occasioned or at least accompanied, the great extinctions, which, at the end of the Pleistocene, decimated the mammals over three-fifths of the earth's land surface (372:75).

No one has been able to explain these widespread extinctions. I shall attempt to explain them as consequences of the last displacement of the crust, but, since the extinctions took place both in North America and in Asia—that is, both in the area presumably moved southward and in the area presumably moved northward—I shall concentrate first on Asia. There we shall find no difficulty in producing evidence to show that the climate of eastern Siberia grew colder as North America grew warmer, just as the theory requires.

Among all the animals that became extinct in Asia, the mammoth has been the most studied. This is because of its size; because of the great range of its distribution, all the way from the New Siberian Islands in the Arctic Ocean, across Siberia and Europe, to North America; because pictures of it drawn by primitive man have been found in the caves of southern France and Spain; but most of all, perhaps, because some well-preserved bodies of mammoths have been found frozen in the mud of Siberia and Alaska. Ivory from these frozen remains has provided a supply for the ivory trade of China and Central Europe since ancient times.

A study of the reports on the frozen mammoths reveals some very remarkable facts. In the first place, they increase in numbers the farther north one goes, and are most numerous in the New Siberian Islands, which lie between the Arctic coast of Siberia and the pole. Secondly, they are accompanied by many other kinds of animals.

Thirdly, although ivory is easily ruined by exposure to the weather, uncounted thousands of pairs of tusks have been preserved in good enough conditions for the ivory trade. A fourth point is that the bodies of many mammoths and a few other animals have been preserved so perfectly (in the frozen ground) as to be edible today. Finally, astonishing as it may seem, it is not true that the mammoth was adapted to a very cold climate. I shall first take up this question of the mammoth's alleged adaptation to cold.

2. THE MAMMOTH'S ADAPTATION TO COLD

It has long been taken for granted, without really careful consideration, that the mammoth was an Arctic animal. The opinion has been based on the mammoth's thick skin, on its hairy coat, and on the deposit of fat usually found under the skin. Yet it can be shown that none of these features mean any special adaptation to cold.

To begin with the skin and the hair, we have a clear presentation of the facts by the French zoologist and dermatologist H. Neuville. His report was published as long ago as 1919 (325). He performed a comparative microscopic study of sections of the skin of a mammoth and that of an Indian elephant, and showed that they were identical in thickness and in structure. They were not merely similar; they were exactly the same. Then he showed that the lack of oil glands in the skin of both animals made their hair less resistant to cold and damp than the hair of the average mammal. In other words, the hair and fur of the mammoth showed a *negative* adaptation to cold. It turns out that the common, ordinary sheep is better adapted to Arctic conditions:

We have . . . two animals very nearly related zoologically, the mammoth and the elephant, one of which lived in severe climates while the other is now confined to certain parts of the torrid zone. The mammoth, it is said, was protected from the cold by its fur and by the thickness of its dermis. But the dermis, as I have said, and as the illustrations prove, is identical in the two instances; it would therefore be hard to attribute a specially adaptive function to the skin of the mammoth. The fur, much more dense, it is true, on the mammoths than on any of the living elephants, nevertheless is present only in a very special condition which is fundamentally identical in all of these animals. Let us examine the consequences of this special condition, consisting, I may repeat, in the absence of cutaneous glands. The physiological function of these glands is very important. [Neuville's footnote here: It is merely necessary to mention that according to the opinion now accepted, that of Unna, the effect of the sebum is to lubricate the fur, thus protecting it against disintegration, and that of the sweat is to soak the epidermis with an oily liquid, protecting it also against desiccation and disintegration . . . the absence of the glandular secretions puts the skin in a condition of less resistance well known in dermatology. It is superfluous to recall that the sebaceous impregnation gives the fur in general its isolating properties and imparts to each of its elements, the hairs, its impermeability, thanks to which they resist with a well-known strength all disintegrating agents, and notably those which are atmospheric. Everyone knows to what degree the presence of grease produced by the sebaceous glands renders wool resistant and isolating, and to what degree the total lack of this fatty matter lessens the value of woolen goods. . . .] (325:331–33).

Neuville points out both that the mammoth lacks sebaceous glands and that the oil from these glands is an important factor in the protection of an animal against cold. It is probable that protection from damp is more important than protection from low temperature. Oil in the hair must certainly impede the penetration of damp. The hair of the mammoth, deprived of oil, would seem to offer poor protection against the dampness of an Arctic blizzard. Sanderson has pointed out that thick fur by itself means nothing: Many animals of the equatorial jungles, such as tigers, have thick fur (365). Fur by itself is not a feature of adaptation to cold, and fur without oil, as Neuville points out so lucidly, is a feature of adaptation to warmth, not cold.

The question of the importance of oily secretions from the skin for the effectiveness of resistance of fur or hair to cold and damp is, however, highly involved. Very many inquiries directed to specialists in universities, medical schools, and research institutes over a period of more than five years failed to elicit sufficiently clear and definite answers until, finally, Dr. Thomas S. Argyris, Professor of Zoology at Brown University, referred me to the Headquarters Research and Development Command of the United States Army. This agency, in turn, very kindly referred me to the British Wool Industries Research Association. I addressed an inquiry to them regarding the effects of natural oil secretions from the skin on the preservation of wool. They replied in general confirmation of Neuville:

. . . Those interested in wool assume that the function of the wool wax is to protect the wool fibres from the weather and to maintain the animal in a dry and warm condition. Arguments in this direction are of course mainly speculative. We do know, however, that shorn wool in its natural state can be stored and transported without entanglement (or felting) of the fibres, while scoured wool becomes entangled so that, during subsequent processing, fibre breakage at the card is significantly increased. It seems reasonable, therefore, to assume that the wool wax is responsible not only for conferring protection against the weather but also for the maintenance of the fleece in an orderly and hence more efficacious state (447).

It appears that there has been no scientific study of the precise points at issue here; no one has measured in any scientific way the quantitative effect of oily secretions in keeping heat in or moisture out. Despite this fact, however, we are at least justified, on the basis of the facts cited above, in rejecting the claims advanced for the hair of the mammoth as an adaptive feature to a very cold climate.

Neuville goes on to destroy one or two other arguments in favor of the mammoth's adaptation to cold:

. . . It has been thought that the reduction of the ears, thick and very small relatively to those of the existing elephants, might be so understood in this sense; such large and thin ears as those of the elephants would probably be very sensitive to the action of cold. But it has also been suggested that the fattiness and peculiar form of the tail of the mammoth was an adaptive character of the same kind; however, it is to the fat rumped sheep, animals of the hot regions, whose range extends to the center of Africa, that we must go for an analogue to the last character.

It is therefore, only thanks to entirely superficial comparisons which do not stand a somewhat detailed analysis, that it has been possible to regard the mammoth as adapted to the cold. On account of the peculiar character of the pelage the animal was, on the contrary, at a disadvantage in this respect (325:331–33).

There remains the question of the layer of fat, about three inches thick, which is found under the skin of the mammoth. This fat is thought to have provided insulation against the bitter cold of the Siberian winter.

The best opinion of physiologists is opposed to the view that the storage of fat by animals is a measure of self-protection against cold. The consensus is, on the contrary, that large fat accumulation testifies chiefly to ample food supply, obtainable without much effort, as indeed is the case with human beings. Physiologists agree that resistance to cold is mainly a question of the metabolic rate, rather than of insulation by fat. Since the length of capillaries in a cubic inch of fat is less than the length of capillaries in a cubic inch of muscle, blood circulation would be better in a thin animal. We might ask the question, Which would be more likely to survive through a Siberian winter, a man burdened with fifty or a hundred pounds of surplus fat or a man of normal build who was all solid muscle, assuming that winter conditions would mean a hard struggle to obtain food? Dr. Charles P. Lyman, Professor of Zoology at Harvard, remarked regarding this question of fat:

It is true that many animals become obese before the winter sets in, but for the most part it seems likely that they become obese because they have an ample food supply in the fall, rather, than that they are stimulated by cold to lay down a supply of fat. Cold will ordinarily increase the metabolic rate of any animal which means that it burns up more fuel in order to maintain its ordinary weight, to say nothing of adding weight in the form of fat. The amount of muscular activity in the daily life of either type of elephant is certainly just as important as the stimulus of cold as far as laying down a supply of fat is concerned (284).

This statement suggests that there is no basis for the assumption that the fat of the mammoths adapted them to an Arctic climate. On the other hand, it is quite true that the storage of fat in the fall may help animals to get through the winter when food is scarce. The winter does not, however, have to be an Arctic winter. A winter such as we have in temperate climates is quite cold enough to cut the available food supply for herbivorous animals. It seems that under favorable circumstances even the African and Indian elephants accumulate quite a lot of fat. F. G. Benedict, in his comprehensive work on the physiology of the elephant, considers it a fatty animal (26).

The resemblances between the mammoth and the Indian elephant extend further than the identity of their skins in thickness and structure, and the fact that they were both fatty animals. Bell suggests that they were only two varieties of the same species:

Falconer insists on the importance of the fact that throughout the whole geological history of each species of elephant there is a great persistence in the structure and mode of growth of each of the teeth, and that this is the best single character by which to distinguish the species from one another. He finds, after a critical examination of a great number of specimens, that in the mammoth each of the molars is subject to the same history and same variation as the corresponding molar in the living Indian elephant (25).

It is clear that the similarities in the life histories of each of the teeth of these two animals were more important than the differences in the shapes of the teeth, which were such as might easily occur in two varieties of the same species. It cannot be denied that two varieties of the same species may be adapted to different climates, but it must be conceded that the adaptation of two varieties of the same species, one to tropical jungles and the other to Arctic conditions, is against the probabilities.

3. THE PRESENT CLIMATE OF SIBERIA

The people who lay the greatest stress on the adaptation of the mammoth to cold ignore the other animals that lived with the mammoths. Yet we know that along with the millions of mammoths, the northern Siberian plains supported vast numbers of rhinoceroses, antelope, horses, bison, and other herbivorous creatures, while a variety of carnivores, including the sabertooth cat, preyed upon them. What good does it do to argue that the mammoth was adapted to cold when it is impossible to use the argument in the case of several of the other animals?

Like the mammoths, these other animals ranged to the far north, to the extreme north of Siberia, to the shores of the Arctic Ocean, and yet farther north to the Lyakhov and New Siberian Islands, only a very short distance from the pole. It has been claimed that all the remains on the islands may have been washed there from the mouths of the Siberian rivers by spring floods; I shall consider this suggestion a little later.

So far as the present climate of Siberia itself is concerned, Nordenskjöld made the following observations of monthly averages of daily Centigrade temperatures during the year along the Lena River (334):

January	—48.9	(—56 F.)	July	+15.4	(+60 F.)
February	—47.2	(—52 F.)	August	+11.9	(+53 F.)
March	—33.9	(—40 F.)	September	+ 2.3	(+36 F.)
April	—14	(+ 7 F.)	October	—13.9	(+ 7 F.)

(Continued)

| May | — 0.4 | (+32 F.) | November | —39.1 | (—36 F.) |
| June | +13.4 | (+56 F.) | December | —45.1 | (—49 F.) |

The average for the whole year was —16.7 (+2° F.). It appears that only three months out of the year are reasonably free from frost. Even so there must be frequent frosts in July, notwithstanding the occasional high midday temperatures. High temperatures on some days would bring the monthly mean down, even if night frosts continued through July.

No doubt it was knowledge of these conditions that caused the great founder of modern geology, Sir Charles Lyell, to remark that it would doubtless be impossible for herds of mammoths and rhinoceroses to subsist throughout the year, even in the southern part of Siberia.

If this is the case with Siberia, what are we to think when we contemplate the New Siberian Islands? There the remains of mammoths and other animals are most numerous of all. There Baron Toll, the Arctic explorer, found remains of a sabertooth tiger, and a fruit tree that had been ninety feet tall when it was standing. The tree was well preserved in the permafrost, with its roots and seeds (113:151). Toll claimed that green leaves and ripe fruit still clung to its branches. Yet, at the present time, the only representative of tree vegetation on the islands is a willow that grows one inch high.

Now let us return to the question of whether all these remains were floated out to the islands on spring floods. Let us begin with a backward view at the history of these islands. Saks, Belov, and Lapina point to evidence that there were luxuriant forests growing on the New Siberian Islands in Miocene and perhaps Pliocene times (364). At the beginning of the Pleistocene the islands were connected with the mainland, and the mammoths ranged over them. In the opinion of these writers the vast numbers of mammoth remains on Great Lyakhov Island indicate that they took refuge on the island when the land was sinking (364:4, note). There is no evidence that they were washed across the intervening sea.

The improbabilities in this suggestion of transportation of these hundreds of thousands of animal bodies across the entire width of the Nordenskjöld Sea, for a distance of more than 200 miles from the mouth of the Lena River, are simply out of all reason. Let us see exactly what is involved.

First, we should have to explain why the hundreds of thousands of animals fell into the river. To be sure, they did not fall in all at once; nevertheless they must have had the habit of falling into the river in very large numbers, because only one body in a very great many could possibly float across 200 miles of ocean. Of those that floated at all only a few would be likely to float in precisely the correct direction to reach the islands. Islands, even large ones, are amazingly easy to miss even in a boat equipped with a rudder and charts. The Lena River has three mouths, one of which points in a direction away from the islands. The two other mouths face the islands across these 200 miles of ocean. Occasionally a piece of driftwood might float across the intervening sea. Occasionally perhaps an animal—if for some reason it did not happen to sink, if it were not eaten by fishes—might be washed up on the shore of one of the islands. It seems probable that only a very powerful current could transport the body of a mammoth across 200 miles of ocean.

But let us suppose that somehow the animals are transported across the ocean. What then? The greatest of the New Siberian Islands is about 150 miles long and about half as wide. Not one single account of the explorations on these islands has mentioned that the animal remains are found only along the beaches. They are obviously found also in the interior. Are we to suppose that the floods of the Lena River were so immense that they could inundate the New Siberian Islands, 200 miles at sea? It is safe to say that all the rivers of Europe and Asia put together, at full flood, would fail to raise the ocean level 200 miles off the coast by more than a few inches at most.

But, again, let us suppose that the remains were merely washed to the present coasts and not into the interior. How then were they preserved? How were hundreds of thousands of mammoths placed above high-water mark? Storms, no doubt, but whatever storms can wash up, other storms can wash away. No accumulation of anything occurs along the coasts because of storms. All that storms can do is to destroy; they can grind up and destroy anything. And they would have ground up and destroyed all the bodies, including, of course, the 90-foot fruit tree with its branches, roots, seeds, green leaves, and ripe fruit.

I think it is plain that the only reason suggestions of this kind are advanced is that there is need to support some theory that has been developed to explain some other part of the evidence, some local problem. Moreover there is need, always need, to discredit the evidence that argues for drastic climatic changes.

Naturally the knowledge that the Arctic islands, though they are now in polar darkness much of the year, were in very recent geologic times able to grow the flourishing forests of a temperate climate eliminates any need to insist that they were always as cold as they are today. Thus it is not a question at all of whether the climate grew colder but merely a question of when the change occurred.

Campbell has contributed a suggestion with regard to the alleged floating of hundreds of thousands of bodies across the Nordenskjöld Sea. He notes that bodies ordinarily float because of gas produced by decomposition. Decomposition is at a minimum in very cold water, and therefore bodies ordinarily do not float in very cold water. As an example of this he points to a peculiarity of Lake Superior. The waters of this lake are very cold. There is an old saying in the lake region that "Lake Superior never gives up its dead." The Arctic Ocean is much colder than the waters of Lake Superior. The water of the Lena would not be warm even in midsummer, but during the spring floods—when the Lena would be swollen with the melt water of the winter snows—the water would be frigid, and the bodies of animals drowned in it would not decompose, nor would they float. They would tend to sink, instead, into the nearest hole and perhaps never come to the surface.

4. A SUDDEN CHANGE OF CLIMATE?

We may reasonably conclude that the climate of Siberia changed at the end of the Pleistocene and that it grew colder. Our problem is to discover what process of change was involved. On the one hand, our theory of displacement of the crust involves a considerable period of time and a gradual movement; on the other hand, the discovery of complete bodies of mammoths and other animals in Siberia, so well preserved in the frozen ground as to be in some cases still edible, seems to argue a cataclysmic change.

To those who, in the past, have argued for a very sudden catastrophe, the specialists in the field have offered opposing theories to explain the preservation of the bodies. One of these was that as the mammoths walked over the frozen ground, over the snowfields, they may have fallen into pits or crevasses and been swallowed up and permanently frozen. Or, again, they might either have broken through river ice and been drowned or have got bogged while feeding along the banks.

There is no doubt that a certain number of animals could have been put into the frozen ground in just the manner suggested above. That this is the explanation for the preservation of the mammoths' bodies generally, however, is unlikely for a number of reasons.

It is not generally realized, in the first place, that it is not merely a matter of the accidental preservation of eighty-odd mammoths and half a dozen rhinoceroses that have been found in the permafrost. These few could perhaps be accounted for by individual accidents, provided, of course, that we agreed that the animals concerned were Arctic animals. The sudden freezing and consequent preservation of the flesh of these animals might be thus explained. But there is another factor of great importance, which has been consistently neglected. It has been overlooked that meat is not the only thing that has to be frozen quickly in order to be preserved. The same is true of ivory. Ivory, it appears, spoils very quickly when it dries out.

Tens of thousands of skeletons and individual bones of many kinds of animals have been discovered in the permafrost. Among them have been found the enormous numbers of mammoths' tusks already mentioned. To be of any use for carving, tusks must either come from freshly killed animals or have been frozen very quickly after the deaths of the animals and kept frozen. Ivory experts testify that if tusks are exposed to the weather they dry out, lose their animal matter, and become useless for carving (280:361–66).

According to Lydekker, about 20,000 pairs of tusks, in perfect condition, were exported for the ivory trade in the few decades preceding 1899, yet even now there is no end in sight. According to Digby, about a quarter of all the mammoth tusks found in Siberia are in good enough condition for ivory turning (113:177). This means that hundreds of thousands of individuals, not merely eighty or so, must have been frozen immediately after death and remained frozen. Obviously it is unreasonable to attempt to account for these hundreds of thousands of individuals by the assumption of such rare individual accidents as have been suggested above. Some powerful general force was certainly at work. Lydekker gives many hints of the nature of this force in the following passage:

. . . In many instances, as is well known, entire carcasses of the mammoth have been found thus buried, with the hair, skin and flesh as fresh as in frozen New Zealand sheep in the hold of a steamer. And sleigh dogs, as well as Yakuts themselves, have often made a hearty meal on mammoth flesh thousands of years old. In instances like these it is evident that mammoths must have been buried and frozen almost immediately after death; but as the majority of the tusks appear to be met with in an isolated condition, often heaped one atop another, it would seem that the carcasses were often broken up by being carried down the rivers before their final entombment. Even then, however, the burial, or at least the freezing, must. have taken place comparatively quickly as exposure in their ordinary condition would speedily deteriorate the quality of the ivory (280:363).

He continues:

How the mammoths were enabled to exist in a region where their remains became so speedily frozen, and how such vast quantities of them became accumulated at certain spots, are questions that do not at present seem capable of being satisfactorily answered; and their discussion would accordingly be useless. . . . (280:363).

Lydekker was not alone in feeling the futility of considering these mysterious facts. For many years, in this field as in others, there has been a tendency to put away questions that could not be answered. However, we shall return to his statement. I shall try to show later on how all the details of the phenomena he describes can be made understandable. For the moment, I would like to point out simply that some sort of abrupt climatic change is required. This conclusion is reinforced by the results of recent research in the frozen-foods industry. This has produced evidence that throws additional doubt on the theory of the preservation of the bodies of mammoths by individual accidents. It seems that the preservation of meat by freezing requires some rather special conditions. Herbert Harris, in an article on Birdseye in *Science Digest*, writes:

What Birdseye had proved was that the faster a food can be frozen at "deep" temperatures of around minus 40 degrees Fahrenheit, the less chance there is of forming the large ice crystals that tear down cellular walls and tissues leaving gaps through which escape the natural juices, nutriment and flavor (202:3).

Harris quotes one of Birdseye's engineers as saying:

. . . take poultry giblets; they can last eight months at 10 below zero, but "turn" in four weeks above it. Or lobster. It lasts 24 months at 10 below but less than twenty days at anything above. . . . (202:5).

In the light of these statements the description of the frozen mammoth flesh given by F. F. Herz is very illuminating. Quoted by Bassett Digby in his book on the mammoth, Herz said that "the flesh is fibrous and marbled with fat." It "looks as fresh as well frozen beef." And this remark is made about flesh known to have been frozen for thousands of years! Some people have reported that they have been made ill by eating this preserved meat, but occasionally, at least, it is really perfectly edible. Thus Joseph Barnes, former correspondent of the New York *Herald Tribune*, remarked on the delicious flavor of some mammoth meat served to him at a dinner at the Academy of Sciences in Moscow in the 1930's (24).

What Birdseye proved was that meat to remain in edible condition must be kept very cold—not merely frozen, but at a temperature far below the freezing point. What the edible mammoth steaks proved was that meat had been so kept in at least a few cases for perhaps 10,000 to 15,000 years in the Siberian tundra. It is reasonable to suppose that the same cause that was responsible for the preservation of the meat also preserved the ivory and therefore that tens or hundreds of thousands of animals were killed in the same way.

How can such low temperatures for the original freeze be reconciled with the idea of individual accidents unless at least the animals died in the middle of the winter? It is quite certain that such temperatures could never have prevailed at the surface or in mudholes during "spring freshets." Ripe seeds and buttercups found in the stomach of one of the mammoths, to be discussed later, showed that his death took place in the middle of the summer. It is obvious that during the summer the temperature at the top of the permafrost zone was and is 32° F. or 0° Centigrade, neither more nor less, since by definition that is where melting begins. And from that point down there would be only a relatively gradual fall in the prevailing temperature of the permafrost.

Even if mammoths died in the winter, it is difficult to see how very many of them could have become well enough buried to escape the warming effects of the thaws of thousands of springs and summers, which would have rotted both the meat and the ivory unless there was a change of climate.

The theory that mammoths may have been preserved by falls into pits or into rivers encounters further objections. Tolmachev, the Russian authority, pointed out that the remains are often found at high points—on the highest points of the tundra (412:51). He notes that the bodies are found in frozen ground, not in ice, and that they must have been buried in mud before freezing. This presents a serious problem because, he says,

> . . . As a matter of fact, the swamps and bogs of a moderate climate with their treacherous pits, in northern Siberia, owing to the permanently frozen ground, could exist only in quite exceptional conditions (412:57).

Howorth remarked on this same problem:

> While it is on the one hand clear that the ground in which tlfe bodies are found has been hard frozen since the carcasses were entombed, it is no less inevitable that when these same carcasses were originally entombed, the ground must have been soft and unfrozen. You cannot thrust flesh into hard frozen earth without destroying it (225a:313).

Since Tolmachev can think of no other solution to this problem, he finds himself forced to conclude that the mammoths got trapped in mud when feeding on river terraces. We have seen that this conflicts seriously with the conditions of temperature required for the preservation of the meat, whether they were feeding on the terraces during the summer, when presumably the fresh-grass supply would be available there, or whether they were shoving aside the heavy snowdrifts during the winter to attempt to get at the dead grass below. For in either case they would fall into unfrozen water, the temperature of which could not be lower than 32° Fahrenheit. Furthermore, if this is the way it happened, why are the animals often found on the highest point of the tundra?

Thus we see that the further we get into this question the thornier it becomes. We shall have, for one thing, to face the problem of the apparently sudden original freeze. How sudden, indeed, must it have been? How can we account for it on the assumption of a comparatively slow displacement of the earth's crust? So far as the first question is concerned, recent research has contributed interesting new data.

Research on the mechanics of the freezing process and its effects on animal tissues has been carried forward considerably since the experiments conducted by Birdseye's engineers. In an article in *Science*, Meryman summarizes the more recent findings. These are based on extremely thorough laboratory research, and they modify, to some extent, the Birdseye findings.

Meryman shows that initial freezing at deep temperatures is not required for the preservation of meat. On the contrary, such sudden deep freeze may destroy the cells. He remarks, "Lovelock considers —5° C. as the lowest temperature to which mammalian cells may be slowly frozen and still survive." Furthermore the tissues survive gradual freezing very well:

> In most, if not all, soft tissue cells there is no gross membrane rupture by slow freezing. Even though it is frozen for long periods of time, upon thawing the water is reimbibed by the cells, and their immediate histological appearance is often indistinguishable from the normal (304: 518–19).

It appears that what damages the cells is dehydration, caused by the withdrawal of water from them to be incorporated in the ice. This process goes on after the initial freezing:

> . . . The principal cause of injury from slow freezing is not the physical presence of extracellular ice crystals, but the denaturation incurred by the dehydration resulting from the incorporation of all free water into ice (ibid.).

There are only two known ways, according to Meryman, to prevent this damage. First, ". . . the temperature may be reduced immediately after freezing to very low, stabilizing temperatures." The other way is artificial; it consists of using glycerine to bind water in the liquid state, preventing freezing.

Meryman shows that once the temperature has fallen to a very low point, it must remain at that point if the frozen product is to escape serious damage. The reason for this is that except at these low temperatures, a recrystallization process may take place in ice, in which numerous small crystals are combined into large ones. The growth of the large crystals may disrupt cells and membranes. He remarks:

> At very low temperatures, recrystallization is relatively slow, and equilibrium is approached while the crystals are quite small. At temperatures near the melting point, recrystallization is rapid, and the crystals may grow to nearly visible size in less than an hour (ibid.).

I am reminded, in writing these lines, of my experience in truck gardening. In trying to reduce damage from frost, I often resorted to a method that was effective but mysterious, for I could not understand why it worked. I learned that if the vegetables got frosted—even heavily frosted—they would not be seriously damaged if I could manage to get out before sunrise and thoroughly hose them off, washing away the frost. If, however, the sun should rise before I was finished, the unwashed vegetables would be damaged. It would seem, according to the explanation given by Meryman, that the frost damage was the result of recrystallization of the ice that had formed within the vegetable fibers. Small crystals, growing into large ones in the hour or so before the sun was up far enough to melt them may have caused the damage.

It follows, from this analysis of the mechanics of freezing, that the preservation of mammoth meat for thousands of years may be accounted for by normal initial freezing, followed by a sharp fall in temperature. Whenever the meat was preserved in an edible condition the deep freeze must have been uninterrupted; there must have been no thaws sufficient to bring the temperature near the freezing point.*

Let us now take a closer look at one of these preserved mammoths and see what it may have to tell us.

5. THE BERESOVKA MAMMOTH

Perhaps the most famous individual mammoth found preserved in the permafrost was the so-called Beresovka mammoth. This mam-

---

* A number of additional considerations are pertinent to the question of the process of preservation of the mammoths.

It appears that the size of the meat unit is an important factor. According to a Science Service dispatch from Washington, published in the *Boston Globe* November 11, 1957:

"Both cut-up and precooked poultry are less stable than whole, uncooked poultry when it comes to undesirable changes. . . ."

It seems that 50° F. would stop the digestive juices.

At 0 F. freezing would be at the rate of one inch an hour with direct contact of the snow with the skin. Here the amount of protective fat on the stomach might be a factor.

It is clear that very low temperatures are required for long-term preservation of meat. However, it is also true that protection of the carcasses from oxygen is essential. The exteriors of many carcasses may spoil, and yet the interiors may last a long time.

moth was discovered sticking out of the ground not far from the bank of the Beresovka River in Siberia about 1901. Word of it reached the capital, St. Petersburg. It so happened that, a long time before, word of another mammoth had come to the ears of Tsar Peter the Great. With his strong interest in natural science, the Tsar had issued a ukase ordering that whenever thereafter another mammoth was discovered, an expedition should be sent out by his Imperial Academy of Sciences to study it.

In accordance with this standing order, a group of distinguished academicians entrained at St. Petersburg and proceeded to the remote district of Siberia where the creature had been reported. When they arrived they found that the wolves had chewed off such parts of the mammoth as projected aboveground, but most of the carcass was still intact. They erected a structure over the body and built fires so as to thaw the ground and permit the removal of the remains. This process was hardly agreeable, since, the moment the meat began to thaw, the stench became terrific. However, several academicians remarked that after a little exposure to the stench, they became used to it. They ended by hardly noticing it.

Eventually the body of the entire mammoth was removed from the ground. The academicians, meantime, made careful observations of its original position. They saw evidence that, in their opinion, the mammoth had been mired in the mud. It looked as if its last struggles had been to get out of the mud, and as if it had frozen to death in a half-standing position. Strangely enough, the animal's penis was fully erect. Two major bones, a leg bone and the pelvic bone, had been broken as if by a fall. There was still some food on the animal's tongue and between his teeth, indicating an abrupt interruption of his last meal. The preliminary conclusion suggested by these facts was that the animal met his death by falling into the river.*

Very special interest attached to the analysis of the contents of this animal's stomach. These consisted of about fifty pounds of material, largely undigested and remarkably well preserved. While the foregoing data were obtained from a translation of parts of the report of the academicians, published by the Smithsonian Institution, the section dealing with the stomach contents was specially translated for this work by my aunt, Mrs. Norman Hapgood. Since there are

---

* This turned out to be incorrect (see below).

many interesting points essential to an understanding of the question, which can be noted only by a reading of the report itself and which do not figure in the published accounts, I reproduce the stomach analysis by V. N. Sukachev, with omission of technical botanical terms where possible, and with omission of bibliographical references to Russian, German, and Latin sources, and some shortening of the comment (400).

We can definitely establish the following types of plants in the food in the stomach and among the teeth of the Beresovka mammoth [Latin names are those of the Russian text]:

a. *Alopecurus alpinus sin.* The remains of this grass are numerous in the contents of the stomach. A significant portion of it consists of stems, with occasional remnants of leaves, usually mixed in with other vegetable remains. . . . All these remains are so little destroyed that one is able to establish with exactitude to what species they belong. . . .

Measurements of the individual parts of these plants, when compared with the varieties of the existing species, showed that the variety contained in the food was more closely related to that now found in the forest regions to the south of the tundra than to the varieties now found in the tundra. Nevertheless, this is an Arctic variety and is widely spread over the Arctic regions, in North America and Eurasia. However, in the forested regions it runs far to the south.

b. *Beckmannia eruciformis* (L.) *Host.* The florets of this plant are numerous in the contents of the stomach and usually are excellently preserved. [The detailed description of the remains (with precise measurements in millimeters) shows the species to be the same as that of the present day, although a little smaller, which may be the result of compaction in the stomach. At the present time the species is widely prevalent in Siberia and in the Arctic generally. It grows in flooded meadows or marshes. It is also found in North America, the south of Europe, and a major part of European Russia (although it has not been reported from northern Russia), almost all of Siberia, Japan, North China, and Mongolia.]

c. *Agropyrum cristatum* (L.) *Bess.* Remains of this plant are very numerous in the contents of the stomach. [They are so well preserved that there is no doubt as to the exact species. The individual specimens are slightly smaller than those of the typical more southern variety growing today, but this could be the result of some reduction of size because of pressure in the stomach, which is noted in other cases.]

The finding of these plants is of very great interest. Not only are they scarcely known anywhere in the Arctic regions, they are even, so far as I have been able to discover, very rare also in the Yakutsk district. . . . Generally speaking the *Agropyrum cristatum L. Bess* is a plant of the plains (steppes) and is widespread in the plains of Dauria. . . . The general range of this plant includes southern Europe (in European Russia it is adapted to the plains belt), southern Siberia, Turkestan, Djungaria, Tian-Shan, and Mongolia.

Nevertheless, the variety found in the stomach differs slightly from both the European and Oriental-Siberian varieties found today.

d. *Hordeum violaceum Boiss. et Huet.* [After a detailed anatomical description of the remains of this plant in the stomach contents, the writer continues.] Our specimens are in no particular different from the specimens of this species from the Yakutsk, Irkutsk, and Transbaikal districts. [The plant is, apparently, no longer found along the Lena River, except south of its junction with the Aldan River. It is found in dry, grassy areas. It is not found in the Arctic regions.] Its northernmost point is apparently Turochansk. . . . Generally speaking, in Siberia this plant is a meadow plant and is also found in moister places in the plains.

e. *Agrostis sp.* . . . it does not appear possible to identify the species positively. [Apparently, no plant precisely similar is known at the present day. Thus it may represent an extinct form.]

f. *Gramina gen. et sp.* A grass, but preservation is not good enough to allow any more precise identification.

g. *Carex lagopina Wahlenb.* The remains of this sedge are numerous in the contents of the stomach. [The specimens exactly resemble varieties growing today. The measurements show no reduction in size. Its range extends to the shores of the Arctic Ocean. It is found in mountainous regions, including the Carpathians, Alps, and Pyrenees. It is also found in the peat bogs of western Prussia, in Siberia as far south as Transbaikalia and Kamchatka, in eastern India, North America, and the southern island of New Zealand.]

h. [Omitted—apparently a numbering error in the text.]

i. *Ranunculus acris L.* [The specimens in the stomach did not permit identification of the precise variety of this buttercup, though pods equally large are occasionally found.] The general range of this plant is very great. It includes all Europe and Siberia, it stretches to the extreme north, spreads to China, Japan, Mongolia, and North America. However, over this area this species very much deteriorates into many varieties which are considered by some to be independent species. [This plant grows in rather dry places. It is not at present found growing together with the *Beckmannia Eruciformis*, although it is found with it in the stomach.]

j. *Oxytropis sordida (Willd) Trantv.* In the contents of the stomach were found several fragments of these beans. . . . In the fragments taken from the teeth there were found eight whole bean pods in a very good state of preservation; they even in places retained five beans. . . . [The plant is now found in Arctic and sub-Arctic regions, but also in the northern forests. It grows in rather dry places.]

In addition to the nine species mentioned above, and described in the report, with numerous measurements, the author reports that two kinds of mosses were identified in the stomach contents by Professor Broterus, of Finland. There were five sprigs of *Hypnum fluitans* (*Dill.*) *L.* and one sprig of *Aulacomnium turgidum* (*Wahlenb.*) *Schwaegr.* The first is common in Siberia north of the 61st parallel of latitude and to the marshlands of northern Europe. Both of them "belong to species widely distributed over both the wooded and the tundra regions."

The report states, further, that another scientist, F. F. Herz, brought back several fragments of woody substances and bark from beneath the mammoth, and of the species of vegetation among which it was lying. Very surprisingly, these were found to differ in a marked degree from the contents of the stomach. A larch (*Larix sp.*) was finally identified, but the genus only, not the species.

Another tree identified in a general way was *Betula Alba L.s.I.*, but the exact species could not be determined. The same was true of a third tree, *Alnus sp.* "All three of these kinds grow at present in the Kolyma River basin, and along the Beresovka, as they are widespread in general from the northern limits of the wooded belt to the southern plains."

The general conclusions reached in the report are as follows:

a. The remains of plants in the mammoth's mouth, between its teeth, were the same as the stomach contents, and represented food the mammoth had not yet swallowed when it was killed.

b. The food consisted preponderantly of grasses and sedge. "No remains at all of conifers were found." Therefore, "One may conclude that the Beresovka mammoth did not, as was previously thought, feed mainly on coniferous vegetation but mainly on meadow grasses." Evidently he wandered into low, moist places and also into higher, drier places such as are now found in the same region.

c. "The finding of the wood remains under the mammoth, and even the cliff itself where the mammoth was lying, suggest that he was not feeding in the place where he died. The majority of the vegetation in his food did not grow along cliffs or in conjunction with species of trees."

d. The discovery of the ripe fruits of sedges, grasses, and other plants suggests that "the mammoth died during the second half of July or the beginning of August."

6. THE INTERPRETATION OF THE REPORT

A vital prerequisite for any correct interpretation of the facts in this case is information on the age of the mammoth. This information was not available when the first edition of this work went to press, but is now at hand. It comes as a great surprise to those, like myself, who assumed that the mammoth must have died during the time that Siberia was moving northward in accordance with the crust-displacement hypothesis; that is, between approximately 12,000 and 18,000 years ago. The age of the mammoth, however, turns out to be at least 39,000 years, and possibly as much as 47,500 years.

Where does this leave us? Can we fit this into our scheme? It appears from this timing (which there is no reason to doubt) that the Beresovka Mammoth died when the climate in Siberia was warming up—after the pole had left the Greenland Sea and migrated to America. His death occurred at a time when we would assume that there was a high turbulence of climatic conditions, and when the level of earthquakes and volcanic eruptions would be at a peak. Since the warming of the climate had probably been going on for several thousand years herds of mammoths and other animals would have been moving northward into areas where the grasslands and forests had been reestablished.

And it was in the middle of this warming trend in Siberia, when the climate was warmer there than it is now, and right in the middle of the summer, that the mammoth died, and his body was immediately frozen! And somehow or other it remained frozen all through the period of about 30,000 years when we have shown through much evidence that the Arctic Ocean was warm and luxuriant forests were growing along the Arctic coasts.

The evidence for the warm Arctic that we have presented in earlier chapters is overwhelming, and it ties in with the evidence we have produced for a warm Antarctic at the same time. It cannot be dismissed just because one mammoth (and a few other animals—see the table below) wanted to stay frozen for the whole period that the Arctic was warm. But it certainly is not easy to see how those bodies could have been kept in deep freeze for such a length of time when the climate of the region where they lay entombed was warm. Off-hand one would be tempted to shout that the thing was impossible.

Of course there has to be a way out. Three or four bodies are not

going to bulldoze us into giving up the assumption of the warm
Arctic that is supported by so much evidence.

But if we are going to hold to our assumption of a warm Arctic,
how are we going to explain the Beresovka Mammoth? Perhaps we
can do it this way:

The evidence shows that the animal suffered a very severe fall,
severe enough to break his pelvis and leg. We learn also that the
food in his stomach and mouth did not match the vegetation around
him at the spot where he was found. He did not fall into water,
because, as was ascertained by another investigator,* large masses of
his blood were found under him. The blood would, of course, have
been washed away had he tumbled into a river. The fact that his
penis was found to be erect indicates that he was not instantly killed
by his fall, but that he froze to death. He was certainly plunged suddenly into extreme cold.

I think we can see how this might have happened. With a high
level of earthquake activity large fissures could be opening in the
crust in considerable numbers, as they commonly do in many
earthquakes. Let us assume that the mammoth fell into one of these.

We must remember that according to our theory a long period of
intense cold had gripped the Siberian coast until only a short while
before. This would have been the time of the pole position in the
Greenland Sea. The situation of the pole just north of Norway
would have logically involved an ice age in the region of the Beresovka River. The frozen ground, or permafrost, of this ice age might
have extended down thousands of feet, as it does today in some
places in the Arctic. When the pole moved to Hudson Bay the
climate in the Beresovka region would have become about like that
of Minnesota today, where the winters are severe enough to prevent,
or greatly delay, the deep melting of a permafrost extending down
thousands of feet.

We may suggest, then, that the Beresovka Mammoth fell into a
deep crevasse or fracture in the earth's crust, perhaps several hundreds

---

* One of my correspondents, Alf H. Hostmark, sent me a quotation from the
German edition of a work by Wegener and Koppen, which I translate as follows:

"The mammoth excavated by Herz and Pfizenmayer in the middle Kolyma
region in 1901 lay in a depression in . . . fossil ice which was evidently an old
glacial remnant into which it had been thrust; great masses of frozen blood had
flowed from the severe wounds it had suffered in its fall."

of feet deep. He might have tumbled down a sloping wall of the crevasse a long way without actually killing himself, but of course at the bottom loose earth dislodged by his fall could have cascaded down upon him and buried him alive. According to biologists I have consulted the erection of the penis could have resulted from the poor animal's emotions of terror and from his pain.

The mammoth might have frozen to death and afterwards been gradually frozen through in the manner I have suggested in the preceding pages. The fissure would very likely have been largely filled in as the result of continuing earth shocks, landslides and the like, and then gradually the temperature of the body would have been reduced to the low temperatures prevailing deep in the permafrost.

And what of that great fissure? What is the existing evidence of it? Why, the valley of the Beresovka River itself! The valley, or channel, of the present river may have been created by the filling in of the fissure.

But how then, you may ask, did the mammoth come to the surface? The answer may be that erosion in the valley was rapid during the ensuing warm period because the river must have been much better fed by its tributaries then than it is now. Moreover it is generally thought that the coast stood higher then, than now, so that the New Siberian and Liakov Islands were connected with the mainland. The result of these factors would probably have been that the river was much larger and flowed much faster than now, and consequently in the 30,000 years or so of the warm period could have eroded the valley to a very considerable depth.

The Beresovka Mammoth, and the other bodies we have of about the same age, might thus have been brought nearer the surface but not actually uncovered until after the climate again grew cold with another poleward shift of Siberia.

The following table shows that warming periods after glaciations—that is, after crust displacements—have been just as fatal to species of animals as periods of increasing cold. I feel that this is because they succumbed to the turbulence of the climate, to the furious storms, to the abrupt changes of temperature caused by massive volcanism, to hurricanes, dust storms, torrential rains and unseasonable snows that probably decimated their food supplies. In America, after all, the horse, mastodon, mammoth and other ice-age animals died out as the climate was warming up. We shall see in the next chapter that cold climate had nothing whatever to do with the massive extinctions of animals in South America at the same time.

Of course, most of the animals found frozen in the Arctic do date, as Table 20 shows, from the end of the Wisconsin glaciation to the time when we assume Siberia was moving northward, and the refrigeration of the climate does account for the good preservation of the mammoth ivory. At the same time, the turbulence of the climatic conditions accounts for the fact that few entire bodies are found. The remains are for the most part just bones scattered about and piled in great heaps, together with heaps of frozen trees. These contribute an air of violence and tragedy to the endless reaches of the desolate tundra.

TABLE 20
Radiocarbon Datings of Mammoths and Other Animals

1. T-299;351a:IV,180. Beresovka River Valley, Siberia .... 39,000+
   or 44,000±3500

   The famous Beresovka Mammoth, discovered in 1901. The finite date (with range from 40,500 to 47,500 years ago) is based on a slight trace of radioactivity left in the sample, but is not considered highly dependable.

2. P-426;351a:V,82. Arlington, Mass., USA .... 41,000+
   42,060±4305

   Ivory tusk (apparently mammoth or mastodon) from an animal that was perhaps roughly contemporary with the Beresovka Mammoth.

3. T-170;351a:IV,179. Sanga-Jurjak River Valley, Yakut ASSR .... 39,000+
   31,500±2000
   44,000±3500

   Fat from a mammoth. Comment suggests that the last date is the most probable.

4. T-172;351a:IV,179. River Elga, Siberia .... 38,000+
   The Nochnoj woolly rhinoceros, portion of skin.

5. T-169;351a:IV,179. Yenisei River Valley, Yakut ASSR .. 32,500+
   35,800±2700

   Mammoth skin. Of three samples dated separately the first gave the "infinite" date, the other two averaged the finite date, but contamination was suspected with these, and therefore all three samples may be much older, or much younger.

6. T-171;351a:IV,179. Lena Delta, Yakut ASSR .... 33,000+
35,800±1200
Mammoth skin. Comment, "A measurable activity was found in the sample (35,800±1200) perhaps due to contamination."

7. T-298;351a:IV,180. Gyda River Valley, Yakut ASSR 33,500±1000
Skin of mammoth.

8. TA-121;351a:X,379. Byzovaya, Komi ASSR .... 18,320±280
Bones of a mammoth.

9. A-375;351a:VI,101. Blackwater Draw, New Mexico,
USA .... 15,750±760
Carbon containing sand with remains of mammoth, horse, bison, camel and sloth. No human remains.

10. M-639;351a:I,173. Kalamazoo County, Michigan,
USA .... 13,200±600
Bone from woodland musk-ox.

11. CX-145;351a:VII,48. Missouri, USA .... 13,170±600
Missouri dire wolf.

12. M-1464;351a:X,102. Lamb Spring, Colorado, USA 13,140±1000
Mammoth bone.

13. UCLA-522;351a:VI,322. Tule Springs, Nevada, USA 13,100±200
Charcoal lumps associated with mammoth, antelope and camel.

14. I(TTC)-246;351a:IV,41. Lubbock, Texas, USA .... 12,650±250
Pelecypod shells associated with Pleistocene horse, mammoth, camel. Comment estimates time of extinction of these animals in Texas between 10,000 and 11,000 B.P. Mammoth not directly dated.

15. UCLA-604;351a:VI,322. Tule Springs, Nevada, USA 12,400±200
Charcoal associated with bone tool and teeth of horse, mammoth and camel.

16. S-246;351a:X,369. Kyle, Saskatchewan .... 12,000±200
Mammoth bone.

17. MO-3;351a:VIII,320. Shore of Lake Taimyr, USSR 11,700±300
Wood of fossil tree from mammoth horizon.

18. SA-53;351a:VI,246. Vailly-sur-Ainse, France .... 11,550±450
Mammoth tooth from gravel pit.

19. T-297;351a:IV,179. Taimyr Peninsula, USSR ...... 11,450±250
Taimyr Mammoth, found in 1948; sinews, skin and hair.

20. M-1361;351a:VIII,256. Genesee County, Michigan,
USA ....11,400±400
Tusk of mammoth, from Clayton Township.

21. M-811;351a:I,190. Cochise County, Arizona, USA ... 11,290±500
Mixed pine, ash and charcoal from Lehner mammoth site.
Bones of nine young mammoths, horse, bison, tapir. (Bones
not directly dated.)

22. I-449;351a:VIII,172. Rawlings, Wyoming, USA .... 11,280±350
Mammoth tusk associated with human artifacts.

23. I-622;351a:VIII,172. Milliken, Colorado, USA ..... 11,200±500
Bone and tusk fragments from a young mammoth.

24. M-1254;351a:VII,123. Gratiot County, Michigan,
USA .... 10,700±400
Tooth from a mastodon.

25. W-1358;351a:VII,374. Big Bone Lick, Kentucky,
USA ....10,600±250
Wood with tusk, and remains of horses, sloth, etc.

26. M-1739;351a:X,62. Gratiot County, Michigan, USA .. 9910±350
Mastodon bones.

27. M-1778;351a:X,63. Lapeer County, Michigan, USA .. 9900±400
Parts of mastodon mandibles.

28. M-774;351a:II,43. Vallejo, Mexico D.F., Mexico .... 9670±400
Charcoal in direct association with mammoth finds.

29. WIS-267;351a:X,476. Dane County, Wisconsin, USA 9630±110
Bone scraps from a mastodon.

30. WIS-265;351a:X,476. Dane County, Wisconsin, USA 9480±100
Bone scraps from a second mastodon.

31. M-694;351a:III,106. Elkhart, Indiana, USA .... 9320±400
Bone from American mastodon, found in marl below 16 ft.
of muck and peat.

32. M-1783;351a:X,64. Lapeer County, Michigan, USA .. 9250±350
Section of left tusk of Rappuhn Mastodon. Wood sticks from
a platform made by men who hunted mastodon averaged
10,500 yrs.

33. M-490;351a:I,174. Chautauqua County, New York,
USA .... 9200±500
Rib of young mastodon. "The tusks, also found, had evidently
been driven into the sand under the muck and quicksand a
number of feet."

34. TX-127;351a:VII,300. Helsinki, Finland .... 9030±165
Mammoth bones. Comment: date considered "much too
young." No reason given.

35. GSC-614;351a:X,216. Tupperville, Ontario .... 8910±150
Mastodon bones.

36. M-1400;351a:VII,123. Berrien Springs, Michigan, USA   8260±300
    Cross section of tusk of nearly complete specimen of *mammuthus jeffersoni*.

37. UCLA-705;351a:VII,338. Santa Rosa Island, California,
    USA .... 8,000±250
    Portion of ilium of dwarf mammoth. Date on associated charcoal is 12,500±250. Comment: "discrepancy between dates is not understood."

38. I-2244;351a:X,263. Peace River District, B.C., Canada   7670±170
    Elephant tusk. Comment: "date indicates that some elephants survived longer than previously believed."

39. M-347;351a:I,173. Lapeer County, Michigan, USA ... 5950±300
    Mastodon tusk fragments.

40. TBN-311;351a:V,59. Caddo County, Oklahoma ...... 4952±304
    Mammoth tusk.

41. M-774;351a:II,43. Santa Isabel Iztapan, State of Mexico,
    Mexico .... 2640±200
    Mammoth bone, associated with stone implements.

7. STORM!

I have referred to the possibility that the extinction of animals and preservation of their bodies may be accounted for in part by violent atmospheric disturbances, and I have offered some evidence that such disturbances did accompany the last displacement of the crust and therefore, presumably, earlier displacements.

It may be hard to distinguish between the effects on animal life of ice action (that is, of being melted out of glaciers and subjected to the action of glacial streams) and the effects of atmospheric factors. Nevertheless perhaps some evidence of the operation of the atmospheric factors is available.

The evidence is presented, in part, by Professor Frank C. Hibben in *The Lost Americans*, and since his description of the evidence is firsthand and is presented so clearly, I have asked his permission to reproduce the pertinent passages.

He begins with a general description of the Alaskan muck, in which enormous quantities of bones (and even parts of bodies) are found:

In many places the Alaskan muck is packed with animal bones and debris in trainload lots. Bones of mammoth, mastodon, several kinds of

bison, horses, wolves, bears, and lions tell a story of a faunal population. . . .

The Alaskan muck is like a fine, dark gray sand. . . . Within this mass,
frozen solid, lie the twisted parts of animals and trees intermingled with
lenses of ice and layers of peat and mosses. It looks as though in the
midst of some cataclysmic catastrophe of ten thousand years ago the
whole Alaskan world of living animals and plants was suddenly frozen
in midmotion in a grim charade. . . .

Throughout the Yukon and its tributaries, the gnawing currents of
the river had eaten into many a frozen bank of muck to reveal bones
and tusks of these animals protruding at all levels. Whole gravel bars
in the muddy river were formed of the jumbled fragments of animal
remains. . . . (212:90–92).

In a later chapter Hibben writes:

The Pleistocene period ended in death. This is no ordinary extinction
of a vague geological period which fizzled to an uncertain end. This
death was catastrophic and all-inclusive. . . . The large animals that
had given their name to the period became extinct. Their death marked
the end of an era.

But how did they die? What caused the extinction of forty million
animals? This mystery forms one of the oldest detective stories in the
world. A good detective story involves humans and death. These conditions are met at the end of the Pleistocene. In this particular case,
the death was of such colossal proportions as to be staggering to contemplate. . . .

The "corpus delicti" of the deceased in this mystery may be found
almost everywhere . . . the animals of the period wandered into every
corner of the New World not actually covered by the ice sheets. Their
bones lie bleaching on the sands of Florida and in the gravels of New
Jersey. They weather out of the dry terraces of Texas and protrude from
the sticky ooze of the tar pits of Wilshire Boulevard in Los Angeles.
Thousands of these remains have been encountered in Mexico and even
in South America. The bodies lie as articulated skeletons revealed by
dust storms, or as isolated bones and fragments in ditches or canals.
The bodies of the victims are everywhere in evidence.

It might at first appear that many of these great animals died natural
deaths; that is, that the remains that we find in the Pleistocene strata
over the continent represent the normal death that ends the ordinary
life cycle. However, where we can study these animals in some detail,
such as in the great bone pits of Nebraska, we find literally thousands
of these remains together. The young lie with the old, foal with dam
and calf with cow. Whole herds of animals were apparently killed together, overcome by some common power.

We have already seen that the muck pits of Alaska are filled with the
evidences of universal death. Mingled in these frozen masses are the

remains of many thousands of animals killed in their prime. The best evidence we could have that this Pleistocene death was not simply a case of the bison and the mammoth dying after their normal span of years is found in the Alaskan muck. In this dark gray frozen stuff is preserved, quite commonly, fragments of ligaments, skin, hair, and even flesh. We have gained from the muck pits of the Yukon Valley a picture of quick extinction. The evidences of violence there are as obvious as in the horror camps of Germany. Such piles of bodies of animals or men simply do not occur by any ordinary natural means. . . . (212:168–70).

It is evident that the animals that were killed far to the south, in Florida, Texas, Mexico, and South America, cannot have been contained in any ice cap, whether thin or thick. Hibben suggests that other factors were at work:

One of the most interesting of the theories of the Pleistocene end is that which explains this ancient tragedy by world-wide, earthshaking volcanic eruptions of catastrophic violence. This bizarre idea, queerly enough, has considerable support, especially in the Alaskan and Siberian regions. Interspersed in the muck depths and sometimes through the very piles of bones and tusks themselves are layers of volcanic ash. There is no doubt that coincidental with the end of the Pleistocene animals, at least in Alaska, there were volcanic eruptions of tremendous proportions. It stands to reason that animals whose flesh is still preserved must have been killed and buried quickly to be preserved at all. Bodies that die and lie on the surface soon disintegrate and the bones are scattered. A volcanic eruption would explain the end of the Alaskan animals all at one time, and in a manner that would satisfy the evidences there as we know them. The herds would be killed in their tracks either by the blanket of volcanic ash covering them and causing death by heat or suffocation, or, indirectly, by volcanic gases. Toxic clouds of gas from volcanic upheavals could well cause death on a gigantic scale. . . .

Throughout the Alaskan mucks, too, there is evidence of atmospheric disturbances of unparalleled violence. Mammoth and bison alike were torn and twisted as though by a cosmic hand in Godly rage. In one place, we can find the foreleg and shoulder of a mammoth with portions of the flesh and the toenails and the hair still clinging to the blackened bones. Close by is the neck and skull of a bison with the vertebrae clinging together with tendons and ligaments and the chitinous covering of the horns intact. There is no mark of a knife or cutting instrument. The animals were simply torn apart and scattered over the landscape like things of straw and string, even though some of them weighed several tons. Mixed with the piles of bones are trees, also twisted and torn and piled in tangled groups; and the whole is covered with fine sifting muck, then frozen solid.

Storms, too, accompany volcanic disturbances of the proportions indicated here. Differences in temperature and the influence of the cubic miles of ash and pumice thrown into the air by eruptions of this sort might well produce winds and blasts of inconceivable violence. If this is the explanation of the end of all this animal life, the Pleistocene period was terminated by a very exciting time indeed (212:176–78).

In Chapter IX we saw that volcanic eruptions, possibly on a great scale, are a corollary of any displacement of the crust; therefore our theory strongly supports and reinforces the suggestions advanced by Hibben, and at the same time his evidence strongly supports our theory. But Hibben points out certain consequences that would flow from our theory, which I have not stressed. Wherever volcanism is very intensive, toxic gases could locally be very effective in destroying life. This is also true of violent local windstorms. Massive volcanic eruptions might, of course, occur anywhere on earth during a movement of the crust, and we saw, in Chapter IX, that they apparently occurred in a good many places, some of them far removed from the ice sheets themselves.

Despite the unquestionable importance of these locally acting factors, it seems that we must give much greater importance to the meteorological results of the universally acting volcanic dust. As we have noted, this dust has a powerful effect in reducing the average temperatures of the earth's surface. A sufficient fall in temperature could easily wipe out large numbers of animals, either directly or by killing their food or even by favoring the spread of epizootic diseases. Then the dust could greatly increase rainfall, which, in certain circumstances would produce extensive floods, thus drowning numbers of animals and perhaps piling their bodies in certain spots. As already mentioned, the dust would also act to increase the temperature differences between the climatic zones (the temperature gradient), thereby increasing, perhaps very noticeably, the average wind velocities everywhere. Violent gales, lasting for days at a time and recurring frequently throughout the year, might raise great dust storms in which animals might be caught and killed by thirst or suffocation. It must not be forgotten that, at the same time, changes in land elevations would be in progress, and these also would be affecting the climate and the availability of food supplies. The gradual character of these changes would be punctuated at times by the abrupt release of accumulating tensions in the crust, accompanied by terrific earthquakes and by sudden changes of elevation locally amounting perhaps to a good many feet, which also could be the cause of floods either inland (by the sudden damming of rivers) or along the coasts. There is, as a matter of fact, as already mentioned, much evidence of turbulence throughout the world during the last North American ice age, not only in the air but in the sea.

It is little wonder that, faced by all these unpleasant conditions, a good many species in all parts of the world, even very far from the ice caps, gave up the struggle for existence.

In conclusion, it appears to me that the whole mass of the evidence relative to the animal and plant remains in the Siberian tundra, interpreted in the light of the evidence from North America, sufficiently confirms the conclusion that there was a northward displacement of Siberia coincident with the southward displacement of North America at the end of the last North American ice age.

chapter **11**

THE EVIDENCE OF VIOLENT EXTINCTION IN SOUTH AMERICA*

by J. B. Delair and E. F. Oppé

In the foregoing chapters the whole of North America and Siberia have testified to violent physical changes and to the destructive effects of unidentified forces upon a widespread animal population at the end of the Pleistocene Epoch. The evidence from South America will be found as strong or even stronger. In this chapter we shall present a number of aspects of this evidence. They include evidence of a geological revolution having to do not so much with ice caps and ice ages as with the upheaval of half a continent in which the deaths of millions of animals resulted from extensive volcanic eruptions

* Since this chapter is a separate contribution the bibliographical references will be found at the end of the chapter.

and vast floods. To begin with, we consider the evidence for great changes in the elevation of the high plateau of Peru and Bolivia in very recent time.

1. THE LOST SEA OF THE ANDES

Outstanding among the unsolved problems of the recent geological history of South America are those connected with that part of the Cordillera where Bolivia and Peru meet. There, in the heart of the Andes at an average elevation of 12,300 feet, extends the highest lacustrine basin in the world, the Meseta or Altiplano, on the floor of which occurs a succession of remarkable lakes.

The largest of these, Lake Titicaca, is navigable, being some 110 miles long, 35 miles wide and 890 feet deep at the maximum. Its waters are only slightly brackish and support the only species of seahorse (*Hippocampus*) known to live in a land-locked body of water. *Hippocampus* is a typically marine creature and, with *Allorchestes* and a few other oceanic forms inhabiting this lake, strongly suggests that the present fauna of Lake Titicaca has survived from a time when the lake communicated directly with the ocean.

Lake Poöpó, some 180 miles southeast of Titicaca and 12,051 feet above sea level, receives its water from Lake Titicaca via the sluggish Desaguadero River; despite the fact that it is about 50 miles long and 20 miles wide, its greatest depth is a mere 9 feet and its water so salt that fishes reaching it from Titicaca seem unable to propagate in it.

The waters of Poöpó seep seasonally southward through the Lacahahuira River into the shallow, marshy, and very briny Lake Coipasa—12,031 feet above sea level—which has no outlet. It is of very uncertain extent, much of its southern portion forming a vast salt desert some 50 miles by 35 miles in area. Still farther south is the immense salt plain of Uyuni, which, at slightly over 12,000 feet above sea level, is about 80 by 70 miles in area. It is joined in the southwest by a long chain of small salt, saltpeter, and borax lakes and marshes lying on the floor of a winding valley nearly 100 miles in length but only 5 to 8 miles wide (6:15).

The sequence is further defined and its strangeness enhanced by continuing south over the Bolivian border to northwestern Argentina. There another series of salt deserts and large saline marshes reaches southward as far as the southern extremity of Atacama province while in the valley between the eastern slopes of the Cordillera and the Sierra de Cordoba is another succession of enormous salt lakes, the largest of which are Salinas Grandes, Sal de la Rioja, and Pampa de la Salina.

Discussing the salinity of the lakes of the high plateau, Professor Arthur Posnansky of La Paz observed:

> Titicaca and Poöpó, lake and salt-bed of Coipasa, salt beds of Uyuni —several of these lakes and salt-beds have chemical compositions similar to those of the ocean (46:23). He pointed out that Lake Titicaca is . . . full of characteristic [saltwater] molluscs, such as Paludestrina and Ancylus, which shows that it is, geologically speaking, of relatively modern origin (ibid.).

Hans S. Bellamy, who gave the problem of the salinity of this region very considerable thought, had the following to say:

> The region in which the feeders of Lake Titicaca rise consist almost exclusively of old crystalline, and younger volcanic rocks; Triassic formations, from which salt is usually derived through extraction, are markedly absent.
>
> Hence the presence of so much salt in the Bolivian Tableland can only be accounted for by postulating a former connection of the great lacustrine basin with the Ocean, and by assuming the eventual evaporation of this body of water when the connection with the Ocean was at last severed (6:16).

The modern oceanic character of the faunas of these lakes* and the chemical composition of the salt deserts support this conclusion. Additional confirmation is to be found in the *recent* age of the strand-lines left by this ancient sea on the slopes of the mountains enclosing the Altiplano. Bellamy called this body of water the Inter-

---

* Agassiz wrote in 1876 that he had found the molluscs in Lake Titicaca of freshwater species, but that the crustacea were marine: "The mollusca are all species of eminently freshwater genera, showing nothing very special. The crustacea, on the other hand, belong mainly to the Orchestriadae forms which have thus far not been found in fresh water at all: their nearest allies are all marine. . . ." (1a:287).

Agassiz was surprised to find Lake Titicaca very poor in both flora and fauna. He had expected that the lake, from its great size and its great presumed age, would be very rich in species. He found the opposite to be the fact. The poverty of the species in the lake and the presence of the saltwater species of molluscs might be considered evidence of the youthfulness of the lake. On the other hand, the higher former level of the lake (Agassiz found evidence that it had formerly stood 300 feet higher) might be regarded as evidence of pluvial conditions during the ice age rather than of a former connection with the ocean.—C.H.

Andean Sea. Indeed, when H. P. Moon wrote his account of the geology of the region he put great stress on the ". . . freshness of many of the strand-lines and the modern character of such fossils as occur (41:32)."

A few miles south of Lake Titicaca lies the celebrated ruin site of Tiahuanaco, a collection of shattered edifices of some ancient civilization, itself outside the present inquiry but bearing very definitely upon the radical changes which have occurred throughout the Altiplano within geologically very recent times. Of these ruins A. Hyatt Verrill wrote:

> Although the ruins are now over thirteen miles from Lake Titicaca there are reasons to think that in the days when the city was occupied it stood on the shores of the Lake itself or on an arm, or bay, for traces of what was apparently a dock or mole are to be seen just north of the principal ruins. If so the lake has receded . . . (52:260).

Bellamy refers to a "canal" which appears to have surrounded the principal group of ruins at Tiahuanaco, including the structure referred to hereafter as the "fortress" (6:51) and adds:

> Some explorers of the site of Tiahuanaco are of the opinion that the "canal" was, at most, only a "dry-moat," and hence will not concede that the peculiar rectangular depressions near the ruins were once actual docks or harbour basins.
>
> But the proofs in favour of our assertion that Tiahuanaco was once a harbour-town are stronger than any of the objections put forward by more superficial observers.
>
> Firstly: there is a rapid fall in level from the edge of the territory which bears culture-remains to the floor of the territory which we say was covered by the waters of the Inter-Andean Sea. . . . The difference in level is about 35 feet north of Tiahuanaco proper. . . .
>
> Secondly: while the soil of the territory which we say was above the water-level contains numerous ceramic fragments and other remains, the former sea-bottom yields practically nothing but the stone-rings with which the fishermen of that time used to weight their nets.
>
> Thirdly: the "dumps" of roughly squared stone blocks [with which the edifices at Tiahuanaco were built] are found only on territory which formerly was sea-bottom (op. cit.: 177).

Bellamy concluded from this last fact that the builders of Tiahuanaco, who obtained their material from quarries many miles distant—for structures which in their skilled and accurate masonry alone remain a mystery—floated their stone blocks in a roughly squared condition on large rafts and that the foundering of these occasionally would leave "dumps" of, in effect, raw material where now found. He made another observation of like force:

> Moreover, the "dry-moat" must have been a water-bearing canal because the great sewer, which drained the overflow of the pond on the platform of the "fortress" of Akapana discharged into it (ibid.).

The salient proof, and one wholly relevant in present review, that Tiahuanaco possessed a waterfront rests upon discernible traces of alkaline incrustations on the sides of the huge stone blocks forming a part of the above-described mole, harbor-basin, or canal wall.

The line of these incrustations corresponds closely with that of the strand-line on the slopes of the surrounding mountains, about which Bellamy wrote:

> It was carefully surveyed for a length of about 375 miles.
> And then it was established that it is not "straight." It was found that the Inter-Andean Sea . . . was not merely a Lake Titicaca of higher level extending far to the south, but that its level showed a slant of a most peculiar character in relation to the present ocean-level, or, which amounts to the same, relative to the present level of Lake Titicaca.
> The level of the Inter-Andean Sea revealed by the ancient . . . strand-line was higher to the north of Tiahuanaco and lower to the south.
> The actuality of this peculiarity cannot be doubted, for it was established independently by different persons at different times, using different methods of surveying.
> The northernmost point at which the former strand-line of the Inter-Andean Sea . . . has been surveyed is on the mountain-slopes near Sillustani and to the west of Lake Umayo in the Peruvian department of Puno.
> There the former littoral is about 295 feet above the present level of Lake Titicaca, whose surface is 12,506 feet above sea-level.
> At Tiahuanaco, at the southern end of Lake Titicaca, the same strand-line is 90 feet above the level of that great sheet of water, and 4 feet below the coping stones of the parapets of the long-dry harbours and docks and canals of that mysterious metropolis. The ancient strand-line and the ruined prehistoric city are linked beyond any doubt.
> The height of the strand-line relative to the ocean-level decreases the further south we go. At the northern end of Lake Poöpó on the mountain slopes south of Oruro it is 12,232 feet above sea-level, or 181 feet above the level of Lake Poöpó, or 274 feet below the level of Lake Titicaca, or 364 feet below the level of the same ancient strand-line in the latitude of Tiahuanaco.
> Still further south, it is discernible just a few feet above the level of

Lake Coipasa. It becomes lost in the Salt Desert of Uyuni some 12,300 feet above sea-level.

From Sillustani to beyond Lake Coipasa, a distance of about 375 miles, the strand-line dips about 800 feet.

A peculiarity of the dip is that it seems to be progressive. In the first quarter of the distance it is only about a foot and a quarter per mile, while in the last fourth it increases to more than two feet per mile. . . .

The strand-line . . . is very distinct. It consists not only of notches cut into the rock by the prolonged action of shore waves, and of fan-like delta deposits of mud and gravel which former streams dropped on meeting the ancient water's edge, but chiefly of conspicuous deposits of white lime, of a thickness of many feet, upon the red sandstone, or brown porphyry and amorphous slate, or grey granite and andesite.

This white streak, which is drawn along the slopes of the mountain-chains surrounding the Altiplano, and visible on the islands of Lake Titicaca like a chalk-line, is the residue of certain calcareous algae, chiefly of *alga characea.*

This lowly organized plant, which contains about 80 per cent of lime, is still found growing in certain shallow shore parts of Lake Titicaca. It only thrives in slightly muddy water down to a depth not exceeding three feet (*op. cit.,* 58–60).

The phenomenon of this slanting strand-line is generally thought to be due to an "imbalanced rise" of South America out of the waters of the ocean. These forces, it has been argued, lifted the continent to a greater height in the north than in the south, thus explaining why the level of the former Inter-Andean Sea is not parallel with that of either Lake Titicaca or the present ocean.

On the basis of paleontological and hydrological evidence, Bellamy believed that in geologically recent time the whole Cordillera was violently upheaved, and the Inter-Andean Sea thereby caused to vanish, the remnants of which have, over long periods of time, shrunk to their present vestigial condition.

Remarkable confirmation of the immensity of this uplift is represented by the ancient agricultural stone terraces surrounding the Titicaca basin. These structures, belonging to some bygone civilization, occur at altitudes far too high to support the growth of crops for which they were originally built. Some rise to 15,000 feet above sea level, or about 2,500 feet above the ruins of Tiahuanaco, and on Mt. Illimani they occur up to 18,400 feet above sea level; that is, above the line of eternal snow (46:39).

Posnansky, who described these terraces as *practically endless,* concluded that the entire Altiplano region was formerly at a much lower level than at present (46:39).* It is clear, however that other areas of the Cordillera underwent profound changes also; Dr. E. Huntington noted from aerial survey photographs of arid and desert regions in Peru: . . . an unexpected number of old ruins, and an almost incredible number of terraces for cultivation (30:578), showing how some ancient race had cultivated formerly fertile tracts, now absolutely desiccated.

2. THE PLEISTOCENE GRAVEYARDS OF SOUTH AMERICA

The discoveries of vast quantities of animal remains in almost every part of South America have invariably been made in recent formations. As long ago as 1887 Sir Henry H. Howorth in his monumental work, *The Mammoth and the Flood*, an enlarged version of an earlier paper dated 1881 (225a), summarized our knowledge of these beds as follows:

In South America the Pleistocene beds are developed on a very large scale. They cover plains of the Argentine Republic, in the form of modified lehm or loess, to which the name Pampas mud was given by Darwin and "formation Pampeene" by D'Orbigny.

In other places they exist in the form of beds of gravel and clay, and occasionally as beds of tufa. As in Europe and North America, we also meet with caverns of Pleistocene age, many of which have been explored in Brazil by Lund, Claussen, Bravard and Liais. The distribution of these beds is exceedingly widespread over South America.

According to Burmeister, they are richest in organic remains in the province of Buenos Aires, becoming less rich as we travel westward and northward. Rich deposits of this age have also been found in the Banda Oriental, at various points on the river Parana, and at Berrero in Patagonia.

* Radiocarbon datings of materials from Tiahuanaco indicate the site is much younger than expected. The early classic style there is dated to about the fifth century B.C., and the following cultural period to about the time of Christ. (Ref. *Radiocarbon*, Vol. IV, 1962, p. 91) The city continued to be occupied as late as the eighth century A.D. (Ref. *Radiocarbon*, Vol. 1, 1961, pp. 54–57) It would seem from these dates that the geological upheavals indicated in this chapter came very late. This is in agreement with the impression of Darwin, for example, that the uplift of the South American coast was very recent. (See p. 365). However, the destructions of most of the animal populations, to be described below, may have resulted from much earlier events.—C.H.

Burmeister says, "the diluvial deposit containing bones of animals of this age extends over the whole Brazilian plain, from the flanks of the Cordilleras to the borders of the Atlantic." They have also been found abundantly in Bolivia on the great plateau; and also west of the mountains both in Peru and Chili.

From Caracas in the north, to the sierra of Tandel in Patagonia in the south, they have, in fact, occurred in more or less abundance over the whole continent.

In the great Argentine plain they are found close to the sea-level, while in Bolivia they occur, according to D'Orbigny, at a height of 4000 metres, and they are found with a singular similarity if not uniformity of contents in all latitudes.

That the surface beds of the Pampas and the deposits in the caves were synchronous, is admitted by all explorers. The same creatures are found in both, of course in different proportions, as is the case elsewhere.

Nor is there any doubt that both sets of beds date from the same horizon as the Mammoth beds of other countries.

The fauna of the Pleistocene beds of the Southern States of North America is, in fact, largely identical with that from the beds we are now discussing; the megatherium and mylodon, the tapir and capybara, the mastodon and horse, &c., &c.; being found in both, and every observer, from Darwin to Burmeister, is agreed in assigning them to the same horizon (29:325–6).

Historically the bones of Pleistocene mammals, especially those of the larger genera, were noticed in South America soon after the Spanish Conquest. Curious theories were advanced by the early discoverers to explain the presence of these bones, usually by reference to a race of giants who were supposed to have anciently inhabited various parts of the New World. Among the earliest reports is that of Pedro de Cieza de León, who wrote:

. . . . when the most illustrious Don Antonio de Mendoza was viceroy and governor of New Spain, he found certain bones of men who must have been even larger than these giants (11:191).

Cieza traveled through Peru and the adjacent lands from 1532 to 1550. A contemporary, Augustín de Zarate, probably referred to the same discovery when he mentioned that Juan de Holmos, a native of Truxillo, excavating near that place, exhumed enormous teeth, a huge rib, and other bones, all of which were, of course, assigned to the legendary giants (56:ch.iv). Joseph de Acosta recorded, only a little later, the discovery of similarly large bones at Puerto Viego and Manta in Ecuador (1:56).

Some very large mammalian bones, among other finds recorded throughout the seventeenth and eighteenth centuries, were described by Father Guevarra in 1770 (24:8) as occurring in the Paraguayan districts of Argentina and in Paraguay itself. Four years later the Jesuit Father Thomas Falkner referred to the discovery on the banks of the Carcaranan or Tercero River of numerous large bones, some of which evidently belonged to a gigantic species of fossil armadillo. Falkner's description is the earliest we have of this creature's existence.

About 1789 the greater part of the skeleton of an enormous unknown animal, which science was later to identify as the *Megatherium*, was found near Lujan, some 9 miles west of Buenos Aires, Argentina. This skeleton was sent to Madrid—where incidentally it still is; the Danish scholar, M. Abildgaard, published the first scientific notice of it in 1793. In 1795 M. Roume published a longer account (48), greatly amplified in 1796 as part of the text of the first memoir on the skeleton by Garriga and Bru. This memoir was translated from the Spanish original in 1804 by the French comparative anatomist, Baron Georges Cuvier (14), who, although he never personally examined the Madrid skeleton, later wrote a detailed account of its osteology from copies of the engravings illustrating Garriga and Bru's monograph (15:vol.iv).

During the years around the dawn of the nineteenth century when the great Alexander von Humboldt was exploring the Orinoco valley, he found elephant bones embedded in gravel near Cumanacoa in Venezuela (36:547). A little later the same traveler found numerous fossil bones of *mastodons* near Santa Fe de Bogotá, in Colombia, these being especially abundant at a locality known as *Camp des Geants* ("Field of the Giants"). Humboldt also found elephant teeth near Concepción, in Chile, and other fossil bones in the Cordillera de Chiquitos near Santa Cruz de la Sierra (19:13). Some of these remains and an elephant molar found by the same traveler on the volcano of Ibambura, at an elevation of 7,200 feet above sea level, were described by Cuvier in 1812 (15).

With these notices and the discovery during 1795 of another *Megatherium* skeleton in Argentina, the world of natural science was introduced to a hitherto entirely unsuspected mammalian fauna which had flourished in South America during geologically very recent times. Later discoveries were to show that man had been contemporaneous with it.

Not long after the appearance of Cuvier's great work, Professor Charles Lyell was shown in the Museum of the American Philosophical Society at Philadelphia a block of limestone from Santas in Brazil, obtained by Captain Elliot of the U.S. Navy about 1827.

The block contained a human skull, teeth, and other bones, together with fragments of shells, some of which *still retained traces of their original colors*. Remains of several hundred other human skeletons were dug out of similar calcareous tufa at the same place, where the presence of *serpulae* in the rock suggested that all the remains were deposited through marine action, for as Lyell observed (38:ii, 200–1), the shell would not have been brought so far inland by natives for food. Dr. C. D. Meigs, who wrote an account of this discovery, said:

Captain Elliot, while riding along the banks of the river Santas on his way from the port of Santas to the town of St. Paul, found a mound three acres in extent and 14 feet high, about 10 miles from the sea and 4 from Santas.

The bones he took with him to America . . . were dug from the face of the hill, where it was cut by the wash of the stream, and are parts of one skeleton out of many hundreds that are still lying in their bed of tufa.

They were lying on the rock in an oblique direction, the heads uppermost, and the lower extremities dipping at an angle of from 20° to 25° below the horizon.

Portions of the bones were invested externally with a stalactitic deposit of carbonate of lime, looking very much like a mummified skin. Close to one of the teeth was a *serpula* and a piece of oyster-shell. The rock in which the skeleton was embedded consisted of fragments of shells united by a stalactitic matter, and contained nodules of carbonaceous matter. . . .

A question naturally arises as to the date of that catastrophe which enclosed several hundred individuals in that tufa of the Rio Santas. . . .

It seems unlikely that these remains were formally buried by sorrowing friends. It is unlikely that so solid a stone should have been formed at so great a distance from the sea. . . . No doubt they are co-existent with the emerged land; they are not to be considered as the results of human industry.

The shore of the Atlantic must have formerly swept nearly in a line with these remarkable deposits. . . . Within this bed, or nearer than it to the sea, are found fossil bones of elephants, &c., which cannot be so old as the unfossilized oyster-shells, since they could not have been fossilized anterior to the existence of the soil out of which they are dug, unless you consider them as boulders, which is inadmissible. . . . (29:355–6).

In a limestone cavern on the borders of the Lagoa do Sumidouro, some three leagues from Santa Lucia, Dr. P. W. Lund excavated the bones of more than thirty individuals (human) of both sexes and various ages. The skeletons lay buried in hard clay overlying the original red soil forming the floor of the cave and were found mixed together in such great confusion—not only with one another but with the remains of the *Megatherium* and other Pleistocene mammals—as to preclude the idea that they had been entombed by the hand of man. All the bones, whether human or animal, showed evidence of having been contemporary with one another.

In other caves investigated by Lund, bones of ancient men were found alongside those of the formidable *Smilodon*, a giant feline which became extinct during the last Pleistocene times. Referring to the evidence from these and other Brazilian fossiliferous caves, the Marquis de Nadaillac wrote:

. . . Doubtless these men and animals lived together and perished together, common victims of catastrophes, the time and cause of which are alike unknown (42:25).

Two further cases are of particular interest. The first of these concerns the discovery, by Savage-Landor, of the remains of primitive humanoid mammals, associated with the bones of creatures regarded by him as gigantic saurians, in volcanic ash and lava deposits encountered in Matto Grosso State (34:vol.i,371–4).

The second case relates to the occurrence of the remains of mastodons, *camels*, and an extinct species of horse in beds of volcanic ash high in the Andes near Punin in Ecuador. Associated with these mammalian bones was the fossilized skull of a woman of *Australoid* type (33:311–2). This cranium, which is dolichocephalous (9:145), was scientifically described in 1925 by Drs. Louis R. Sullivan and Milo Hellman (51) and has since become generally known as the "Punin" skull.

The presence of an Australoid type in Ecuadorean South America during geologically recent times poses questions about prehistoric human populations in the continent, to the solution or partial solution of which the different configuration of South America before, or up to, late Pleistocene times, although conjectural in many respects, may well contribute. The critical importance of the Punin and Matto Grosso discoveries in the present context, however, lies in their stark demonstration that in South America human and animal denizens of the late Pleistocene world were exposed to, and perished by, geological upheavals of inconceivable violence and extent.

3. "FRESHNESS" OF FOSSIL REMAINS

In abundance of Pleistocene animal fossils South America compares very well with Siberia and North America. Authorities are as one in stressing the freshness of a high proportion of the skeletal remains and associated substances found on a continent mostly within the tropical zone.

Darwin observed, in *Voyage of the Beagle* (1876 edition), that some remains of a large unknown mammal exhumed from Pleistocene deposits in the Banda Oriental district of Uruguay appeared so fresh that:

> . . . it is difficult to believe that they have lain buried for ages underground. The bone contains so much animal matter, that when heated in the flame of a spirit-lamp it not only exhales a very strong animal odour, but likewise burns with a slight flame (op. cit., iii, 181).

Fossil bones when subjected to heat or fire do not ordinarily burn with a flame. Elsewhere Darwin referred to the perfect preservation of even the minutest details of fossil bones dug up at Bahía Blanca. In his description of the skeleton of the giant sloth *Mylodon*, Professor Richard Owen noted a similarly perfect state of preservation as regards the individual bones and concluded that the individual represented must have been buried *almost immediately* upon death. Dr. P. W. Lund also recorded how he found, in a Brazilian cavern, a skeleton of *Scelidotherium oweni*, in which not only did all the bones lie in correct relative position, but they were covered with a cellulose tissue of calcareous matter evidently derived from the petrifaction of the soft parts of the animal (35:377–383).

From the extensive literature on the subject it is impracticable here either to acknowledge further sources or to extract the many additional examples known of similar burials. Special reference, however, should be made to the undermentioned specimens derived from the Pampas areas, which suggest flood action:

a. Mammalian skeletons in natural, or upright, positions. These are embedded respectively in undisturbed and largely unstratified beds of gravel, loam, or mud.

b. A skeleton of an adult female *Mylodon* found in a natural posture alongside that of its young one embedded in gravel.

In addition, a markedly high percentage of *Megatherium* skeletons apparently had lost only their left extremities, suggesting human agency.

In these and other cases it is generally agreed that the individuals perished *before* their carcasses were subjected to the eventful last stages of entombment. Included in this category are the discoveries in the La Pumilla Valley of the carapaces of giant armadillos standing *upright on their edges* in the Pampas mud (8:ii,85). Burmeister, who noted these occurrences, also mentioned that the cuirasses of the great glyptodonts are generally found *reversed*, and that many instances are known of large, though usually incomplete, mammalian skeletons being found *upside down*, or with their bones manifestly disjointed and scattered within a small space (*ibid.*).

These extraordinary modes of burial are further exemplified by the groups or caches of animal fossils unearthed at widely separated South American localities, in which incongruous animal types (carnivores and herbivores) are mixed promiscuously with human bones. These are found not only in the Pampas formation but also in Brazilian caves and in volcanic ash at Punin and elsewhere. No less significant is the association—over truly widespread areas—of fossilized land and sea creatures mingled in no order and yet entombed in the same geological horizon, and also the occurrence of *mastodon* remains in the Cordilleras at altitudes impossibly high for their ordinary existence. Clearly these varied, but apparently contemporaneous, burials all over the South American continent are the results of different and relatively localized effects of a single tremendous upheaval, the numerous ramifications of which operated synchronously. In seeking to explain one of these effects, one must explain them all.

4. THEORIES AND CONSIDERATIONS

The aforecited facts, in their full accumulative presentation, are inconsistent with nature's normal disposal of bodies. This has long been recognized, and the mysterious character of the problem has been admitted by the highest authorities. But no comprehensive explanation has been advanced to date which accounts for all the phenomena. Theories ascribing mass animal extermination variously to glacial cold (29:343), colossal sand and wind storms (Bravard), immense floods (Lund and others), volcanic gas (27:47), local river flooding (Burmeister), pestilence and parasitical poisoning (*ibid.*) are not lacking, but none of these has been able to win general acceptance.

Now, however, that it is possible, with the aid of the theory developed in this book, to evaluate the position, who will doubt that South America exhibits part and counterpart of the global convulsion which terminated the Pleistocene world? The reader will not fail to be impressed by the conclusions reached by Alcide d'Orbigny, to whose standing and accuracy tribute must still be paid; he wrote:

It would seem that one cause destroyed the terrestrial animals of South America, and that this cause is to be found in great dislocations of the ground caused by the upheaval of the Cordilleras.

If not, it is difficult to conceive on the one hand the sudden and fortuitous destruction of the great animals which inhabited the American continents, and on the other the vast deposit of Pampan mud.

I argue that this destruction was caused by an invasion of the continent by water, a view which is completely *en rapport* with the facts presented by the great Pampan deposit, which was clearly laid down by water.

How otherwise can we account for this complete destruction and the homogeneity of the Pampas deposits containing bones? I find an evident proof of this in the immense number of bones and of entire animals whose numbers are greatest at the outlets of the valleys, as Mr. Darwin shows.

He found the greatest number of the remains at Bahía Blanca, at Bajada, also on the coast, and on affluents of the Río Negro, also at the outlet of the valley. This proves that the animals were floated, and hence were chiefly carried to the coast.

This hypothesis necessitates that the Pampas mud was deposited suddenly as the result of violent floods of water, which carried off the soil and other superfluous debris, and mingled them together. This homogeneity of the soil in all parts of the Pampas, even in places 200 leagues apart, is very remarkable.

These are not different strata differently coloured, but a homogeneous mass, which is more or less porous, and shows no signs of distinct stratification. The deposit is also of one uniform colour, as if it had been mixed in one muddy flood slightly tinted by oxide of iron.

The bones, again, are only found isolated in the lower strata, while entire animals occur on the circumference or the upper part of the basin.

Thus they are very rare at Buenos Aires, while they abound in the Banda Oriental and in the White Bay. Mr. Darwin says they are heaped up in the latter place, which again supports the contention.

Another argument may be drawn from the fact that the Pampas mud is identical in colour and appearance with the earth in which the fossil remains occur in the caverns and fissures of Minaes Geraes in Brazil, and the fragments brought by M. Claussen are completely like the others in colour and texture.

My final conclusion from the geological facts I have observed in America is, that there was a perfect coincidence between the upheaval of the Cordilleras, the destruction of the great race of animals, and the great deposit of Pampas mud (21:iii, 3, 82, 85, 86).

Nor will a researcher today withhold agreement with Professor Thomas Henry Huxley, quoted by Howorth, who said in 1869:

To my mind there appears to be no sort of theoretical antagonism between Catastrophism and Uniformitarianism; on the contrary, it is very conceivable that catastrophes may be part and parcel of uniformity.

Let me illustrate my case by analogy. The working of a clock is a model of uniform action. Good timekeeping means uniformity of action. But the striking of a clock is essentially a catastrophe.

The hammer might be made to blow up a barrel of gunpowder, or turn on a deluge of water, and by proper arrangement the clock, instead of marking the hours, might strike all sorts of irregular intervals, never twice alike in the intervals, force or numbers of its blows.

Nevertheless, all these irregular and apparently lawless catastrophes would be the result of an absolutely Uniformitarian action, and we might have two schools of clock theorists, one studying the hammer and the other the pendulum. (Address to the Geological Society, 1869)

Bibliography

1. Acosta, J. de, "Historia y moral de las Indias, en que se tratan las cosas." (Seville) 1590
1a. Agassiz, Alexander, "Hydrographic Sketch of Lake Titicaca. Proceedings of the American Academy of Arts and Sciences," Vol. XI, 1876, 283–292.

2. Alvim, G. F., "Jazigos brasileiros de mamiferos fosseis." *Not. prel. no. 18. Div. Geol. Min.* 1939
3. Ameghino, F., "Antiguedad del Hombre en el Plata." 2 vols. (Paris) 1881
4. Ameghino, F., "Notes sobre una pequena colecion de huesos mamiferos procedentes de las grutas calcareas de Iporanga, en el Estado de Sao Paulo, Brasil." *Rev. Museu Paulista, vii (1907), 1908.*
5. Badariotti, P. N., "Exploracao do Norte de Matto Grosso: regiao do Alto Paraguay e Planalto fos Parecis." (Sao Paulo) 1898
6. Bellamy, H. S., "Built before the Flood." (London) 1947
7. Branner, J. C., "On the occurrence of fossil remains of mammals in the interior of the States of Pernambuco and Alagoas, Brazil." *Amer. Journ. Sci., vol. xiii,* pp. 133–7, Feb. 1902.
8. Burmeister, C., "Déscription Physique de la République Argentina." 2 vols. 1876
9. Bushnell, C. H. C., "Archaeology of the Santa Elena Peninsula in South-West Ecuador." (Cambridge) 1951
10. Cazal, M. A. de, "Corographia Brasilica." 1817
11. Cieza de León, P. de, "The Second Part of the Chronicle of Peru." *Hakluyt Society, vol. lxviii.* (1560) (London edition) 1883.
12. Clift, W., "Notice on the Megatherium." *Geol. Trans. 2nd Series, vol. iii.* 1832
13. Couto, C. de P., "Peter Wilhelm Lund; Memorias sobre a Paleontologia Brasileira, revistas e commentadas." (Rio de Janeiro) 1950
14. Cuvier, G., *Annales du Museum.* tome v. 1804
15. Cuvier, G., "Recherches sur les Ossemans Fossiles." 4 vols. (Paris) 1812
16. d'Alton, E., "Uber die von Herrn Sellow mitgebrachten fossilen Panzerfragmente aus der Banda Oriental und die dazu gehorigen Knochen-Uberreste." (Berlin) 1834
17. d'Archiac, E. J., "Geologie et Palaeontologie." (Paris) 1866
18. Darwin, C., "Geological Observations on the Volcanic Islands and Parts of South America." (London) Part II. 1846
19. Darwin, C., (Editor) "The Zoology of the Voyage of H.M.S. Beagle, under the Command of Captain Fitzroy, during the years 1832 to 1836." (London) 1840
20. Darwin, C., "Journal of Researches into the Natural History and Geology of the Countries Visited During the Voyage of H.M.S. Beagle Round the World." (New York edition) 1878
21. d'Orbigny, A., "Voyage dans l'Amérique méridionale." 3 vols. (Paris) 1842
22. Evans, J. W., "The Geology of Matto Grosso." *Quart. Journ. Geol. Soc. London, vol. L,* Feb. 1894, pp. 86–104
23. Gervais, P., *Journal de Zoologie, vol. ii.* 1872
24. Guevarra, Father, "Historia del Paraguay, Rio de la Plata y Tucuman." 1770

25. Hartt, F., "Geology and Physical Geography of Brazil." (Boston) 1870
26. Hatcher, J. B., "On the Geology of Southern Patagonia." *Amer. Journ. Sci., vol. IV, pp.* 327–354. 1897
27. Hibben, F. C., "Treasure in the Dust." (London edition) 1953
28. Holland, W. J., "Fossil Mammals collected at pedra Vermelha Bahia, Brazil, by Gerald A. Waring." *Ann. Carnegie Museum, vol. xiii, nos. i and ii.* 1920
29. Howorth, H. H., "The Mammoth and the Flood." (London) 1887
30. Huntington, E., "Climatic Pulsations, in *Hylluings+skrift*" (dedicated to Sven Hedin) 1935
31. Ihering, H. von, "Conchas marinas da formacao pampeana de la Plata." *Revista do Museo Paulista, vol. i, pp.* 223–231. 1895
32. Joyce, T. A., "South American Archaeology: An Introduction to the Archaeology of the South American Continent with Special Reference to the Early History of Peru." (London) 1912
33. Keith, A., "New Discoveries Relating to the Antiquity of Man." (London) 1931
34. Landor, A. H. Savage-, "Across Unknown South America." 2 vols. (London) 1913
35. Liais, E., "Climats, géologie, faune et géographie botanique du Bresil." (Paris) 1872
36. Liddle, R. A., "The Geology of Venezuela and Trinidad." (Revised edition) (London) 1946
37. Lund,. P. W., *Ann. Nat. History, vol. III,* July 1839, pp. 235–6.
38. Lyell, C., "Travels in North America." 2 vols. 1852
39. Maria, H. N., "Los Grandes Mamiferos Fosiles de la Region de Barquisimeto." *Bol. Geol. y Min., tome i, nos. 2, 3, and 4.* (Caracas) 1937
40. Maury, C., "New Genera and new species of fossil terrestrial mollusca from Brazil." *Amer. Mus. Nat. Hist. science paper 764,* Nov. 1935.
41. Moon, H. P., "The Geology and Physiography of the Altiplano of Peru and Bolivia." *Trans. Linn. Soc. London, 3rd series, vol. i, pt. 1.* 1939
42. Nadaillac, Marquis de, "Prehistoric America." (English Translation) (London) 1885
43. Neu-Wied, M., "Reise." 2 vols. 1817
44. Oakenfull, J. C., "Brazil: Past, Present and Future." (London) 1919
45. Oliveira, A. J., and Leonardos, O. H., "Geologia do Brazil." 2nd edition. (Rio de Janeiro) 1943
46. Posnansky, A., "Tiahuanaco: The Cradle of American Man." (La Paz) 1945
47. Rheinhardt, J., "Bone Caves of Brazil and their Animal Remains." *Amer. Journ. Sci., vol. xcvi, pp.* 264–5. 1868
48. Roume, M., *Bulletin de la Société Philomathique.* 1795

49. Roxo, M. G. de O., "Note on a New Species of Toxodon, Owen; T. Lopesi, Roxo." (Rio de Janeiro) 1921
50. Spix, J. B. von, and von Martius, C. F. P., "Reise in Brazilien." 2 vols. (Munich) 1823–4
51. Sullivan, L. R., and Hellman, M. *Anthropological papers of the Amer. Mus. Nat. Hist., vol. xxiii.* (New York) 1925
52. Verrill, A. H., "Old Civilizations of the New World." (N\. v York) 1929
53. Walther, H. V., "Extinct Bear, Arctotherium brasiliensis, from Lagoa Funda Cave, Minas Geraes." *Grafica Guarani.* (Rio de Janeiro) 1940
54. Weiss, C. S., "Geological Memoir on the Provinces of S. Pedro do Sul and the Banda Oriental." *Abhandl. der. Kon. Akad. der Wissenschaften zu Berlin.* 1827
55. Woodward, A. S., "Notes on the Geology and Palaeontology of Argentina." *Geol. Mag. dec. iv, vol. IV,* pp. 4f. 1897
56. Zarate, A. de, "Histoire de la Découverte et de la Conquête du Perou." (1st edition) (Antwerp) 1555

chapter **12**

SOME PROBLEMS
OF EVOLUTION

IN THE PRECEDING chapters we have reviewed a mass of evidence that suggests displacements of the earth's crust, at comparatively short intervals, during the earth's history. We shall now see that this assumption throws some light on the process of evolution.

1. THE CAUSE OF EVOLUTION

A century ago, in the *Origin of Species*, Darwin suggested natural selection as the mechanism to account for evolution. The combination of the occurrence of natural variations with elimination of the unfit individuals in the competitive struggle for existence helped to explain a process of unending, gradual change in the forms of life.

Darwin did not consider that this was the whole answer. He admitted, for example, that he could not explain the numerous instances of the worldwide extinction of many forms of life simultaneously, especially in those cases where, apparently, there were no competitors and no successors to the extinct forms. Biologists today are in agreement that evolution has occurred, but they also feel that the process has not been satisfactorily explained. Thus Doctor Barghoorn, of Harvard, has recently referred to "our limited understanding of the actual *causes* of evolution," while quoting Dr. George Gaylord Simpson, author of the widely read *Meaning of Evolution*, as remarking, ". . . search for *the* cause of evolution has been abandoned" (375:238). There is a tendency at the present time for specialists to recognize a large number of interacting factors that may, together, conceivably account for evolution, though their relative importance is not agreed upon. This situation does not exclude the possibility that the confusion may indeed arise because one factor is still missing—a factor which, when added, will bring the others into proper focus.

2. THE PROBLEM OF TIME

While no biologists since Darwin's time have questioned the basic fact of evolution, numerous difficulties have developed with natural selection. In the first place, while Darwin could present evidence of changes produced in varieties of plants and animals by artificial selective breeding, he was not able to show how, even under artificial conditions, such changes could lead to the establishment of new species. Recently some progress may have been made in solving this problem, but by the end of the nineteenth century, Darwin's explanation of the mechanism of evolution had been largely abandoned. Natural selection had come to be considered, by many biologists, as chiefly a negative factor, capable of eliminating unadapted variations but not of producing new species.

Around the turn of the century the attention of evolutionists was turned to mutation, the sudden change in hereditary characteristics produced by an alteration of the basic genetic factors, genes and chromosomes. One of the early mutationists, Hugo de Vries, believed that a large-scale mutation might produce a new kind of plant or animal in a single step (115:96). Many evolutionists then adopted mutation and gave up natural selection as the explanation of evolution.

This did not, however, end the controversy. A neo-Darwinian school, clinging to natural selection, raised damaging objections to the theory of evolution through massive mutations. They insisted, for one thing, that different plants or animals differed by a great many minor traits rather than by a few major ones. This would mean that a great many mutations would be required and that these mutations would have to take place in the same individual or in the same line of descent. The fact that mutation is apparently an entirely accidental process rendered the mathematical chances against the coincidence of many mutations in one individual or in one line of individuals completely overwhelming.

But this was by no means the only difficulty. The antimutationists could argue that since mutations were purely accidental changes in the hereditary factors and did not occur in response to needs created by the environment, most mutations would be positively harmful, or at least negative, and would have no effect on the adaptation of the organism to its environment. Only a chance mutation now and then could help an organism to survive. Mutationists were unable to show the existence of any principle by which mutations would be adaptive; that is, brought about as a part of an effort of an organism to adapt to the environment. Some recent experiments indicate that such adaptive mutation may occur, perhaps under special and rare conditions, but it still cannot be shown that adaptation by mutation has been an important factor in evolution.

The mutationists did establish, of course, that minor mutations were of frequent occurrence and might even be induced artificially; therefore evolutionists accepted them, but they recognized them as just another way of accounting for the occurrence of variations. The law of natural selection would still be required in order to eliminate the harmful mutations, which would constitute the great majority of all mutations. For a while it seemed that, in this way, the basic question of evolution was answered.

It soon appeared that this was very far from being the case. The acceptance of mutations by the Darwinians as a factor in evolution did not solve the problem. It became clear, as time passed, that a major difficulty remained. Attention was concentrated on the rate at which mutation and natural selection could be effective in changing life forms. Mathematical studies showed that such changes would take place, according to the theory, at rates so slow that even long geological eras would provide insufficient time for evolution. Professor Dodson wrote:

In nature, neither mutation nor selection will ordinarily occur alone, and so the two will act simultaneously, perhaps in the same direction, perhaps in opposite directions. . . . Most frequently, selection will work against mutation, as the majority of possible mutations are deleterious. This will result in very slow change, if any. . . . (115:298).

He emphasized:

It appears that it is extremely difficult for mild selection pressures, unaided by any other factor, to establish a new dominant gene in a species. . . . (115:298).

By "mild selection pressures," Dodson means those conditions of competition between life forms pointed out by Darwin; that is, the competition that goes on at all times. What he suggests here is that some more drastic influence must have operated to produce evolutionary change.

After discussing Haldane's mathematical calculations indicating the astronomical numbers of generations that might be required to change a plant or animal under the influence of mild selection pressures, Dodson quotes Dobzhansky (the leader of the neo-Darwinian school) on their implications:

. . . The number of generations needed for the change may, however, be so tremendous that the efficiency of selection alone as an evolutionary agent may be open to doubt, and this even if time on a geological scale is involved (115:298).

Thus the problem is clearly posed: It is the problem of time. It is necessary to find some way of explaining how natural selection can have operated at rate sufficiently rapid to account for evolution. A factor of acceleration is required.

Some writers, when they saw that evolution could not be explained even with the enormous amounts of time available under the current concepts of the lengths of the geological periods, felt compelled to revert to mystical explanations. Writers such as Du Noüy (119) concluded that evolution was totally inconceivable unless its course had been indicated in advance by the reigning influence of cosmic purpose. For these writers, the end or final purpose of evolution must be the active controlling force of the whole process. The process, at basis, could be understood only as the direct effect and evidence of the will of God.

Another solution was proposed by Richard Goldschmidt, who became the leader of the anti-Darwinians. He renewed the emphasis on major or macromutations. As Dodson puts it:

. . . Goldschmidt believes that the neo-Darwinian theory places too great a burden upon natural selection, and hence that the work of selection must be shortened by some other process, namely systematic mutation (115:299).

By "systematic mutation" is meant a mutation that changes not merely an individual trait of an organism but a whole complex of traits; that is, that changes a basic principle of the biological system. The great advantage of Goldschmidt's theory is that it may greatly reduce the number of "genes" required to account for the traits of a single individual. Under present concepts of genetics, for example, from 5,000 to 15,000 "genes" may be called for to account for all the traits of the fruit fly, Drosophila melanogaster, while as many as 120,000 may be required for man (115:245).

The majority of writers on evolution today seem to feel that Goldschmidt's specific arguments for macromutations have been refuted. I can contribute no opinion on this technical question. But from my point of view the most significant thing about the Goldschmidt theory is that he produced it in an effort to gain time for the process of evolution, to accelerate it, so that the amount of evolutionary change in life forms could be brought into rough agreement with the available amount of geological time. The rejection of his theory, if the rejection is indeed based upon sound considerations, means that another factor must be found to account for the tempo of evolution.

3. CLIMATE AND EVOLUTION

Evolutionists in general agree that climatic change must have had a powerful influence on evolution. Geologists have, as I have pointed out, found a correspondence between periods of climatic change and changes in the forms of life. It is evident that as long as the general environment remains roughly the same, there can be only gentle selection pressures such as, apparently, are inadequate to account for evolution. With static environmental conditions, forms of life may continue virtually unchanged for tens or hundreds of millions of years. There are any number of organisms living today whose very similar ancestors lived in remote geological periods. To name merely a few, there is the newly discovered coelacanth, a fish whose ancestors, one hundred or more million years ago, looked as he does today; the recently discovered Dawn Redwood, found growing in China after having been regarded as extinct since its close relatives disappeared in Alaska about 20,000,000 years ago; the sphenodon, a reptile of New Zealand, whose ancestors, very closely resembling him, were contemporaries of *Tyrannosaurus rex*; horseshoe crabs, whose time span may amount to half a billion years; palm trees, whose age has just been "jumped" another 10,000,000 years (261); sharks; scorpions, and so on. Sanderson has pointed out that "living fossils" are simply too numerous to list (365). We can take it that, if external conditions are stable, or if animals and plants can migrate around to find the conditions they are used to, they may continue to exist indefinitely.

At the same time, it is equally true that any kind of animal or plant may succumb in the course of the usual and continuous competition between life forms and of the local or transitory climatic variations that are always occurring. To forget this fact would distort the picture. Furthermore recent studies have shown that new varieties of plants and animals can appear within very short periods of time, on the order of a century or less, if they live in conditions of isolation (115:365). But these rapidly produced varieties are not the same, of course, as established species.

A factor which undeniably must produce pressure for profound change in the forms of life is major climatic change. Clearly this will apply what evolutionists call "strong selection pressure." In this case life forms will have but three alternatives: to migrate, to adapt, or to die. Geologists and biologists have never denied the truth of this: Coleman, for example, recognized the importance of the glacial periods in "hastening and intensifying" the process of evolution (87:62). Lull recognizes the importance of basic climatic change thus:

. . . For changes of climate react directly upon plant life, and hence both directly and indirectly upon that of animals, while restriction or amplification of habitat and the severance and formation of land-bridges provide the essential isolation, or by the introduction of new forms increase competition, both of which stimulate evolutionary progress (278:84).

The problem has been, until now, that major climatic changes, and concomitant changes in the distribution of land and sea, could not be explained by any acceptable theory. They were inexplicable events in themselves; their coincidence in time was inexplicable. Even more serious, they were assumed to have happened only at such extremely long intervals that the total number of such major climatic "revolutions" was too small to account for more than a very insignificant portion of evolutionary history.

To recapitulate what has already been said, if drastic climatic and geographical change is the most obvious factor to which to look for changes in life forms, then it is to the acceleration of that factor that we must look for the acceleration of evolution. In the previous chapters we have been led again and again by the force of the evidence to the concept of displacements of the earth's crust. There is no reasonable doubt as to the effect that such displacements, at relatively short intervals, would have on the tempo of evolution. They could not fail enormously to accelerate the several aspects of the evolutionary process. Let us now examine some of these special aspects in more detail.

Wright has pointed out that the rate of evolutionary change may have been accelerated at various times through the mass transformation of one kind of plant or animal into another (115:314). This requires that all over the area of distribution of the life form in question there must be strong pressure for change in the same direction. This means that similar new varieties would appear simultaneously and independently in countless localities or that well-adapted new varieties would spread and become established rapidly. Quite obviously this would tend to accelerate evolution.

But how would such mass transformation be brought about? It could result only from profound transformation of the environment. The required change would have to be general and would have to tend in the same direction for a considerable period of time. No short-range fluctuations and, above all, no merely local climatic changes would suffice. A displacement of the crust appears to meet all these requirements. For a period of many thousands of years, some areas, moving toward the equator, would be growing warmer; others, moving toward the poles, would be growing colder. In the areas moving toward the equator (not necessarily reaching the equator, however, or even the tropics) the increase of sunlight would mean more luxuriant life conditions; for many species this might mean increased food supplies and an extended distribution. It would also be likely to mean increased competition with other forms. Many effects would depend upon whether the displacement carried the area in question into the wet tropics or into the dry horse latitudes, or merely from an arctic into a temperate climate. Meanwhile, of course, in areas displaced poleward, opposite trends would exist; here the forms of life would have to adapt to diminishing light, to increased cold, to decreased food supplies.

What is important is that these changes of climate would apply over great areas of the earth. In one movement of the crust, two opposite quarters of the earth's surface would be moving equatorward while two others were moving poleward. Thus the climatic changes would be in the same direction over very great areas: the entire distribution, perhaps, of many plants and animals. Mass transformation of life forms might therefore be expected to occur; not mass transformations of all life forms, of course, but merely one or two short steps in the mass transformation of one or a few kinds of plants or animals. New varieties might be established in great numbers during a single movement of the crust; but by this I do not mean to imply that many new "species" would be. The latter may be the end results of a considerable number of displacements of the crust. I hope that the reader will not ask me to define "species." In this book I use the term simply to denote forms of life that are reasonably distinct and relatively permanent.

We must remember that the different areas of the earth's surface would be unequally shifted in a crust displacement. I have explained (Introduction) that the amount of the displacement would depend on whether an area was near to, or distant from, the meridian of displacement. Selection pressures would vary accordingly.

Since we consider displacements to have taken place in short periods of the order of 5 or 10 thousand years, it seems likely that most plants and animals in areas radically displaced by a given movement would be unlikely to succeed in adapting. Some would migrate into areas having climates similar to their accustomed climates. Some would disappear. Some would develop varieties adapted to changed conditions. Even though there would be no wholesale creation of new plants and animals, the age-long process of change would have received an acceleration.

Another important, generally accepted requirement for evolution, as already suggested, besides climatic change, is geographical isolation to permit the development of new varieties. Geneticists agree that the larger the population of a given sort of plant or animal, the harder it is for a new variety to get established, because cross-breeding tends to destroy the new variety. If, however, populations are cut off from each other, and are reduced in numbers, a new variant has a much better chance to become dominant and establish itself as a variety in that locality. As already pointed out, crust displacements can account for the alternation of conditions of geographical isolation and intercommunication at the tempo required to account for evolution, because they can account for rapid, recurrent changes of sea level. Let us now visualize the consequences of a displacement of the crust resulting in a subsidence of a continental area displaced equatorward. Let us suppose a moderate subsidence of a few hundred feet only, over a period of a few thousand years. The result, of course, would be the deep intrusion of the sea into the continent. The sea would invade valleys, cutting off one part of the mainland from another and creating islands and island groups. Many populations of the same kind of plant or animal would thus be isolated and left for many thousands of years to develop and establish new variant forms.

Let us suppose many new varieties to have become established in the islands and in areas of the mainland separated from each other by tongues of the sea. The next requirement of evolution is that these new varieties be brought into competition and that the best adapted of them be disseminated into more varied habitats. This might be brought about by a new movement of the crust, such as would displace this area poleward. The area will now be uplifted, the sea will withdraw, and the life forms formerly isolated will mingle and enter a phase of competition.

The situation that compels the adaptation of the forms of life to colder, drier climates (poleward displacement) also will adapt the forms of life to higher elevations, to mountain heights. Thus, if we consider all the effects of crust displacement, both toward the equator and toward the poles, we can see that crust displacement constitutes the most powerful engine imaginable for forcing life forms to adapt to all possible habitats.

4. THE DISTRIBUTION OF SPECIES

Another important question is the problem of the origin of the present and past distribution of species over the face of the earth. Darwin and Wallace attempted to explain the numerous difficulties in this field, but their explanations have, in general, become less and less satisfactory with the passing years. These are the questions:

a. How did certain species cross wide oceans to become established on different continents?

b. What accounts for the richness of some islands, and the impoverishment of others, with respect to their fauna and flora?

c. How did many kinds of animals and plants get distributed from the north temperate to the south temperate zones, or from one polar zone to another, across the tropics?

d. Why are certain species of freshwater fish, inhabiting the lakes and rivers of Europe, also found in the lakes and rivers of North America?

Some of the answers to these puzzling questions will already be clear from what has been said about land bridges. Land bridges, or sunken continents, are obviously necessary to explain many of these distributions between continents and between continents and islands. Sunken continents have already been discussed (Chapter IX). Here I would like to discuss the situation that confronts us if we are not allowed to postulate sunken continents or land bridges.

If we cannot find an acceptable mechanism to account for the creation and destruction of land bridges (or sunken continents) we are forced back upon the ingenious "sweepstakes" idea, which has been much overworked, as an explanation of the distribution of species. The idea arose because it was observed that seabirds, or migratory birds, may carry the seeds of plants or the eggs of insects from continent to continent, and that some species manage to cross, by chance, bodies of water on floating objects such as logs or even ice. Although it conveniently ignores about nine tenths of the evidence, this idea has gained considerable importance. Even though many species have migrated in this way, the idea is no substitute for land bridges. Nor, it may be added, is one land bridge, at Bering Strait, able to do the work of explaining the infinite number of plant and animal migrations in all climatic zones in all geological periods. Many land bridges are required, and for these an explanation is necessary. The theory presented in this book, however, can explain the creation and destruction of land bridges (and sunken continents), and therefore it can explain the distribution of species across large bodies of water.

The impoverishment of certain island faunas and floras as compared with others may be understood as follows. Some of these islands may have rich faunas and floras because, in recent time, they have had land connections with adjacent continents. This would be true of the Philippines, of Java, of Sumatra, and of numerous other islands in that area, whose former continental connections with either Asia or Australia have been argued for by Wallace (435) and others. It is not a question of showing that the species in these islands came from the continents; it is simply true that there were land connections and that the species wandered back and forth; we do not know where they originated.

An island like Java can have a rich fauna and flora not only because of having had rather recent connections with the continent of Asia, but also because it is mountainous. This makes it possible, supposing at some time an equatorward displacement of the island into a warmer latitude, for temperate climate species to ascend into the mountains and so survive. Such variety of climate conditions, due to differences of altitude of different parts of the island, would favor the preservation of a rich flora and fauna.

Let us contrast with Java the situation of a small island or island group, such as the Bermudas, the Azores, or the Canaries, where in general we find the life forms to be impoverished. These islands, often far from the nearest continent, may have been separated from it, of course, for long periods of time. Now let us suppose one of them, say the Azores, to be displaced through about 2,000 miles of latitude in one movement of the crust in either direction. Where will the indigenous species go? Obviously there will be no refuge for them; therefore many of them will succumb. Subsequently the sea will be an effective barrier to the repopulation of the islands from the mainland.

As to the distribution of life forms across the climatic zones, referred to as "bipolar mirrorism," Darwin proposed an explanation in Chapter 12 of the *Origin of Species* that can no longer be accepted. He supposed, first, that glacial periods alternated in the northern and southern hemispheres. This idea has long since been given up. Then Darwin reasoned that when there was an ice age in the northern hemisphere, the climatic zones would be displaced southward, and the temperate-zone species would migrate southward. When that ice age ended and the climate warmed up, the temperate species that had migrated southward would now ascend into the mountains, where they would survive, in the tropic zone. There are, of course, mountains in the tropics high enough to be snow-capped the year round; on these even arctic plants might exist.

The next step, according to Darwin, would be the onset of an ice age in the southern hemisphere. Now the temperature in the southern tropics would fall and become temperate, and the temperate species would descend from their mountains and migrate across the valleys southward to the south temperate zone. In this way the migration of the species from the northern to the southern temperate zone would be accomplished.

Now this idea of the species clambering up and down the mountainsides in response to the changing weather is a good one and gives us one key to the problem. Where Darwin went wrong was in his alternating ice age theory; he could hardly be blamed in view of the prevailing ignorance about ice ages. Darwin, of course, lived at a time when people were first getting used to the idea of ice ages. But if Darwin was wrong, if ice ages do not regularly alternate in the northern and southern hemispheres, how do we explain bipolar mirrorism? For some decades now, glaciologists have been holding grimly to the theory that ice ages were always simultaneous in the two hemispheres. In maintaining this view, they have ignored the fact that they have made mincemeat of Darwin's explanation of bipolar mirrorism. But this does not concern them. They are concerned with explaining ice ages, not with the distribution of species. They have suggested no alternative explanation for the migration of species across the climatic zones. Instead they have constructed a theory that puts the migration of species, and even the survival of tropical species, into the realm of sheer impossibility.

They insist, we remember, that the temperature of the whole earth was simultaneously lowered in glacial periods. We have seen that at various times in the past great continental ice caps have existed at sea level within the tropics, and even on the equator itself. I have already pointed out that if the world temperature had been lowered enough to permit a continental ice cap in the Congo, there would have been no place of refuge for tropical species of plants and animals. Nowhere along the circle of the equator around the earth would any tropical species have survived. This would be equally true of land and sea forms of life.

Bipolar mirrorism, however, presents no problems if we reconsider it in terms of displacements of the crust. One movement, let us suppose, takes the Rocky Mountains 2,000 miles to the south. The species climb higher. Later, in a series of movements of the crust (not always, of course, in the same direction), the Rockies finally end up south of the equator in a temperate climate. Now the species climb down and occupy the temperate valleys of the southern hemisphere. The mountain chain has functioned as a ferryboat, simply transporting species back and forth.

At this point it is interesting to reflect on how useful it is to have these high mountain ranges. A low mountain range would never do. It could never ferry a load of species across the tropical zone.

5. THE PERIODS OF REVOLUTIONARY CHANGES IN LIFE FORMS

The reader may have gained the impression that, while certain aspects of evolution have escaped satisfactory explanation, at least the process itself has continued evenly through all time. To this reader it may come as a shock, as it did to me, to learn that this is not at all the case. There have been remarkable variations in the rate of evolution. For long periods it has marked time, and then some force, hitherto unidentified, has initiated a phase of rapid change, a revolution changing so many forms of plant and animal life as to alter the general complexion of life on the earth. All paleontologists appear to agree on this point. Doctor Simpson uses the term "Virenzperiod" to define the periods of rapid change. Others refer to "explosive" phases of evolution or to "quantum evolution." It must be understood that development during these periods is rapid only relatively; new forms are still not created overnight.

One phenomenon that frequently occurs during these periods is termed "adaptive radiation." This is a kind of explosion in which one form (or species) rapidly gives rise to dozens, scores, or even hundreds of new forms apparently at one and the same time. How is this phenomenon accounted for?

We must distinguish between the biological process and the circumstances that cause it to occur. The process is easily explained. Let us suppose that a form of plant or animal is spread over a considerable area. Its total population may include some millions of individuals; over its whole distribution there will be local variations in the environment, and consequently there will be selection pressures operating simultaneously but in different directions on different parts of the population in different habitats. New varieties of the plant or animal will tend to appear to take advantage of special opportunities offered by particular local environments. This sort of thing is always going on, but it does not, by itself, produce explosions of adaptive radiation.

Something more is required. Normally a new variety of any form has to compete with other forms already in possession of the necessary supplies of food, light, and water. The situation that has the particular combination of these things required by a given plant or animal is referred to as its life, or ecological, niche. Naturally, if this niche is already effectively occupied the spread of the new variety is restricted. As an analogy, think of a garden in which you have set out one hundred expensive strawberry plants of a totally new variety just before being called away for two months on urgent business requiring your presence in a foreign country. What now happens? Weeds immediately take over the niche you had hoped to preserve (artificially) for the spread of the strawberry plants. Their spread is restricted, and their survival may be threatened.

In nature what seem to be required to permit the very rapid dissemination of many new variant forms of the original plant or animal is an absence of competition. Empty life niches are required. The question is, How is an empty life niche produced? Occasionally, of course, it may have been there from the beginning; it may never have been occupied, because, presumably, there never was any form of life that could utilize it, but after two billion or more years of evolution, such primeval biological vacuums are few indeed. Life niches have, in general, been very well occupied for a very long time. Something is required, therefore, to empty them.

This is where our theory comes in. The effects of a displacement can be visualized in two stages. In the first a movement of a large continental area through many degrees of latitude might well cause a very general extermination of the inhabitants. We have seen how, in several instances, this occurred during the late Pleistocene (Chapter VIII). The consequence of the extermination of many kinds of plants and animals (which is not to say their extinction, for many of them might survive in other areas) would be to leave their life niches empty.

The second stage, initiated by a new movement of the crust, would be marked by the opening up of avenues for the immigration of life forms from other land areas. Life forms entering the continent would now enjoy a field day. They would multiply; they would occupy rapidly a tremendous area and all manner of habitats; they would produce variant forms, and the variant forms would occupy appropriate niches. Thus explosive evolution would take place. The new forms need not always be immigrants; they could equally well be local survivors of the period of depopulation, of the displacement, who had somehow managed to hold their own under unfavorable conditions. It seems highly probable, indeed, that displacements of the earth's crust are the explanation of explosive evolution.

We have already made mention of the fact that an interrelationship between the revolutionary periods in evolution and the critical phases of change in other geological areas has been noted by many observers. Lull, for example, says,

. . . There are times of quickening, tne expression points of evolution, which are almost invariably coincident with some great geological change, and the correspondence is so exact and so frequent that the laws of chance may not be invoked as an explanation (278:687).

Umbgrove mentions two specific examples of this phenomenon:

The most important point of all, as far as we are concerned, is that the two major periods of strong differentiation of plant life correspond with two major periods of mountain-building and glaciation of the Upper Paleozoic and Pleistocene (419:292).

The same thing is described by Professor Erling Dorf, of Princeton (349:575–91). We need not take too seriously the small number of turning points mentioned by them for the reason that everything, after all, is relative. The turning points mentioned by Umbgrove might turn out to have been, in some respects, the most important turning points in the history of life, and yet there may have been a hundred lesser, but still very important, turning points.

Geologists who have sought an explanation of the relationship between biological and geological change have, in some cases, favored the idea that geological change, such as the formation of new mountain ranges, might have caused both ice ages and biological change. We have seen that this will not account for ice ages. We have also seen that geologists now generally admit their failure to explain mountain building. It is unsatisfactory to attempt to explain the known by the unknown; it will not do to drag in mountain building as the cause of evolution when the former also is unexplained.

Displacements of the earth's crust appear to be the connecting link between these different processes: they explain, at one and the same time, ice ages, mountain formation, and the significant turning points of evolution.

6. THE EXTINCTION OF SPECIES

It has already been shown that periodical displacements of the lithosphere can provide an explanation for the extinction of species. Some further discussion of this problem is, however, required.

It has been suggested that the history of any particular species can be compared with the life of an individual, with its phases of youth, maturity, and old age. Thus the explosive period is the youth of a species, the period of quiet and prosperous enjoyment of its life niche is maturity, and its degenerative phase is its old age. Finally extinction results from the exhaustion of the vital force of the species. This theory assumes an innate cause and a natural order for the succession of the phases.

This idea has been widely disseminated, and in one form or another it has served to confuse all the issues and obscure the known facts. It is one more of those philosophical abstractions that people resort to who come up against an unsolved problem and cannot stand the psychological tension of persevering in the search for truth. It is important that the essentials of this matter should be made clear.

In the first place, the idea that a species is analogous to an individual, and must go through similar phases, is a modern revival of the Scholastic logic of the Middle Ages, like the microcosm-macrocosm analogy (according to which some people have recently argued that since planets are satellites of the sun, and electrons are satellites of the nucleus of the atom, then planets are exactly like electrons and must obey the same laws of physics). The alleged vital force, which is supposed to set a preordained limit to the life of a species, completely escapes scientific observation and experiment. It is not only a mere assumption, it is also an unjustified assumption.

The facts of paleontology do not agree with the analogy between the life phases of a species and those of an individual. In very many cases the same phase may be repeated several times in the life of a species, and other phases may be omitted altogether, as we shall see below. For this reason the theory brings caustic comment from Doctor Simpson. After discussing the two phases of adaptive radiation (youth) and "intrazonal adaptation" (establishment in a stable but limited environment), which is analogous to maturity—these often do follow each other in this order—he explains their relationship thus:

> The sequence radiation-intrazonal evolution is usual, simply because radiation does not occur unless there are diverse zones within which evolution will follow. Occasionally, nevertheless, something happens to close the zones so soon that radiation is curtailed, or the intrazonal phase is even shorter than the radiation. The camariate crinoids, for instance, seem to have been in the full swing of a radiation when they all became extinct in the Carboniferous. . . . (380:232).

We note that Doctor Simpson says, "Something happens." What happens? He does not care to suggest what might happen to close the zones, to curtail the radiation, to destroy the species. No one has ever suggested a reasonable explanation of these things, but they can be understood as effects of repeated displacements of the crust.

Doctor Simpson has remarked elsewhere that he does not object to the analogy between the species and the individual, provided it may be allowed that youth may follow maturity, and may occur more than once!

Not only may phases occur in the wrong order and be repeated, but also some may be omitted. This seems particularly true of the last, or so-called senile, period. No concept has had so wide a currency with so little support in evidence as that of the alleged

degeneration of species. The reasoning behind it is essentially speci-
ous: If a form of life becomes extinct, and if some "exaggerated"
trait can be pointed to, which might have produced this extinction,
then it is claimed that the species was degenerate. This is, of course,
merely hindsight, because it ignores the fact that some of the oddest
creatures in the world have lasted for millions of years and still exist.
It is true that some kinds of plants and animals become adapted to
very narrowly specialized environments, so that an almost imper-
ceptible change in the environment may destroy them. These forms
may, if you like, be called overspecialized, but they cannot be called
degenerate. No inner process of decay has taken place in the organism. Its extinction results from the external circumstance that
destroys its relationship with its environment. Is the specialist who
has spent his entire life in the study of the pre-Cambrian and therefore is incapable of making his living in any field outside of geology,
or even outside pre-Cambrian geology, degenerate? If he starves to
death, is his extinction due to degeneration? The reasoning is analogous.

But, supposing that we allow a phenomenon of degeneration in
species, it is still true that most species disappear without showing
any indication whatever of a decline of their "vital force." The
majority of them are cut off in the vigor of maturity, or in "youth,"
as in the case of the camariate crinoids. Moreover there is no rule as
to the relative length of the different periods. Doctor Simpson remarks:

Diversification may be brief or prolonged, and may be of limited scope
or may ramify into the most extraordinarily varied zones covering a
breadth of total adaptation that would have been totally unpredictable
and incredible if we were aware only of the beginning of the process
(380:222–23).

Again, he says,

. . . Episodes of proliferation may come early, middle or late in the
history of a group. This confirms the conclusion that adaptive radiation
is episodic but not cyclic (380:235).

We have already noted that Darwin recognized that the ordinary
competition of species could not account for the mass extinction of
whole groups, of which, even then, there were many known instances
in the fossil record. Since his day, paleontologists have found very

many more cases of apparently well-adapted species which in some cases had flourished for tens of millions of years and yet suddenly disappeared, sometimes leaving their life niches empty and at other times giving way to inferior species as their successors. For the Pleistocene alone, the last million years, as we have seen, the examples of this include the mammoth, the mastodon, the sabertooth cat, the giant beaver, the giant sloth, the giant bison, and countless extinct varieties of still existing forms like horses, deer, camels, peccaries, armadillos, wolves, bears, etc. Doctor Simpson, in discussing the extinction of the dinosaurs, remarks:

> It should be emphasized that these mass extinctions are not instantaneous, or even brief, events. They extend over periods of tens of millions of years. . . . This makes the phenomenon all the more mysterious, because we have to think of environmental changes that not only affected a great many different groups in different environments, but also did so very slowly and very persistently. The only general and true statement that can now be made about, say, the extinction of the dinosaurs is that they all lost adaptation in the course of some long environmental change the nature of which is entirely unknown (380:302).

If the dinosaurs lost adaptation, it was not because they changed. The same is true of the sabertooth cat, which existed for 40,000,000 years and, according to Simpson, was apparently as well adapted at the end of that period as at the beginning (382:43–44). The gradual elimination of the dinosaurs can be understood as the result of constant shiftings of the earth's crust, which eliminated these reptiles first in one area and then in another. No doubt, dinosaurs repeatedly reoccupied areas from which they had previously been eliminated, but eventually—perhaps much more recently than some people think —they were destroyed. Being cold-blooded creatures, of course, they would find it quite intolerable to be shifted into the cold zones. but there is not the slightest reason to think they were degenerate. Simpson attacks the entire idea of degeneration of species (382:72, 81). He quotes Rensch:

> In innumerable cases lineages become extinct without there being recognizable in the last forms any sort of morphological or pathological degenerative phenomena (380:292).

Dodson gives a good example of the piecemeal extinction of species. He cites the case of the mastodons, relatives of the elephants, which became extinct first in the old world and then in the new

(115:371). Other examples could be cited from the Pleistocene, when many species became extinct in the Americas, while their close relatives, such as horses, camels, and various kinds of elephants, survived in the eastern hemisphere. Now one might ask the question, If a species becomes extinct on one continent but continues to flourish on another, is it or is it not senile? What stage is it in then? We can understand all these events as the results of piecemeal destructions of animal populations in crust displacements. We can see in them the process of the creation of empty environments, preparing the way for a new stage of explosive evolution. Simpson directly suggests the connection between these two things:

> . . . Opportunity may come as an inheritance from the dead, the extinct, who bequeath adaptive zones free from competitors. Jurassic Virenz for ammonites followed extinction of all but one family, perhaps all but one genus, of Triassic ammonites; early Tertiary mammalian Virenz followed mysterious decimation of the Cretaceous reptiles. . . . (382:73).

There is another question regarding the extinction of species that should be answered. Perhaps it will be asked, If crust displacements killed off the dinosaurs, why did they not eliminate also the very numerous other reptiles that still survive? If the last displacement at the close of the Pleistocene eliminated the mammoth and certain other mammals from America, why did other animals survive? The answer is, essentially, that it is a question of the mathematical chances of survival. It is a question of the numbers of the animals, the geographical extent and variety of their habitats, their particular individual aptitudes, and the ever-present factor of sheer accident. It may be true that size militated against some species, but it may have worked in favor of others. The very largest animal of all—the whale—still survives. Elephants compare favorably with all but the very largest extinct mammals.

7. THE GAPS IN THE FOSSIL RECORD

One further point remains for our consideration. A feature of the fossil record that greatly impressed Darwin was the curious way in which species appear, full-blown, with no indication of transition forms, much like the mythical birth of Venus. The paleontologist suddenly comes upon a species, or a whole group of them, which have not been found before. They are all fully evolved; they obviously have had long histories; there must have been hundreds or even thousands of ancestral forms for them; but absolutely no trace of the preceding forms can be found. It happened this way with the dinosaurs, which appeared in Africa, with a great many species already fully developed, at the beginning of the Mesozoic Era. They seem to have come, literally, out of nothing. Sometimes ancestral forms of a particular plant or animal will be found at a great distance—on another continent perhaps—but always there appear to have existed many intermediary links which have been lost. Even in the case of the horse, where an unusually good record exists, there are many missing links.

A part of the reason for this situation is, of course, the imperfect preservation of the fossil record. There appear to be several reasons for this. Fossilization itself is a very rare event; very few individuals of any species are preserved, and the great majority of all the species that have existed have disappeared without a trace. Then, of the fossils that were preserved in the rocks since the beginning of the sedimentary record, about 95 percent have been destroyed, since about that percentage of all the sedimentary rocks of the older periods has been eroded away and redeposited, with consequent destruction of all fossils. Finally, of the fossils that have been preserved, it is very unlikely that paleontologists can have seen and studied more than a very insignificant proportion—let us say, to put the matter as liberally as possible, that they may have seen one millionth of the existing fossils. Many of the latter, of course, are buried deep in the earth or under the numerous shallow seas and will never be seen.

But true as this is, it does not quite satisfy. Relatively few and scattered as fossils may be, it is still to be wondered at that we do not have a respectable handful of reasonably complete life histories. The light cast on this matter by the theory of crust displacement is quite startling. We have seen that such movements would necessitate frequent migrations of whole faunas and floras. It would necessarily follow, from the theory of crust displacement, that species would as a rule be separated by considerable geographical distances from the places of their origin. This would be all the more certain since the rate of developement of new forms is probably very slow as compared with the rate at which crust displacements may occur.

It could actually be rather seldom that one plant or animal would complete much of its life history in the same place. The "missing links" would usually have been separated by great geographical distances from the homes of their decendants. Moreover the successive movements of the crust, with the resulting changes in the distribution of land and sea, would leave much of the fossil record under the present shallow (or even deep) seas and out of our reach.

8. SUMMARY

To sum up: It would seem that in crust displacement we have the missing factor that can bring all the other evolutionary factors into proper focus and correct perspective. By crust displacements we may accelerate the tempo of natural selection, provide the conditions of isolation and competition required for change in life forms, and account for periods of revolutionary change, for the distribution of species across oceans and climatic zones, and for the extinction of species. We may also account for the significant association of turning points in evolution with geological and climatic changes, presenting them as different results of the same cause. But for crust displacements to have had these effects, and if they are, indeed, to account for the evolution of species, they must have occurred very often throughout the history of the earth.

CONCLUSION

1. A GENERAL SUMMARY

In this book I have presented a highly detailed mass of material, and I have sought to relate it to a single, essentially simple hypothesis. It now remains to summarize the evidence and the argument for the hypothesis.

We have seen that the problem of the geographical stability of the poles has long been a vexatious matter for science. From time to time theories of polar shift have been advanced, supported by large quantities of evidence, but the proposed mechanisms have been found defective, and in consequence the theories have been rejected. The failure of the theories has led, in the following years, to neglect of the evidence or to its analysis in accordance with theories conforming to the doctrine of the permanence of the poles. Although all the older theories of polar change, including that of Wegener, have been discredited, the evidence in favor of polar change has constantly increased. As a consequence, many writers at the present time are discussing polar shift, but none of them has as yet suggested an acceptable mechanism.

The general evidence for displacements of the lithosphere is exceedingly rich. In turn, the assumption of such displacements serves to solve a wide range of problems such as the causes of ice ages, warm polar climates, mountain building; it provides a mechanism that may account for changes in the elevations of land areas and in the topography of the ocean floors; it also provides a basis for the resolution of conflicts in isostatic theory. For the period of the late Pleistocene, the theory permits the construction of a chronology of polar shifts, with three successive tentative polar positions in the Yukon, the Greenland Sea, and Hudson Bay preceding the present position of the pole. The evidence for the location of the Hudson Bay region at the pole during the last North American ice age is overwhelming, and this fact in itself provides the principal support for the assumption of the earlier shifts. The tempo of change indicated for the late Pleistocene is reflected in evidence from earlier geological periods.

The theory is able to explain not only the general succession of climatic changes in various parts of the world in the late Pleistocene; it can account also for the detailed history of the last North American ice cap. It can explain the fluctuations of that ice cap, its repeated retreats and readvances. It shows that the effects of volcanism may have been responsible for the oscillations. It shows also that these same effects, added to the effects of gradual climatic change, may account for the widespread extinctions of species at the end of the Pleistocene, and from this we may asume that the same cause may have been responsible for numerous extinctions in earlier geological periods. By providing a reasonable basis for the assumptions of rapid climatic change and rapid topographical change, the theory provides solutions for many problems in the evolution and distribution of species.

Our theory of displacement depends upon two assumptions, and on two only. One of these is that an unbalanced mass within the lithosphere is exerting a sufficient centrifugal effect. The other assumption is that at some point below the crust a weak layer exists that will permit the displacement of the crust over it. The body of geological evidence presented in this book provides very strong indirect support for both these assumptions.

As to the mechanics of displacements, Campbell has provided the necessary constructions. To some, the simplicity of his thought may be unnerving, but I feel assured that in the end this simplicity itself will be the justification for reposing wide confidence in this theory. For it appears that no recondite principle can vitiate it. Who can argue with formulas so simple that a high-school student can, and usually does, master them?

In addition to the support provided by evidence from the field, our theory receives support from logic. It has been recognized that one characteristic of sound new theories is the simplicity of their basic assumptions, and another is their capacity to explain a greater number of facts or a greater range of problems than previous theories.

It was the simplicity of this theory that first aroused the interest of Einstein, in whose philosophy of science simplicity was a prime consideration. It appeared to him also that it might explain a far greater number of facts than were explainable by the various theories that have been produced to explain the leading problems of the earth separately.

2. REMAINING PROBLEMS: FUTURE RESEARCH

I am conscious of the many problems that remain to be solved, especially that of the mechanism of displacement. These are intensified by the assumptions I have been forced to make regarding the speed and recent date of some of these displacements. Many difficulties, some real and some fictitious, will confront this theory as they do any new and far-reaching assumption.

An immediate problem which, however, is soluble relates to the geomagnetic evidence for recent displacements. Some of this has been presented in this book but admittedly it is insufficiently precise. I am sure that more evidence will be forthcoming; and if more evidence is to be found it must first be looked for. Perhaps a good place to look for it would be in Iceland, which is situated with reference to the locations of our assumed poles in such a way that maximum changes in magnetic directions would have accompanied the polar shifts.

Other possibilities for research suggest themselves. The following is an incomplete list:

1. Since the Ross Sea cores were lifted in 1949, no attempt has been made to date other cores in the same way, to confirm or revise the results obtained by Urry. This is a surprising oversight that should be remedied. It seems that the controversial implications of these cores have succeeded in arousing absolutely no interest in geological quarters. This reflects the lethargy regarding basic questions of geological theory so much deplored by Daly. It is important to date many more such cores from all around the coasts of Antarctica, not only to establish the facts regarding a possible warm period there but also to determine whether climatic changes affected different parts of the continent in the sequence required by our hypothetical polar succession.

2. A determined effort should be made to establish beyond question the time of the end of the Sangamon Interglacial period in North America.

3. A project should be organized to carry out the dating by radiocarbon of several thousand samples of wood and animal remains from the circumpolar permafrost areas of the Arctic in order to find out whether the climatic changes followed the pattern deducible from our assumed polar succession and perhaps to determine the speed of the displacements more accurately.

4. Studies of the age of the Antarctic ice should be planned to provide evidence confirming or revising the presumed polar succession. In some areas the bottom ice should be much older than in others, and the ages should be in line with our assumptions regarding the various polar shifts.

5. There is at present no world map of gravity anomalies from which it might be possible to calculate the tangential stresses now active in the lithosphere. This should be compiled (though it would be a large task) and the results processed by computers. If the results should indicate a net tangential stress in the same direction as the present indicated movement of the North Pole toward Greenland, one might assume the possibility of a connection between them.

6. Benioff has pointed out that during the twentieth century there has been an ascending curve of major earthquakes of unexplained origin (29). These have apparently tended to increase both in intensity and in frequency. An investigation should be undertaken to determine whether this ascending curve is related to the possible acceleration of the secular displacement of the pole, which I have referred to in Chapter I. If they are connected, it may follow that earthquakes are caused by movements of the lithosphere.

7. Extensive studies are necessary in connection with every observed isostatic anomaly (of significant magnitude) on the earth's surface, whether positive or negative, to determine the tangential component of its gross centrifugal or centripetal effect. A computer should be utilized to determine the direction of any net horizontal stress that might affect the stability of the lithosphere. In this isostatic study the triaxial distortions of the geoid should not be excluded from the calculations.

8. Studies should be directed to the question of whether opposing

centrifugal (or centripetal) effects of isostatic anomalies may not be factors in seismicity. For example it is known that the Hawaiian Islands constitute an enormous mass of basalt superposed on the earth's crust but not compensated isostatically. Their low latitude implies that the centrifugal effect which the earth's rotation exerts on their mass is enormous. It is mathematically certain that the tangential component of this effect, acting in an equatorward direction, must transmit a significant stress to the floor of the Pacific. Results of this stress should be apparent, and it may be possible to identify specific seismic phenomena with this source.

9. The assumption of a displacement of the earth's crust at the end of the ice age carries the corollary of numerous changes in the elevation of various land masses with reference to sea level. Some of these changes of elevation may have been drastic. Evidence exists of such changes, but it is limited to only a few areas, and the observations are incomplete and imprecise. A world survey of such recent geological changes is required, but it is a task for many geologists that will take much time.

Let me say a final word to the students, the young men and women in our colleges and high schools: The mysteries of the earth beckon to you. What man now knows is little enough, and most of his general concepts in every field are vitiated by the artificial concepts he has created to cover his ignorance. These concepts must be destroyed. One tool exists that can accomplish this destruction, and this tool is in your hands. It is simply curiosity—the instinct to ask and to question. It should be kept sharp and used without mercy.

APPENDIX
Note 1.

PREFACE TO THE FRENCH EDITION
Earth's Shifting Crust

By Yves Rocard, Professor in the Faculty of
Sciences at Paris, and Director of the Physics Laboratory
of the *Ecole Normale Supérieure.*

It is not entirely natural that a physicist should take the responsibility
of presenting the French edition of Charles H. Hapgood's remarkable
work. However, I am covered by the authority of the great Albert Einstein, who from the beginning approved the author's ideas.

But I should also say that I have found in this book a theory that
delivers me from painful uncertainties on at least two points:

First of all, being involved with seismology and performing experiments in which at a distance of 200 kilometers we register the effects
of small explosions, I have a number of cases in which so-called "Pn"
seismic waves reach our apparatus after having, so to speak, struck the
terrestrial crust from below. One would expect to find a great regularity
in the propagation of these waves, the floor of the crust (mohorovicic
discontinuity) presumably being uniform. However, these waves evidence
a considerable heterogeneity in azimuth. If we adopt Charles Hapgood's
idea of a crust that turns by gliding on the mantle it is clear that a
slow turbulence is going to affect the interface in much the same way
that fleecy clouds are affected in certain kinds of weather. It therefore
seems to me that the present work contains in general the principle
of an explanation of the facts I was at a loss to explain, and in any
case it is a stimulant for research.

A second point is the following. Our modern seismographs are sensitive to the "noise" of limited agitation at every point in the earth,
even in the absence of any seismic wave. One may in this noise distinguish a man-made vibration (for example, a train four kilometers
away, or a big city ten kilometers off) and also an atmospheric effect
(from changing pressure of the wind on the soil) and sometimes one
registers also the effects of great storms at a distance. Yet there remains

a continued rolling noise of cracklings in the earth which owes nothing to any cause that varies within the human scale of time. We find these cracklings strong enough if we are in France, but twenty to fifty times weaker in the great primary shields such as at the Hoggar. Might this not be an occultation of the crust itself in movement? The order of magnitude here is correct. Doubtless one could say that readjustments of the crust owing to isostasy would produce the same effect. Perhaps that is true. But then a study on a worldwide scale of this noise in the depths of the earth would make possible a map of the cracklings at the surface of the globe, and there is no doubt that one would then see the design of the great axes of deformation—confirming or refuting Hapgood.

In addition, the work attacks very engaging questions—such as the death and sudden freezing of the mammoth of the Beresovka, overtaken with grasses and small summer flowers in his mouth many thousands of years ago—to interest every cultivated reader. One finds specialized technical language only in the citations, and the text itself is accessible to all.

The author somewhere denies being a physicist. A physicist would no doubt have added to his book a chapter treating the past variations of terrestrial magnetism. It would in fact be interesting to see if Hapgood's moving terrestrial crust explains or fails to explain the variations evidenced by rock magnetism, a subject that we do not find treated in the book.

Finally the principal point remains—that of the physical possibility, with an off-center polar ice cap, of a centrifugal force great enough to cause the hypothetical displacement, with its enormous consequences. Without taking sides finally as to this possibility I would like to cite, just the same, the frosting of telegraph wires: The ice accumulates asymmetrically in the direction of the freezing wind; it is heavy and tends to descend, thus twisting the wires; the new ice that is deposited accentuates this torsion, and if the freezing is continued long enough this results in breakage of the wires. It seems to me that there we have a picture of Hapgood's effect.

For all these reasons, small and great, it seemed to be interesting to call attention to this French translation, which is most accurate, thanks to Mme. Anne Frauger, and which will now profit from the wide circulation of the Payot publications.

Note 2.

LETTERS FROM ALBERT EINSTEIN AND GEORGE SARTON

A. Einstein
112 Mercer Street
Princeton, N. J.

November 24, 1952

Mr. Charles Hapgood
2 Allerton Street
Provincetown, Mass.

Dear Sir:

I have read already some years ago in a popular article about the idea that excentric masses of ice, accumulated near a pole, could produce from time to time considerable dislocations of the floating rigid crust of the earth. I have never occupied myself with this problem but my impression is that a careful study of this hypothesis is really desirable.

I think that our factual knowledge of the underlying facts is at present not precise enough for a reliable answer based exclusively on calculations. Knowledge of geological and paleontological facts may be of decisive importance in the matter. In any case, it would not be justified to discard the idea a priori as adventurous.

The question whether high pressure may not be able to produce fusion of nuclei is also quite justified. It is not known to me if a quantitative theory has been worked out by astrophysicists. The action of pressure would not be a static effect as classical mechanics would suggest, but a kinetic effect corresponding not to temperature but to degeneracy of gases of high density. You should correspond about this with an astrophysicist experienced in quantum theory, f.i. Dr. M. Schwarzschild at the Princeton University Observatory.

Sincerely yours,
(Signed)   A. Einstein

Princeton,
May 8, 1953

Dear Mr. Hapgood,

I thank you very much for the manuscript that you sent me on May 3rd. I find your arguments very impressive and have the impression that your hypothesis is correct. One can hardly doubt that significant shifts of the crust of the earth have taken place repeatedly and within a short time. The empirical material you have compiled would hardly permit another interpretation.

It is certainly true, too, that ice is continually deposited in the polar regions. These deposits must lead to instability of the crust when it is not sufficiently weak to constantly keep in balance by the adjustment of the polar regions.

The thickness of the icecap at the polar regions must, if this is the case, constantly increase, at least where a foundation of rock is present. One should be able to estimate empirically the annual increase of the polar icecaps. If there exists at least in one part of the polar regions a rock foundation for the icecap, one should be able to calculate how much time was needed to deposit the whole of the icecap. The amount of the ice that flowed off should be negligible in this calculation. In this way one could almost prove your hypothesis.

Another striking circumstance appears in connection with the ellipticity of the meridians. If according to your hypothesis an approximate folding of the meridional volume takes place, that is, folding of a meridional volume within an equatorial volume (which is considerably larger), this event will have to be accompanied by a fracture of the hard crust of the earth. This also fits in very well with the existing phenomena of the volcanic coastal regions with their mainly north-south extension and the narrowness in the east-west direction. Without your hypothesis one could hardly find a halfway reasonable explanation for these weak spots of the present-day crust of the earth.

Excuse me for not writing in English. My secretary has been away for some time, and "spelling" makes frightful difficulties for me.

With sincere respect and kind regards,

Yours,
(Signed)   A. Einstein

(Translated by Ilse Politzer)

5 Channing Place,
Cambridge, Mass.,
Tuesday '55 06.07

Dear Mr. Hapgood,

I have read your lecture at the AMNH, the discussion which followed and the Einstein documents, with *deep* interest. I really think that you are on the right track but have no authority to express a more definite opinion. It is clear that the only opinions which matter are those which are the results of independent studies by competent specialists.

The combination of ideas is so new that the history of science has nothing to contribute to its understanding. The fact that there have been earlier theories like those of Wegener, Kreichgauer and Vening Meinesz simply proves that some meteorologic and geologic problems had to be solved, and exercised the minds of men of science. What you need is not historical facts, but physical ones, and mathematical developments.

With every good wish
(Signed)
George Sarton

Note 3.

THE LATE JAMES H. CAMPBELL ON THE MECHANICS OF THE DISPLACEMENT OF THE LITHOSPHERE

1. CALCULATING THE CENTRIFUGAL EFFECT OF THE ICE CAP

I have already mentioned how preliminary calculations were made of the possible centrifugal effect of the Antarctic cap. I have mentioned that the calculation was first made by Buker, and later somewhat revised by Campbell. However, Campbell recognized, early in his examination of the theory, that this effect, since it operated at right angles to the axis of rotation and was not tangential to the surface, would not produce a horizontal movement of the crust, even if the magnitude of the effect was sufficient for the purpose. It would be necessary, he felt, to find the tangential component of the total quantity of the centrifugal effect. He accomplished this by the application of the principle of the parallelogram of forces (Fig. 35). However, his use of the principle is not that usually presented in high-school and college textbooks of physics. The definition of the law of the parallelogram of forces is as follows:

> If two forces acting on a point be represented in direction and intensity by the adjacent sides of a parallelogram, their resultant will be represented by that diagonal of the parallelogram which passes through the point (249:489).

In the three parallelograms in Figure 35, the two forces acting on the point a are the force of gravity, represented by the line a–d (a radial line to the center of the earth), and the tangential component of the centrifugal effect of the ice cap, represented by the line a–c, while the "resultant" of these forces is the diagonal a–b, at right angles to the axis of rotation. The reader will note that Campbell has here inverted the

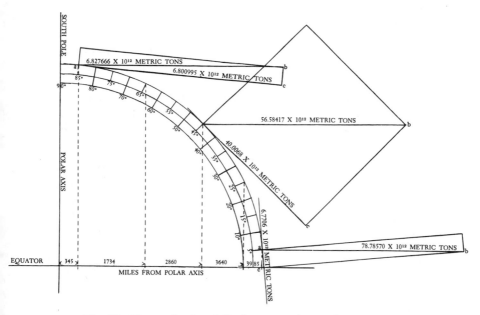

Fig. 35. *The mechanics of displacement: the parallelogram of forces.*

terms of the definition but without changing the quantities of the forces in relationship to each other. The "resultant" in the definition is our starting point; it is the estimated total centrifugal effect of the ice cap. But it is evident that it is unimportant whether the given quantity is the diagonal or the side of the parallelogram; the parallelogram permits the finding of the unknown quantity from the known quantity, whichever the latter is. The parallelogram therefore permits a finding of the

TABLE 21

The Progression of the Centrifugal Effects as the Ice Cap Is Displaced toward the Equator

The figures indicate the relative quantities of the total centrifugal effect at right angles to the axis and of the tangential component acting horizontally and tending to displace the crust, for various latitudes, assuming that the ice cap would maintain its present estimated weight (without melting) until it reached the equator. The calculations are by James Hunter Campbell.

LATI-TUDE	MILES FROM AXIS OF ROTATION	LINEAR VELOCITY Feet per Second	WEIGHT OF ICE Short Tons	CENTRIFUGAL EFFECT Short Tons	Metric Tons	TANGENTIAL COMPONENT Metric Tons
85°	345	132.47	2.5×10^{16}	7.526×10^{12}	6.827×10^{12}	6.800×10^{12}
65°	1,734	665.81	2.5×10^{16}	3.782×10^{13}	3.431×10^{13}	31.099×10^{12}
45°	2,860	1,098.1	2.5×10^{16}	6.237×10^{13}	5.658×10^{13}	40.006×10^{12}
25°	3,640	1,397.7	2.5×10^{16}	7.940×10^{13}	7.203×10^{13}	30.109×10^{12}
5°	3,985	1,529.4	2.5×10^{16}	8.684×10^{13}	7.878×10^{13}	6.770×10^{12}

quantity of the tangential component. The rotating effect of this force exerted on the earth's crust is illustrated by the weight shown in Figure 36.

By definition, the tangential component of the centrifugal effect is exerted horizontally on the earth's crust. Now the question arises as to whether this force will be exerted on the crust itself, or whether it will act on the earth's body as a whole and thus tend to be dissipated in depth. This is the same as the question whether the ice cap will tend to shift the whole planet on its axis or merely to shift the crust. We have already decided that it will tend to shift the crust only, because of the existence of a soft, viscous, and plastic layer under the crust. We may therefore conceive of this force, the tangential component of the total effect, as acting on the crust alone, while recognizing that the displacement of the crust involves frictional effects with the sublayer. We have seen that a special characteristic of the mechanism under discussion is that it provides a constantly growing force that will overcome this friction rather than be absorbed by it.

2. THE WEDGE EFFECT

The problem to be solved by the calculation was to find the quantity of the centrifugal thrust of the ice cap in terms of pressure per square inch on the earth's crust, so that this quantity could be compared with the estimated tensile strength of the crust. If these quantities should be found to be of about the same magnitude, it would follow that the ice cap had the potentiality of bringing about the fracturing of the crust, which, because of the slightly oblate shape of the earth, was necessary to its displacement over the lower layers. Einstein, in a letter received during an early stage of the investigation, suggested that this necessity for the fracturing of the crust was the only serious hindrance to crust displacements in response to centrifugal effects. He wrote:

> For your theory it is only essential that an excentrically situated mass rising above the mean level of the earth-surface is producing a centrifugal momentum acting on the rigid crust of the earth. The earth-crust would change its position through even a very small centrifugal force if the crust would be of spheric symmetry. The only force that I can see which can prevent such sliding motion of the crust is the ellipsoidic form of the crust (and of the fluid core). This form gives to the crust a certain amount of stability which allows the dislocation of the crust only if the centrifugal momentum has a certain magnitude. The dislocation may then occur and be accompanied by a break of the crust. . . . (128).

Campbell, visualizing the sliding of the crust, perceived that a bursting stress would be caused in the crust when parts of it were displaced toward or across the equatorial region, where the diameter of the earth is greater. It became possible to visualize it in the manner suggested by Campbell in the lower left-hand drawing of Figure 36. The drawing shows, in black, two wedge-shaped cross sections representing the earth's equatorial bulge, or that part of it underlying the areas of the crust moving

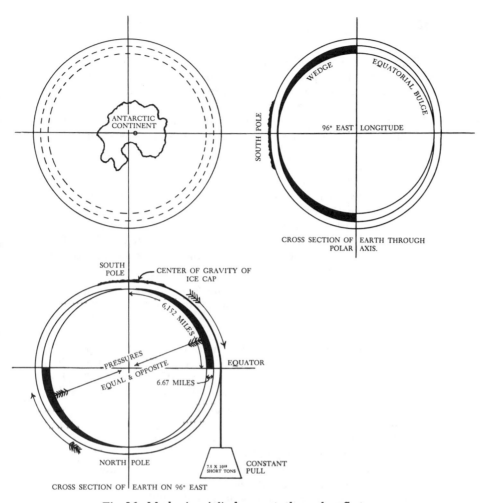

Fig. 36. Mechanics of displacement: the wedge effect.

The Antarctic continent is shown, with center of gravity displaced to the right, on a line representing the 96th degree of East Longitude. The figure at the right represents the continuation of the movement of the ice cap along this meridian, mounting the bulge, which must be visualized three-dimensionally. The lower left-hand figure shows the vertical cross section of the earth under the ice cap, with the two wedges pushing the crust out as it approaches the equator. The proportions of the wedges are shown, and an equilibrium of equal and opposite pressures is indicated. The tangential pull of the ice cap is indicated by the suspended weight.

equatorward in the displacement. Half of the bulge is, in fact, involved; the other half underlies the areas of the crust being simultaneously displaced poleward, but these do not affect the point at issue.

The reader will note that the equatorial bulge of the earth is represented in this drawing as lying underneath the crust, so that the crust is not a part of it. This is a new way of visualizing the bulge, introduced by Campbell, and justified by him on the ground that since the earth's crust is of the same general thickness all over the earth regardless of latitude (even though it may be of differing thicknesses from place to place, and under mountains, continents, and ocean basins), then the differences in the polar and equatorial diameters of the earth are accounted for by differences in the thicknesses of the layers underlying the crust, and the bulge itself represents added matter in the subcrustal layers in the equatorial regions.

The bulge, viewed in this way as lying beneath the crust, appeared to form two wedges against which the crust had to be displaced by the centrifugal thrust of the ice cap. Campbell reached the conclusion that to estimate the bursting stress produced on the crust in the equatorial region, it would be necessary to apply the mechanical principle of the wedge, which has the effect of multiplying the effect of an applied force. This principle is usually given as follows:

> The wedge is a pair of inclined planes united by their bases. In the application of pressure to the head or butt end of the wedge, to cause it to penetrate a resisting body, the applied force is to the resistance as the thickness of the wedge is to its length (249:512).

This statement means that a wedge multiplies the splitting power (or bursting stress) produced by an applied force in the proportion of the length of the wedge to its thickness at the butt end. Figure 37 shows the application of this principle to the earth. The formula for calculating the wedge effect is presented at the extreme left, where P = pressure (as thrust of the ice cap transmitted to the crust), Q = the mutual pressure between crust and bulge, or bursting stress, h = the height (that is, the length) of the wedge, and b = the base or butt end. The bursting stress equals the pressure applied to the butt and multiplied by the ratio of the thickness of the butt end to the length of the wedge. The length of the wedge, in this case, is about 6,000 miles, and its thickness at the butt end is 6.67 miles, so that the ratio is about 1,000:1, and the quantity of the thrust of the ice cap should consequently be multiplied by 1,000; however, there are two wedges, one on each side of the earth, and therefore the thrust is multiplied only 500 times. Nevertheless the significance of such a multiplication of the effect of the ice cap is self-evident.

It was not a simple matter to apply the principle of the wedge to the earth. As in the case of the principle of the parallelogram, the formula could not simply be copied from a textbook; it had to be imaginatively applied. For example, in the diagram P, or pressure, is shown exerted on the butt end, like a sledge hammer hitting the butt end of a wedge to split a log. But obviously, the thrust of the ice cap is not, in the first instance, applied in this way. It takes an act of the imagination to realize that in effect it amounts to the same thing. The ice cap is really pushing or pulling the crust toward the equator on both sides of the earth, but

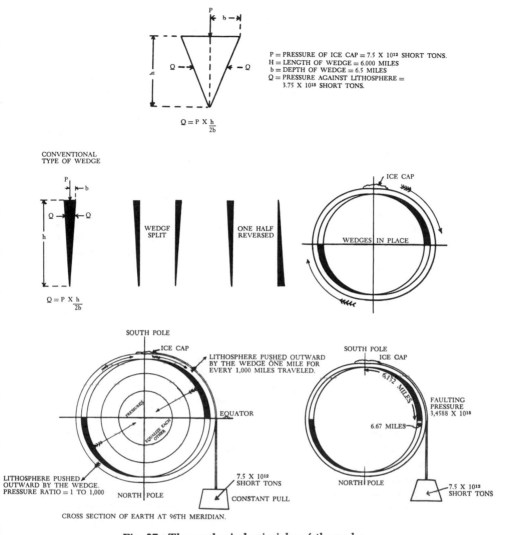

P = PRESSURE OF ICE CAP = 7.5 X 10¹² SHORT TONS.
H = LENGTH OF WEDGE = 6.000 MILES
b = DEPTH OF WEDGE = 6.5 MILES
Q = PRESSURE AGAINST LITHOSPHERE =
   3.75 X 10¹⁵ SHORT TONS.

$$Q = P \times \frac{h}{2b}$$

CONVENTIONAL
TYPE OF WEDGE

$$Q = P \times \frac{h}{2b}$$

WEDGE
SPLIT

ONE HALF
REVERSED

ICE CAP

WEDGES IN PLACE

SOUTH POLE

ICE CAP

LITHOSPHERE PUSHED OUTWARD
BY THE WEDGE ONE MILE FOR
EVERY 1,000 MILES TRAVELED.

EQUATOR

LITHOSPHERE PUSHED
OUTWARD BY THE WEDGE.
PRESSURE RATIO = 1 TO 1,000

NORTH POLE

7.5 X 10¹²
SHORT TONS

CONSTANT PULL

SOUTH POLE
ICE CAP

6.132 MILES

FAULTING
PRESSURE
3,4588 X 10¹⁵

6.67 MILES

NORTH POLE

7.5 X 10¹²
SHORT TONS

CROSS SECTION OF EARTH AT 96TH MERIDIAN.

Fig. 37. *The mechanical principles of the wedge.*

the matter may just as well be looked at in the opposite way, as if the
ice cap and crust stood still or were under no horizontal pressure, but
force was being applied to the butt end of the wedge. Either way, the
mathematics is the same. Campbell's application of this principle to
the problem of estimating the bursting stress on the crust was discussed
with physicists, including Frankland, Bridgman, and Einstein (see below), none of whom questioned its soundness.

After finding the quantity of the total stress on the earth's crust produced by the centrifugal effect transmitted from the ice cap, Campbell

reduced it to pressure per square inch by dividing it into the number of square inches in a cross section of the earth's crust, assuming an average thickness of the crust of about 40 miles. This estimate of the crust's thickness is a liberal one, since some writers, including Umbgrove, suggest that it may be no more than half as much. Since a lesser thickness for the crust would mean a higher figure for the bursting stress per square inch, an error in this direction may serve to counter the effect of the possible partial isostatic compensation of the ice cap, which we have disregarded in the tentative calculation of its centrifugal effect. Thus if half the ice cap is isostatically compensated, but the crust is only 20 miles thick, then Campbell's estimate of the pressure per square inch will be unchanged.

Campbell found that the bursting stress on the crust per square inch amounted to about 1,700 pounds (see p. 339). In comparison with this, I found that the crushing point of basalt at the earth's surface has been estimated, from laboratory experiments, at 2,500 pounds. A number of points must be considered in reaching conclusions regarding the possible significance of these comparative figures. First, the crushing strength of any rock is considered to be higher than its tensile, or breaking, strength. Thus the tensile strength of basalt, the principal constituent of the earth's crust, is probably considerably closer to the estimated quantity of the bursting stress. A second important consideration is that the earth's crust is unequal in thickness and strength from place to place and is everywhere penetrated by deep fractures. It would naturally fail at its weakest point. As we shall see below, Einstein, in view of these facts, said that he would be satisfied as to the plausibility of the mechanics of this theory if the ratio of the bursting stress to the strength of the crust was 1:100. The ratio as shown by Campbell is very much closer than the ratio demanded by Einstein. It seems therefore reasonable to suppose that at some point of the future growth of the ice cap, which is now, it appears, still growing, the crust may respond to the increasing bursting stress by fracturing. When this occurs, it may be expected that a process will begin of gradual fracturing and folding of the crust, accompanied by the beginning of its displacement over the underlying layers.

Campbell has pointed out that no very great force is required to accomplish a widespread fracturing of the crust during a displacement. At the first local failure of the crust in response to the bursting stress, the stress will be relieved at that point, to become effective immediately at an adjacent point. Thus the fracture will travel through the crust without the application of additional force. From this it is clear that the steady application of a small force would suffice to fracture the crust for a great distance. In his conversation with Einstein, an account of which is given below, Campbell gave a convincing illustration of this principle.

The ability of the crust to resist fracture is slight. Jeffreys found that a strain equal to the weight of a layer of rock 2,200 feet in height would fracture it to its full depth (238:202). It is clear that the tensile strength of the crust does not compare with its crushing strength, which, also according to Jeffreys, is sufficient to enable it to transmit mountain-making stresses for any distance. Campbell visualizes the process of crust displacement not as a continuous movement but as a staged movement resulting from an interaction alternately of the direct thrust of the ice cap and of the bursting stress. He writes:

. . . There are two distinctly separate functions performed by the mass of the icecap. . . . The first is the centrifugal momentum causing the lithosphere to change its position in relation to the poles. . . . When the lithosphere comes to a standstill for want of a sufficient force, the second function of the icecap gets busy and builds up a pressure of tremendous potential, five hundred times the force produced by the icecap, and will continue to add to this pressure at the rate of five hundred times the increasing pressure of the growing icecap, until it finally splits the lithosphere. Then the pressure will drop, and the first function will take hold, and once again start to move the lithosphere. This alternate action will continue to take place until the icecap is destroyed. . . .

. . . The wedge does not multiply the power of the centrifugal momentum, as such. The power disposed for the movement of the lithosphere remains the same as it always has been, but the static pressure that will fracture the lithosphere, thereby permitting the centrifugal momentum of the icecap to start moving the lithosphere, will be multiplied by 500.

The wedge has been functioning ever since the first permanent snow fell on the Antarctic continent; it is functioning today. . . . At the same time, the centrifugal momentum of the icecap is just standing by, waiting for the lithosphere to fracture and be released for the journey toward the equator. The tensile strain will fade away with the faulting of the lithosphere, but should the fractures freeze up again from any cause, the tensile strain will tend to build up again, and the same series of actions will repeat themselves (66).

3. THE PROBLEM OF FRICTION

A number of objections have been raised, in the course of consultations with specialists, to the mechanism of displacement suggested by Campbell.

One of these is the problem of friction with the subcrustal layer. It may at first appear that friction would be a powerful brake on any extensive displacement of the crust, and unquestionably it would have an effect. Yet there are several mechanical factors that could aid a displacement. A leading consideration is that the suggested movement is a gliding motion. Gliding is the most economical form of motion. It has been said, in fact, that gliding constitutes an ideal form of motion that utilizes 100 percent of energy, as opposed to the sphere and the cylinder, which, being round, lose 30 percent of their energy in rotation, which reduces their speed considerably. Frankland has suggested that a rise of temperature at the interface of the crust and the lower layer, as the result of friction, could facilitate a displacement. Campbell considers that the underlayer, or asthenophere, would act more like a lubricant than a retardant. He compares the movement to the motion of ice floes: ". . . Observe how vast fields of ice are started in motion just by the friction of the wind on the surface of the ice. . . . Again, you will see the same thing by visiting a pond where they are cutting ice. You will see men pushing around blocks of ice of three or four hundred square feet with the greatest ease as long as the ice is floating in the water. . . ."

4. THE CALCULATIONS

The following are the calculations of the centrifugal effect of the present Antarctic ice cap, and of the resulting bursting stress on the crust, as worked out by Campbell. The phraseology is in part that of Dr. John M. Frankland, of the Federal Bureau of Standards, who was kind enough to review these calculations.

a. Calculation of the Centrifugal Effect of the Rotation of the Antarctic Ice Cap:

Assume isostatic adjustment o, center of gravity of the ice cap 345 miles from the polar axis, and volume of the ice equal to 6,000,000 cubic miles.

W = Weight of the ice cap = 2.500×10^{16} short tons.

F = Centrifugal effect in pounds = $\dfrac{Wv^2}{gR}$, where

v = Velocity of revolving ice cap, 132 feet per second,

R = Distance from the axis of rotation to the center of gravity of the ice cap = 345 miles = 1,821,600 feet,

g = Acceleration due to gravity = 32.

$$F = \frac{Wv^2}{gR} = \frac{2.5 \times 10^{16} \times 132.47^2}{32 \times 1,821,600} =$$

$$\frac{43,870.75 \times 10^{16}}{58,291,200} =$$

7.5×10^{12} short tons = 6.8×10^{12} metric tons,

total centrifugal effect, 6.8×10^{12} metric tons,

radial force tangential to the earth's surface,

6.8×10^{12} metric tons (see p. 331).

This, of course, is an upper estimate, which may be too large by a factor of two or three.

b. Calculation of the Bursting Stress on the Lithosphere:

An approximation of the bursting stress caused by this centrifugal effect can be reached by simple methods as follows. More elaborate approaches hardly seem justified in view of the uncertainty of the magnitude of the centrifugal force.

It is assumed that the entire resistance to the motion of the litho-sphere arises from the fact that the earth is not a perfect sphere, but is an oblate spheroid. The tangential, or shearing, stresses between the lithosphere and the underlying asthenosphere are considered negligible because of the time factor and because of the assumed viscosity of the asthenosphere. If one considers the great circle passing through the center of gravity of the ice cap, at right angles to the meridian of centrifugal thrust of the ice cap, it is evident that the circumference of this great circle will be increased if the ice cap is displaced away from the pole. Of course, any stress system that arises in this way will be two-dimensional, but one will hardly be in error by a factor of more than two, if one neglects the two-dimensional character of the stresses and assumes instead that they are uniaxial. The only purpose of this computation is, of course, to show the order of magnitude of the effect.

With this kind of approximation, one may view the equatorial bulge as a kind of wedge up which the lithosphere is being pushed. There are, of course, two wedges, one on each side of the globe.

The bursting stress is the product of the tangential effect of the ice cap by the ratio of the gradient of the bulge:

(1) Thickness of bulge (wedge) at its butt end $= 6.67$ miles.

(2) Ratio of travel to lift, of bulge wedge $= 6,152: 6.67$.

(3) Stress, on cross section of the lithosphere (taken as 40 miles thick) $= \dfrac{7.5 \times 10^{12} \times 6,152}{6.67 \times 2}$

$= 3.4588 \times 10^{15}$ short tons.

$= \dfrac{3.4588 \times 10^{15}}{990,894} = 3.5 \times 10^{7}$ short tons per sq. in.

$=$ Approximately 1,700 lbs. per sq. in.

5. NOTES OF A CONFERENCE WITH EINSTEIN

In January, 1955, Campbell and I had the privilege of a conference with Einstein at which a number of important questions relating to the theory were discussed. Subsequently I prepared the following statement, which I submitted to him for his approval. He approved it as an accurate report of our discussion, but he desired that it should not be interpreted as an official endorsement on his part of Campbell's calculations in detail, which he had had insufficient opportunity to study. Those present at the meeting included Dr. Einstein, Mr. Campbell, Mrs. Mary G. Grand, and myself.

"After the introductory remarks, Mr. Hapgood explained to Dr. Einstein that while, in the development of the theory, he had himself been concerned mainly with the field evidence in geology and paleontology, Mr. Campbell had contributed the basic concepts in mechanics and geophysics.

"Mr. Hapgood explained further that Mr. Campbell's calculations had now advanced to a point where he felt that a consultation was necessary. The principal question was whether the tangential portion of the centrifugal effect resulting from the rotation of the icecap was of the correct order of magnitude to cause fracturing of the earth's rigid crust. Dr. Einstein had stated in a letter to Mr. Hapgood that, owing to the oblate shape of the earth, the crust could not be displaced without fracturing and that the tensile strength of the crust, opposing such fracturing, was the only force he could see that could prevent a displacement of the crust. He had already suggested, therefore, that it would be necessary to compare the bursting stresses proceeding from the icecap with the available data on the strengths of the crustal rocks.

"It was this problem that now, through the calculations made by Mr. Campbell, seemed to be solved.

"Mr. Campbell explained to Dr. Einstein the principles he had followed in making the calculations. He used photostatic drawings as illustrations. He showed that the crust, in attempting to pass over the equatorial bulge of the earth, would be stretched to a slight degree. A bursting stress would arise that would tend to tear the crust apart. This stress would in all probability exceed the elastic limit of the crustal rocks: that is, they would tend to yield by fracture, if the stress was great enough. Dr. Einstein said, Yes, but he wondered how an equilibrium of force would be created? Mr. Campbell pointed out that two equal and opposite pressures would arise, since, at the same time, on two opposite sides of the globe, two opposite sectors or quadrants of the crust would be attempting to cross the bulge.

"Dr. Einstein agreed that this was reasonable, but raised the question of the behavior of the semiliquid underlayer of the bulge, under pressure from the rigid crust. After some discussion it was agreed that this underlayer, despite its lack of strength, would not be displaced, because of the effect of the centrifugal momentum of the earth.

"Mr. Campbell then explained the application of a principle by which the tangential stress proceeding from the icecap was greatly magnified. He considered that the bulge of the earth, starting with zero thickness at the poles, and approaching 6.67 miles in thickness at the equator, behaved physically as a wedge resisting the movement of the crust. Since the distance from pole to equator is about 6,000 miles, the ratio of this wedge was 1,000:1; but the existence of two wedges on opposite sides of the globe reduced the ratio to 500:1. The icecap's tangential effect, multiplied by 500, and divided by the number of square inches of the cross section of the lithospheric shell at the equator (assuming the crust to be 40 miles thick), produced a bursting stress on that shell of 1,738 pounds per square inch. After examining each step in the argument twice Dr. Einstein had the impression that the principles were right, and that the effects were of the right order of magnitude. He stated that he would be satisfied if the bursting stress and the strength of the crust were in the ratio of not more than 1:100, since the crust varied so greatly in strength from place to place, and would undoubtedly yield at its weakest point.

"Mr. Campbell explained an effect he had often observed, which illustrated the process by which the crust might yield to fracture. A common method of splitting a block of granite is to drill two small holes, about six inches apart, near the center of the long axis of the granite, and insert and drive home a wedge in each hole. A bursting stress of sufficient magnitude is brought to bear to split the rock. However, the rock is not split all at once. Enough stress is brought to bear to start a fracture, but the fracture does not take place instantaneously. If the wedges are put in place in the evening, it will be found next morning that the whole rock has been split evenly along a line extending through the two holes. The fracture has slowly migrated through the rock during the night. The force required to split rock in this way is but a fraction of that required to split it all at once. So far as the earth's crust is concerned, what is required is not a force sufficient to split it all at once, but simply a force sufficient to initiate a fracture or fractures, which will then gradually extend themselves during possibly considerable periods of time.

"Mr. Hapgood next described the geological evidence of world-wide fracture systems extending through the crust, and weakening it, and the remarkable similarity of these patterns to those which, theoretically, would result from a movement of the crust. Dr. Einstein expressed the keenest interest in this evidence.

"Mr. Hapgood referred to the Hough-Urry findings of the dates of climatic change in Antarctica during the Pleistocene. Dr. Einstein stated that the method of radioactive dating developed by W. D. Urry was sound and reliable. As a result, Dr. Einstein was in full agreement that the data from Antarctica, indicating that that continent enjoyed a temperate climate at a time when a continental icecap lay over much of North America, virtually compel the conclusion that a shift of the earth's entire crust must have taken place.

"Dr. Einstein asked Mr. Hapgood what objections geologists had been making to the theory. Mr. Hapgood replied that it was principally a question of the number of such movements. Urry's evidence would imply four such displacements at irregular intervals during the last 50,000 years.* Dr. Einstein replied that this seemed to be a large number. However, he said, if the evidence could not be explained in any other way, even this large number would have to be accepted. The gradualistic notions common in geology were, in his opinion, merely a habit of mind, and were not necessarily justified by the empirical data.

"At this point the discussion turned to astronomy. Mr. Hapgood did not understand why men who would boggle at the rate of change

* This figure was subsequently revised, in the light of much geological evidence (Chs. VII, VIII, IX), to three displacements in the last 130,000 years.

required by the theory of crustal movements thought nothing of accepting the view that the entire universe had been created in half an hour. Dr. Einstein said that, unfortunately, the evidence seemed to point that way. After considerable discussion he added that it was not, however, necessary to take the present state of our knowledge very seriously. Future developments might show us how to reach a different conclusion from the evidence. Much that we regard as knowledge today may someday be regarded as error.

"Toward the end of the interview Dr. Einstein indicated a number of points where further research would be desirable. He suggested the need for a gravitational study of the Antarctic continent, and for a study of the rates of crustal adjustment to increasing or decreasing loads of ice. He commented upon the difficulties that confront those who wish to introduce new theories, and quoted Planck's remark that theories change not because anybody gets converted but because those who hold the old theories eventually die off."

Note 4.

THE MECHANICS
OF CENTRIFUGAL EFFECT

As I have mentioned, there is a possibility of two points of view regarding the particular centrifugal effect postulated by Campbell and myself. It is therefore necessary to provide additional clarification of some of the points at issue. To a certain extent it may be a question of a situation in which new definitions or clearer definitions of accustomed terms are called for, but it also appears to us that in some cases, at least, physicists whom we have consulted in the course of our work are proceeding upon the basis of assumptions that are in conflict with ours. Therefore it is necessary to re-examine these assumptions. A comprehensive discussion of the matter must begin with a review of the broader questions of the mechanics of rotation already briefly referred to in the Introduction.

The existence of a very common misunderstanding regarding the mechanics of the earth's rotation, particularly related to the problem of the stability of the poles, was made clear to me by a difference of opinion that arose at the beginning of my inquiry. Brown, whose work was the starting point of my own, was an engineer, and his concepts of the earth's motions were based upon simple mechanics. He understood gyroscopic action and the stabilizing role of the rim of a rotating flywheel. He also understood the laws of centrifugal effect as applied to weights eccentric to the axes of spin of rotating bodies. It was my good fortune that Campbell, who was to carry the work forward, also was a mechanical engineer.

Brown had made the statement that it was the equatorial bulge of the globe that stabilized it with reference to the axis of rotation; he had compared it to the rim of a flywheel. I found that this statement was disputed by some physicists. The physicists suggested that the stability of the earth on its axis was not owing to the centrifugal effect of the rotation of the equatorial bulge alone, but to that of the rotation of the entire mass of the earth. Later I discovered a passage in Coleman that appeared to express their point of view:

> It may be suggested that the earth is a gyroscope, and, as such, has a very powerful tendency to keep its axis of rotation pointing continuously in the same direction. Any sudden change in the direction would probably wreck the world completely (87:263).

I wished to obtain a clear statement of the rights of this matter. Accordingly I corresponded with specialists, who eventually referred me to the works of James Clerk Maxwell, in one of whose papers I found the following statement in support of Brown's position:

> . . . The permanence of latitude essentially depends on the inequality of the earth's axes, for if they had all been equal, any alteration of the crust of the earth would have produced new principal axes, and the axis of rotation would travel round about those axes, altering the latitudes of all places, and yet not in the least altering the position of the axis of rotation among the stars (296:261).

For the word "axes" in the second line we may read "diameters," and of course Maxwell is referring to the inequality of the polar and equatorial diameters that is, to the existence of the equatorial bulge, to which therefore he directly attributes the stability of the earth on its axis of rotation.

Maxwell, in the foregoing passage, suggests that in the absence of the equatorial bulge, any change in the crust (meaning, it is clear, the creation of any protuberance or excess weight at any point) would change the position of the planet on the axis of rotation. Even before I located this passage in Maxwell, a peculiar device designed by Brown had made this principle clear to me by observation. This device consisted of a globe mounted on three trunnions in such a way that it could rotate in any direction. The globe was a perfect sphere and had no equatorial bulge. It was suspended by a string to an overhead point. To rotate this sphere, all that was necessary was to wind it up and then let it go. Brown had a weight attached to the South Pole of the sphere, and it was observable that, as soon as the sphere began to rotate rapidly, the weight was flung to the equator, where it stabilized the direction of rotation as long as the speed of rotation was maintained. Later Campbell made a larger model of Brown's device, which I rotated unweighted, and I observed that it had no stable axis of spin. Two motions were observable: a rapid rotation and a slow, random drifting. It was evident that the mass of the sphere acted as a stabilizer of the speed of rotation but had no influence on its direction. This experiment, strongly confirming Brown's claim, encouraged me to persist until I could find positive theoretical confirmation of the observation, which eventually I did in the works of Maxwell.

I was amazed and chagrined in this connection to note a phenomenon which, nevertheless, is as old as science itself. The professors—most of them, at any rate—would not come to see the device.

Perhaps I should describe this device in greater detail. A trunnion is like a ring or a hoop, made of metal. A globe is mounted in this trunnion on two pivots set into the ring at points opposite each other (180 degrees apart). Then, if the ring is held (as it often is on a model globe) by a pediment or stand, the globe will rotate. Its axis will be determined by the fixed positions of the pivots set into the ring.

Now, if, instead of fixing the ring into a stand, or pediment, we set it into another, larger ring by inserting two pivots into the larger ring at two points at right angles to those of the inner ring, we have an axis within an axis, and the globe can be made to rotate in either direction. If a third ring is used, then the globe has freedom of action in any direction whatever.

There is still the problem of imparting momentum to this globe. Since it has no fixed axis, this is a difficult problem. Brown solved it by suspending the device from the ceiling by a string attached to the outermost trunnion. This string could be wound up by rotating the outermost trunnion in one direction by hand for a while, just as a boy may wind up the rubber bands used to give momentum to a toy airplane. Then, when the trunnion is released, the string unwinds, putting the globe itself in rapid rotation, but a free rotation, one not confined to a fixed axis.

In view of continued skepticism, I could not be entirely satisfied by the Maxwell statement, supported though it might be by the demonstration. Since I am not myself a physicist, I felt it not unlikely that some persons would conclude that, in the first place, I had misunderstood Maxwell, and that, in the second place, I was incapable of interpreting correctly the evidence of my eyes. I therefore wished to obtain an authoritative interpretation of Maxwell's statement, and accordingly I wrote Dr. Harlow Shapley, the Director of the Harvard Observatory, as follows:

After a year of intensive work with a group of people here, I have concluded that the work we are doing is dependent upon a clear answer to the question as to whether the geographical poles are stabilized by the momentum of rotation of the earth, or solely by that of the equatorial bulge. I have had discussions about this with Dr. Adams of the Coast and Geodetic Survey, and with Dr. Clemence of the Naval Observatory. They have given me references to the work of Clerk Maxwell and others, without quite satisfying me. I am not, of course, equipped to understand all of the technicalities, but I am hoping that you can give me a steer in nontechnical terms on the general concepts.

My hunch is that, contrary to a widespread impression, it is the bulge alone that stabilizes the geographical poles. As I reason it out, if the earth were a perfect sphere, the energy of its rotation, derived from its mass in motion, would "stabilize" the speed of the rotation, but would have no reference to its direction. If we suppose that somebody could reach out from Mars with a pole, and give the earth a strong push at an angle of 90° from the direction of rotation, the earth would be shifted on the axis of rotation to an extent determined by the ratio of the force of the push to the mass of the earth. In fact, if the earth had no bulge, it would never have stable poles, but would rotate every which way. . . .

If my view of the matter is sound, important consequences follow, but I am not quite certain of the validity of my premises.

Dr. Shapley's reply, dated February 2, 1951, was, in part, as follows:

Dr. [Harold] Jeffreys was fortunately here at the Harvard Observatory and I could turn over your inquiry to him. I now have his reply. He says in effect that the fullest discussion of the points mentioned by you is in Routh's Rigid Dynamics, probably in volume I. Most textbooks of rigid dynamics will have something about it. The theory goes back to Euler. Really *both* the rotation and the equatorial bulge are needed to maintain stability. Without rotation the body could be at rest at any position; with rotation but without the equatorial bulge it could rotate permanently about an axis in any direction. . . . (343).

With this statement I decided to rest content. It seemed to me that Brown's position in the matter was correct. Maxwell showed both by the use of his dynamical top and theoretically what Brown showed by his device: that a rotating sphere tends to throw the heaviest weights on its surface to the equator of spin. Maxwell and, after him, George H. Darwin recognized that the equatorial bulge of the globe stabilized the direction of the earth's rotation just as a weight on the surface of a model sphere would do when the sphere was rotated rapidly.

Yet there is a distinct difference between the earth and the model globe. The earth's approximately round shape is not due to the fact that it is a strong, rigid body, for it is not. Its roundness is due primarily to the force of gravity, which in fact holds the earth together. The earth as a whole is a very weak body, and if it were not for the effect of gravity the centrifugal effect of the rotation would disrupt the earth and send all its component masses hurtling outward into interstellar space.

There is also a difference between the equatorial bulge of the earth and a weight attached to the surface of a model globe at its equator of spin. This difference consists in the fact that the earth's equatorial bulge and the flattenings at its poles have been produced by the yielding of the earth's body in response to the centrifugal effect of its rotation. The amount of the yielding has been determined by the ratio of the forces of rotation and gravity. The shape of the earth thus represents a balance of these two forces, a balance that is perfect, theoretically, at every point of the earth's surface. It therefore follows that any unit of material in this balanced surface will be at rest. For this reason, such a surface has been called an equipotential surface.

The balance of the forces of rotation and gravity at every point of the earth's surface can be understood also in his way. The shape of the earth, as we have pointed out, is oblate. This means that as you go toward the equator you are getting farther from the earth's center. In a sense, therefore, you are going uphill. Likewise, when you are going toward the poles you are getting closer to the earth's center and therefore you are going downhill. But we can all see that it takes no more energy to move toward the equator than it does toward a pole. Also, water in the ocean does not run downhill toward the poles. The earth's surface acts as if it were perfectly level. The reason for this is that as you go toward the equator, going uphill, the centrifugal effect of the earth's rotation increases just enough to compensate for the gradient, while, if you move toward the poles, the centrifugal effect declines in proportion. The forces of gravity and rotation are therefore balanced, and no centrifugal effect will tend to propel a mass in this equipotential surface toward the equator, and no gravitational effect will tend to propel it toward the poles. The fact that the force of gravity is absolutely much greater than the centrifugal effect of the rotation is shown by the fact that the flattening of the earth is very slight. The equatorial bulge amounts to 6.7 miles in comparison with the earth's mean radius of 4,000 miles. This is a ratio of only .17 percent.

The past century has been notable for extensive studies of the effects of gravity at the earth's surface. The theory of isostasy has been developed, and the actually existing state of balance of the surface features of the earth's crust has been measured in various ways and for various purposes. As we have seen, there are various difficulties with the theory of isostasy, some of which may be soluble in terms of the theory presented in this book. At the same time, but independently, studies of centrifugal effects at the earth's surface have been undertaken. Eötvös investigated the centrifugal effects that would arise if a given mass had its center of gravity above the equipotential surface. This could occur even with masses in isostatic equilibrium. To visualize this case, we may take the example of a block of ice floating in water.

Ice is lighter than water. When a block of ice falls into a body of water it displaces its own weight of water, and then floats with a tenth of its mass above the water level. It is now in equilibrium, or in isostatic adjustment, even though its upper part projects a considerable distance up out of the water. This upper tenth, in the meantime, has displaced air, not water. It is a solid mass of far greater density than the air it has displaced. Its center of gravity, midway between its summit and the water surface, is farther from the axis of rotation of the earth than was that of the mass of water it has displaced. Since points move faster with the earth's rotation the farther they are from this axis, the mass has now been given added velocity. Added velocity means an increase in the centrifugal effect, and one not compensated by gravity, since the amount of mass is the same as before, and therefore the effect of gravity at that point has not been altered. A tangential component of this added centrifugal momentum will tend to move this ice mass toward the equator.

Eötvös applied this same principle to parts of the earth's crust. We have seen that, according to the theory of isostasy, mountains and continents are elevated above the ocean bottoms because they are composed of lighter materials, and they are considered to be "floating" in an approximate gravitational balance with the heavier crustal formations under the oceans. Eötvös considered the centrifugal effects that might arise from the elevations of the centers of gravity of continental formations above those of the oceanic sectors of the crust, and calculated them mathematically. He found that the effects were comparatively slight. Attempts have been made to account for the drift of continents through these effects, but his calculations show they are too small to have considerable effects. Since Eötvös's time, it has been generally assumed that any centrifugal effects that were to be considered in relationship to the earth's crust must be effects resulting from variations in the vertical position of centers of gravity of masses in gravitational balance, that is, elevations of these centers above the equipotential surface, or depressions of them below it, owing to differences in relative density of the masses involved.

Let us now consider, in connection with this, the effect of departures of given masses from the state of isostatic or gravitational equilibrium. We have already seen that there are remarkable departures from isostatic balance, some resulting from deformities of the crust. In these irregularities in the distribution of matter, resulting from the limited failure of isostatic adjustment, we must recognize the existence of another surface of the earth, in contradistinction to the equipotential or geoidal surface already mentioned. We may call this surface the gravitational surface, or the surface of equal mass. This is a real surface. It is not, however, the visible surface. A high plateau may represent an area of deficient mass, and an ocean basin may represent an area of excess mass. We have seen that there are many oceanic areas that show positive isostatic anomalies, or the existence of local excesses of mass in the earth's crust. We can easily see the distinction between the level equipotential surface of the geoid, represented by sea level, and the surface of mass that may deviate considerably from the level surface.

The mechanism for crust displacement presented in this book depends upon recognition of the fact that distortions of mass on the earth's surface, of whatever type, if they constitute anomalous additions of mass at points on the earth's surface, will give rise to centrifugal effects like the effect of the mass attached to the surface of Brown's rotating model sphere, in accordance with ordinary principles of mechanics, and measurable by the standard formula for calculating centrifugal effects.

The difference between an Eötvös effect and one produced by an uncompensated mass may be illustrated in another way. Let us return to our example of a mass of ice. Campbell has suggested the example of an iceberg before and after its separation from its parent, land-based ice cap. It is assumed that the ice cap is uncompensated. The iceberg, breaking off from the ice cap, falls into the water. Before this event the ice cap, by assumption, is outside the equilibrium surface of the geoid; the rotation of the earth acts upon it precisely as the rotation of Brown's model sphere acts upon the weight fixed to its surface.

But let us see what happens when the iceberg falls into the sea. It now reaches gravitational equilibrium. It sinks and displaces its weight in water. It is now a part of the equipotential surface of the geoid (though the portion projecting above sea level is not, and therefore exerts an Eötvös effect).

Now what is the quantitative relationship between the Eötvös effect and the original centrifugal effect of the iceberg? It is plain that now nine tenths of the ice is within the equilibrium surface. For this nine tenths of the mass the equatorward centrifugal momentum produced by the earth's rotation is precisely canceled by the poleward component of the force of gravity at that point, so that there is no net centrifugal effect. Only one tenth of the ice remains to exert an effect, and the quantity of this effect, furthermore, is determined by the elevation of the center of gravity of this tenth of the iceberg above sea level. But the elevation has been enormously reduced. It has, in fact, been reduced to one tenth of the elevation of the center of gravity before the fall of the iceberg into the sea. Campbell has pointed out that, as a result, the centrifugal momentum not compensated by gravity has now been reduced to one one hundredth of the quantity of the effect of the ice mass when it was totally uncompensated.

It appears, therefore, that the question as to whether a mass is in isostatic adjustment or not is the essence of the matter. The ice cap, if totally uncompensated, may produce a centrifugal effect one hundred times the Eötvös effect for the same mass; furthermore it may be calculated by the usual formula with the reservation that a small poleward component of gravity caused by the oblateness of the earth and proportional to the degree of the oblateness must be taken into consideration.

Let us attempt to define and clarify this poleward component of the force of gravity and to estimate its probable relative magnitude. It applies both to masses in equilibrium but with elevated centers of gravity and to any mass resting on the earth's surface but uncompensated. Its effect will be greater in the latter case than in the former. In both cases it will tend to counteract the equatorward component of the centrifugal effect of any anomalous mass.

The poleward component of the force of gravity results from the oblateness of the earth. It may be visualized as follows: If you should place a marble at the equator, and if the rotation of the earth should be interrupted so that the earth would be at rest, then the marble would tend to roll toward one of the poles, because the poles are closer to the center of the earth and therefore downhill. As I have mentioned, this applies both to masses out of isostatic equilibrium and to those in equilibrium but with elevated centers of gravity (that is, to masses standing higher because of their lesser average density). However, as I have pointed out, there will be a quantitative difference between the poleward effects of gravity in these two cases of about 100:1.

In both cases these effects would tend to counterbalance the equatorward component of the total centrifugal effect of the anomalous mass. The question is: What proportion of the equatorward effect would be thus counterbalanced? This is the crux of the matter.

The answer to this problem may be found in the following consideration. The force with which any object rolls downhill is proportionate not only to its weight but to the gradient of the slope. On a flat surface the marble is at rest. It would develop maximum momentum if it could fall straight down toward the earth's center (if the surface were vertical). Between these extremes of zero and maximum momentum there must be an even curve of increasing momentum with increasing gradient. (It would follow, of course, that a sled would develop twice the momentum if going down a hill twice as steep.)

To apply this principle to the ice cap, we may observe that if there were no oblateness to the earth, there would be no poleward component of gravity. If, on the other hand, the oblateness were increased to the point where the ice cap could fall straight down, it would develop maximum momentum, the product of its velocity and of its weight. Between these extremes, the poleward momentum would be proportional to the gradient. We have seen, however, that this gradient amounts to only .17 percent. It follows from this that the poleward component of gravity acting on the ice cap will be .17 percent of the tangential component of the centrifugal effect of the ice cap. This of course is a relatively negligible quantity.

It may be objected that in this discussion we have offered no mathematical calculations in support of the positions taken, and that therefore we have no quantitative basis for our theory. This is, however, a misunderstanding. It is essential, before mathematical computations are made, to understand the assumptions on which they are based. In our correspondence we have more than once received communications in which the authors have indirectly or directly stated that the question as to whether a given mass was or was not isostatically compensated was irrelevant. It has seemed to us, on the other hand, that the actual balanced surface or shape of the earth as determined by the balance of gravity and the centrifugal effect of the rotation—that is, the geoid, or the equipotential surfaces—while perfectly valid as an assumption for many calculations, was irrelevant for our problem. We feel it must be conceded that if the conformity of the earth's materials in general to the balance of the two forces of gravity and rotation, so as to create the oblate shape of the earth—the geoid—is important, the failure of some of the materials to conform to this shape is also important. By definition, a mass that is not isostatically compensated fails to conform to this shape. Thus the real surface of mass differs from the geoid and cannot be called an equipotential surface. We feel that the real "surface of mass" of the earth cannot be disregarded.

In this situation equations are of no use. They will not help us attain clarity. What is needed instead is a re-examination of the assumptions on which equations have been made. This is an intellectual problem of the logical development of ideas and corresponds to the process advocated by Maxwell as superior, in some situations, to calculations. Discussing the intricacies of the mechanics of rotation before the Royal Society, Maxwell remarked:

. . . If any further progress is to be made in simplifying and arranging the theory, it must be by the method that Poinsot has repeatedly pointed out as the only one that can lead to a true knowledge of the subject—that of proceeding from one distinct idea to another, instead of trusting to symbols and equations (296:248ff).

Let us remember that the author of this remark was one of the greatest mathematical physicists of all time. As such, he understood the limitations of mathematics, of which the most essential is that all calculations must be based in the last analysis on assumptions that consist of clear ideas, logically expressible in words.

Note 5.

A DISCUSSION OF ISOSTASY: VIEWS OF DALY, VENING MEINESZ, JEFFREYS, AND OTHERS

1. DISTURBANCES OF ISOSTASY

One of the troubles with the theory of isostasy is that the failures of the crust to adapt to gravitational balance have been found to be more numerous and more serious than expected. Daly lists and discusses a large number of them. It appears, for example, that the whole chain of the Hawaiian Islands, with their undersea connecting masses of heavy basalt, are uncompensated (97:303). These islands rise from the deep floor of the Pacific, and their peaks tower two and a half miles above sea level. Their gigantic weight rests upon the crust, and under the weight the crust has bent down slightly, but it has not given way. This is the more remarkable since the islands appear to be several million years old. It indicates that at this point the earth's crust is strong enough to bear a very considerable weight without yielding. The Great Rift Valley of Africa, which we have already discussed, is uncompensated, despite its great age (97:221). There are also enormous anomalies in the East Indies. According to Umbgrove, Vening Meinesz found that the negative anomalies (that is, the deficiency of matter) in the great ocean deeps in that area and the positive anomalies on each side caused a total gravity deviation of 400 milligals. One milligal, according to Daly, would amount to about 10 meters of granite (97:394), so the total deflection of the crust from gravitational balance here would amount to 4,000 meters of granite, or roughly three miles of granite, which in turn would be the equivalent of an ice sheet about nine miles thick. And the crust has borne this enormous strain, apparently, for some millions of years. According to Daly, the Nero Deep, near the island of Guam, has deviations from gravitational balance of the same magnitude (97:291). Among uncompensated features on the lands are the Harz Mountains, in Germany (97:349), and the Himalayas, which stand about 864 feet higher than they should (97:235). A particularly interesting case is that of the island of Cyprus, of considerable size, which stands about one kilometer, or

3,000 feet, higher than it should, and yet shows no signs of subsiding. Daly says:

From Mace's table of anomalies and from his map, it appears that we have here a sector of the earth, measuring more than 225 kilometers in length and 100 kilometers in width, and bearing an uncompensated load equal to one kilometer of granite, spread evenly over the sector. . . . (97:212–13).

These facts would appear to argue a very considerable strength of the crust to resist the pressure toward establishment of gravitational, or isostatic, balance. However, in all the cases so far mentioned it is true that the deviations have occurred in comparatively narrow areas. The Hawaiian Islands, for example, represent a long, narrow segment of the crust. Obviously the crust can support loads with small span more easily than loads with a very great span. These deviations, therefore, may not tell us much about the gravitational status of the Antarctic ice cap, which, of course, has an enormous span, since it covers a whole continent. Since they are insignificant quantitatively as compared with the possible effect of the continental ice cap of Antarctica, they will not, of themselves, answer the question raised by Einstein in the last paragraph of his foreword.

Of more importance are isostatic anomalies of broad span, and these are, surprisingly, quite plentiful. Daly mentions one along the Pacific coast. This is a negative anomaly—a deficiency of mass. Daly explains that according to one formula (the "International Formula"), it covers an area 2,100 miles long, and 360 to 660 miles wide; according to another formula (the "Heiskanen"), it is reduced to one half both in intensity and in extent (97:371). Taking the lesser estimate, the deficiency of mass over this large area still amounts to the equivalent of a continuous ice sheet 1,000 to 1,200 feet thick. So it appears that over this large span the crust can bear that amount of negative weight (that is, of pressure from within the earth) without giving way, at least for a short period of time. In other parts of the United States there are positive anomalies of the same magnitude, and these obtain over large areas.

A far more extraordinary case is an enormous area of negative mass that covers part of India and most of the adjacent Arabian Sea. The width of the negative area in India is 780 miles. Daly, after noting the challenge presented by this fact to the whole theory of isostasy, goes on to say:

The situation becomes even more thought provoking when we remember that Vening Meinesz found negative Hayford anomalies all across the Arabian Sea, 2500 kilometers in width. Apparently, therefore, negative anomalies here dominate over a total area much greater than, for example, the huge glaciated tract of Fenno-Scandia [Finland and Scandinavia]. And yet there is no evidence that the lithosphere under India and the Arabian Sea is being upwarped. The fact that Fenno-Scandia, though less (negatively) loaded than the Arabian Sea-India region, is being upwarped, as if by isostatic adjustment, emphasizes the need to examine the Asiatic field with particular care. . . . (97:365).

Let us remember that a negative load means simply pressure from within the earth outward, and positive load pressure from the surface inward. In principle, they are the same insofar as their evidence for the strength of the crust goes. It seems that here the crust is quite able to bear a large load over a great span without yielding. Daly points out that many parts of India are distorted on the positive side; there is an excess of matter over considerable areas, and he remarks:

. . . India, among all the extensive regions with relatively close networks of plumb-bob and gravity stations, is being regarded by some high authorities as departing so far from isostasy that one should no longer recognize a principle of isostasy at all. . . . (97:224–25).

A particularly important aspect of these great deviations from gravitational balance of the crust in India is that they are not local distortions, not the result of local surface features such as hills and valleys. These surface features may well once, and quite recently, have been in good isostatic balance. The distortion lies deeper:

. . . In India practically all the gravity anomalies seem to have no apparent relation to local conditions. Only one explanation seems possible—that is, that they are due to a very deep seated gentle undulation of the lower crustal layers underlying all the superficial rocks; it is evidently a very uniform, broad sweeping feature at a great depth, and must be uncompensated, since if it were compensated it would cause no anomaly at the surface (97:241–42).

Forced to find some way of explaining how the crust could bear such loads (positive and negative) in India and still yield easily to isostatic adjustment in other areas, Daly suggests that the strength of the crust in India might be explained by a recent lateral compression of the whole peninsula, which, he says, is evidenced by the folding there of the young sedimentary rocks (97:391–92).

Daly does not suggest a possible cause for this lateral compression of the whole peninsula; such a compression, part of the process of mountain building, he has already characterized as "utterly mysterious." But it must be clear that it is precisely the type of distortion that might be expected to result from a displacement of the earth's crust. Such a movement could very well account both for the depression of lower India and for the uncompensated elevation of the Himalayas. It can be said, moreover, that no displacement of the crust could possibly take place without creating, at some points on the earth, precisely such deep-lying gentle undulations of the crust.

But still another point may be urged in support of this solution of the problem. We have shown that the last movement of the crust appears to have been approximately along the 83rd meridian, with North America moving southward from the pole. This movement would have subjected India to maximum displacement and to maximum compression. In this last movement India would have been moved across the equator and northward toward the pole, to its present latitude.

Daly's suggestion that compression may increase the tensile strength of the crust opens up most interesting possibilities. We may find here, in connection with the theory of crust displacement, a solution to very puzzling problems of isostatic theory. The crust of the earth shows enormous differences from place to place in its degree of isostatic adjustment and in its sensitivity to the addition or removal of loads. Applying Daly's suggestion, we may infer that the differences may owe their origin to recent displacements of the crust. Areas recently moved poleward, having undergone compression and still retaining compression, would, according to Daly's suggestion, have greater strength to sustain the distortions; areas recently displaced equatorward, having undergone extension, or stretching, would have less strength to resist gravitational adjustment, and, moreover, the widespread fracturing accompanying the movement would facilitate adjustment.

This suggestion of Daly's also has great significance for the understanding of the absence of much volcanism in the polar regions. It has been observed that these regions are relatively quiet with respect to volcanoes. There is only one volcano in the whole continent of Antarctica, so far as we know. What can be the reason for this? It may be thought that this may result from the polar cold, but this cannot be true. The influence of surface temperatures penetrates only a short distance into the crust; volcanoes originate from greater depths. The solution may be found in the fact that, according to our theory, both the present polar areas are areas that were moved poleward in the last movement of the crust and were therefore compressed.

The importance of finding a reasonable solution for the profound contradictions in the theory of isostasy has been emphasized by several recent writers. Professor Bain, of Amherst, writes:

Isostatic adjustment exists only in imagination. I present the existence of peneplains in witness thereof. Establishment of the Rocky Mountain peneplain or the Old Flat Top Peneplain of the western states requires erosion of at least 10,000 feet of the rock over the main arch of the Front Range. The rivers wore the land down slowly to grade equilibrium without observable rise of the unloaded region or subsidence of the loaded region throwing all gravity out of equilibrium. Then in the brief interval of a small part of a geological epoch the land surface rose to re-establish near gravity equilibrium. . . . (19).

Now, as I understand Bain's statement, his point is that in numerous instances erosion has worn away mountain ranges, leaving flat plains (peneplains), and in the instance he cites it seems that during the prolonged period when the erosion was taking place (erosion that resulted in removal of no less than 10,000 feet of rock from one area and the deposition of the resulting sediments in another), the crust did not respond by rising in the first area and sinking in the second. Gravitational balance was thus sadly set askew and remained so for a long time. Then, relatively suddenly, equilibrium was re-established. How do we explain this?

I think it is necessary to take into consideration the fact that just as compression will be at a maximum along the meridian of displacement of the crust in the poleward direction, extension or stretching will likewise be at a maximum along the same meridian in the equatorward direction. But, in both cases, areas removed from this meridian will be displaced proportionately less, and large areas will undergo very little or no displacement and consequently very little or no compression or extension. Since, as we saw in earlier chapters, successive movements of the crust may oscillate along meridians placed close together, it follows that, for long periods, compression may be sustained in particular areas and isostatic adjustment impeded in those areas. Eventually a movement of the crust in a different direction will permit the delayed adjustment to take place.

In this way, too, we may explain the data upon which Jeffreys based his conclusion that isostasy is an exceptional condition of the earth's surface, which is re-established only at long intervals. The theory presented in this book offers a solution for the cause of the geological revolutions which, he supposed, shattered the crust at long intervals, bringing about the formation of mountains and permitting the re-establishment of crustal balance.

With regard to the vast negative or positive distortions of isostasy, the displacement theory has a solution to offer. Let us suppose a movement of the crust causing widespread slight distortion of the earth from its equilibrium shape, distortion such as now prevails across parts of India and all of the Arabian Sea. It is essential to realize that the long persistence of such anomalies, and the apparent lack of any tendency to adjustment, may have no relationship to the strength of the crust. It may be due, quite simply, to the fact that the matter in the sublayer (the asthenosphere) is too viscous to flow rapidly, and that when it has to flow such great distances, and in such great volume as would be required to compensate the sweeping undulations of the geoid caused by a movement of the crust, great periods of time are required, periods so long that our instruments have not been able to detect the progress of isostatic adjustment.

The advantage of the theory of crust displacements is that it can reconcile the data supporting the conviction of geologists that the crust must be too weak to support major loads out of adjustment over great spans of territory, with the observed fact that in some cases it appears to do so. Furthermore we may, with this theory, grant the crust enough strength under certain conditions (of compression) to support heavy loads of narrow span, such as the Hawaiian Islands, and still understand its extreme weakness in areas of extension, where it appears to adjust easily to rather minor loads.

Einstein, in the foreword, referred to the possible centrifugal effects of these distortions within the crust. The following principles apply:

a. A positive load on the crust, like the ice cap, will exert a centrifugal

effect equatorward; correspondingly the effects of negative loads must
be poleward.

b. The effects of positive loads on one side of the equator will be opposed
to the effects of positive loads on the other side of the equator; equal
positive loads in equal longitudes and latitudes will cancel each other
across the equator, and the same is true of negative loads.

c. Despite the fact that such loads may cancel each other wholly or in
part insofar as the transmission of a net centrifugal momentum to
the crust in any given direction is concerned, nevertheless their op-
position will involve the creation of persisting stresses in the crust,
and these may be a cause of seismic activity.

d. Crustal distortions, unlike ice caps, are comparatively permanent features; many may persist through one or more displacements; their
effect will change quantitatively according to their changes of latitude
and longitude.

e. At the termination of each crustal movement, the distortions of the
rock structures of the crust should be approximately balanced across
the equator. In a period of several thousand years following such a
movement, however, the process of isostatic adjustment, proceeding
faster in some areas than in others, may disturb this balance and pre-
dispose the crust to a new displacement.

2. THE TRIAXIAL SHAPE OF THE EARTH

We cannot leave the subject of the gravitational adjustment of the earth's
surface without mentioning the greatest distortion of all, the triaxial
deformation of the earth. It is all the more important to consider this
question since here we shall see, at one and the same time, a solution
for one of the greatest of geological conundrums, and one of the most
powerful arguments in support of the theory of displacements of the
earth's crust.

Not long ago, scientists became aware of the fact that there is a deviation in the shape of the earth from the idealized form of a flattened,
or oblate, spheroid. The increasingly accurate measurements of geodesy
have shown that the earth has bumps and irregular lumps in various
places, which seem to correspond to a third axis running through the
earth. As a result of this, scientists now consider that the true shape of
the earth is that of a "triaxial ellipsoid."

An axis, of course, is not a material thing. It is only a line that somebody imagines running through a sphere to give a dimension to that
sphere in that direction. Three axes of the earth mean one through the
poles, on which the earth rotates (the axis of rotation); one through the
equator, called the equatorial axis, twenty-six miles longer than the polar
axis; and now a third axis, roughly through the equator, at an angle to
the other equatorial axis.

The result of having two axes of different lengths running through the equator is, of course, that the equator itself is a little flattened; it is oval rather than truly circular. The flattening is very slight. According to Daly, one axis through the equator is 2,300 feet longer than the other (97:32); Jeffreys, according to Daly, prefers half that figure. Daly finds that the longer diameter through the equator (the major axis) runs from the Atlantic Ocean, at 25° W. Long., to the Pacific, at 155° E. Long., and the shorter diameter (or minor axis) runs from the western United States, at 115° W. Long., to the Indian Ocean, at 65° E. Long. (97:32). Just as the actual amount of the flattening of the equator is uncertain, so are the precise situations of the major and minor equatorial axes. More recently, determinations by the United States Coast Geodetic Survey have suggested a slightly different position for one of these axes. Moreover the third axis apparently does not run precisely through the equator. The result is that the earth's shape is distorted by protuberances of various sizes and shapes. If we take Jeffrey's estimate of their magnitude, we see that they amount to the equivalent of about 2,000 feet of rock, or over a mile of ice, and of course the anomalies have enormous spans, on the order of thousands of miles.

Despite their vastly greater magnitude, these triaxial protuberances have one thing in common with those in India. Just as Daly observed that the Indian anomalies must result from sweeping undulations of the geoid at some depth in the crust, underlying all the surface features, so do the triaxial protuberances indicate distortion in depth rather than at the surface. In India the surface features would be in fairly good isostatic adjustment if the deep-seated undulations were disregarded, while the geodesist Heiskanen, according to Daly, found that if he disregarded the triaxial protuberances—if he regarded the triaxial ellipsoid as the natural shape of the earth—all his anomalies were reduced to one half, both in extent and in intensity (97:368).

It does not seem reasonable simply to disregard distortions of the shape of the earth of this magnitude unless we have an explanation of them that is convincing. Daly provided an explanation, but for a number of reasons it seems to me unsatisfactory.

It was plain to him that the strength of the crust could not possibly support such enormous distortions over such spans. Therefore he made one or two alternative suggestions, advancing them as possibilities only. He suggested, first, that assuming an original molten condition of the earth, it is possible that the material in the liquid melt was not of uniform density on opposite sides of the earth, and that therefore when the mesosphere (the inner solid shell underlying the asthenosphere) solidified, it was heavier on one side than on the other—that is, lopsided—and the resulting unevenness of gravity at the surface influenced the equilibrium; that is, the elevation from place to place of the surface layers. This is an ingenious suggestion, but it requires the assumption of the cooling of the earth, which is itself doubtful. Thus this particular explanation rests upon speculation, and upon speculation that is not well supported.

The same is true of Daly's second suggestion. He supposes that the lopsidedness of the internal shell may have resulted from the separation of the moon from the earth, at which time the bed of the Pacific may have been created. The arguments that once supported this theory of the origin of the moon have, in recent years, been gradually whittled away until little remains of them. This, then, is also a hazardous speculation.

Daly's fertile mind has produced a third suggestion. He feels that perhaps the triaxiality may have resulted from the effects of continental drift, which he felt compelled to support because there was no other way to explain the innumerable facts of paleontology and geology, many of which have been already cited in this book. We have seen, however, that continental drift will not do.

It seems that all the arguments that Daly uses to support his suggestion that the triaxial protuberances are not supported by the crust, but from below the crust, fail to stand examination. They are supported by no convincing mass of evidence. There is obviously a sort of desperate urgency about them. A strong need impels him to hoist them up. The nature of this need is perfectly clear.

The need is to save the theory of isostasy. It is to smooth the path in front of a theory that has many useful applications and has a great deal to be said for it. The theory is threatened by the unexplained anomalies referred to above; it is still more threatened by these massive distortions of the shape of the planet, the triaxial protuberances. They are wholly and absolutely irreconcilable with the known principles of physics, as opposed to speculations. Either the shape of the earth is established by the balance of the force of gravity and the centrifugal effect of the rotation, or it is not. The geoid, so established, is distorted, and it proves impossible to explain the distortion either by the resistance of the crust to the aforementioned forces or by the (undemonstrated) lopsidedness of the internal shell.

But displacements of the earth's crust may explain the matter, and in the simplest possible fashion.

We have seen that areas displaced poleward in a movement of the crust will be elevated relative to sea level. Two areas will be displaced poleward at the same time, one to each pole, and both will be elevated somewhat with reference to the equilibrium surface. The distance through the earth between these points will be increased slightly. At the same time, two other areas will be displaced equatorward. They will subside, and the diameter through the earth between them will be shortened to some extent. These areas will be centered on the meridian of the movement of the crust. At 90 degrees' remove on each side from this meridian there will be no movement; here are the so-called "pivot areas" that do not change their latitude. They will therefore not change their elevation: A diameter through the earth between them will be unchanged.

As the consequence of this, we see that in one direction the diameter of the earth through the equator is shortened; in the other direction through the equator it is not. The result must inevitably be the ellipticity, or ovalarity, of the equator.

The consequences of the displacement do not end here. As we have stated from time to time, much complicated folding and faulting of the crust, much shifting of matter below the crust, would be inevitable or likely, and these would have effects at the surface, including basining and doming. Hence some of the protuberances now being discovered may have nothing to do with the triaxial distortions and may simply be confused with them.

Now, it is clear that this explanation of the triaxiality requires neither complicated and hazardous speculations about the earth's interior nor an incredible strength in the earth's crust. The protuberances remain because the matter below the crust is too highly viscous to flow the great distances and in the great volume that would be required to re-establish the normal shape of the earth; that is, it is too viscous to have been able to do so in the very short period that has elapsed since the last movement of the crust. But no doubt the readjustment is proceeding slowly; no doubt the triaxial bumps are now the reduced remnants of those that existed at the end of the last movement of the crust.

If all anomalies in the crust cause centrifugal effects, then these vast triaxial protuberances must do so. These must, as I have pointed out, be balanced across the equator or have been so at the termination of the last movement. Since then isostatic adjustment has probably been proceeding, and therefore the balance of forces established when the crust stopped moving may now no longer exist. The instability of the crust may have been thereby increased.

Note 6.

CALCULATION OF THE STABILIZING CENTRIFUGAL EFFECT OF THE EQUATORIAL BULGE OF THE EARTH

Department of Commerce
U. S. Coast and Geodetic Survey
Washington 25

June 28, 1950

Mr. Charles H. Hapgood
Springfield College
Springfield 9, Massachusetts

Dear Sir:

This is in reply to your letter of May 26, 1950.

If it is desired to compute the centrifugal force developed by rotating a shell of variable thickness but constant density about an axis of symmetry, the fundamental definition of centrifugal force can be used for each particle, but an integration must be performed to sum the contributions of each particle. For the details of this computation see the attached sheet.

I mention here an error which occurs in the righthand member of the formula you give for F. A factor R, the mean radius, has been omitted.

The centrifugal force due to the equatorial bulge would seem to have little bearing on the stability of the earth. The centrifugal force at its greatest is only 1/289 part of the gravitational attraction, and so even if the crust had no strength, there would be no disruption. The oblate shape of the earth is a result of the centrifugal force due to the rotation, the resultant mean sea level surface of the earth being an equilibrium surface under the resultant of gravitational and centrifugal forces.*

---

* Doctor Gallen seems to me in error here. He is treating the equatorial bulge as if it were a liquid, whereas it is composed entirely of the rigid crystalline rocks of the lithosphere. Regardless of how this bulge was produced, the angular velocity of its particles has been increased because of their greater distance from the axis of rotation. The principles of rigid dynamics of rotating bodies should apply here, as given by Clerk Maxwell (see p. 344).

Although the centrifugal force itself has no bearing on the stability of the earth, the bulge which it has produced, when subjected to the varying gravitational attractions of the Sun and Moon, gives rise to the precessional and nutational motions of the earth's axis in space. The disturbing force due to the equatorial bulge of the earth is so small a part of the total momentum of rotation that it produces a precessional motion of only 50″ per year. Redistribution of matter within or on the earth would alter the precession, but an enormous redistribution would be required to produce an appreciable change. In this connection, you may wish to consult George H. Darwin, "On the Influence of Geological Changes on the Earth's Axis of Rotation", Philos. Trans. of the Royal Society, Vol. 167 pt. 1.

Very truly yours,

(Signed)   F. L. Gallen
Acting Director

Attachment

June 28, 1950

Computation of Centrifugal Force

Let the equations of the sphere and the ellipsoid of revolution be

(1) $x^2 + y^2 + z^2 = b^2$, and

(2) $\dfrac{x^2}{a^2} + \dfrac{y^2}{b^2} + \dfrac{z^2}{a^2} = 1,$

where the axis of y is the axis of revolution. Take as the element of mass, dM, the ring generated by revolving the rectangle dxdy about the axis of y. We have

(3) $dM = 2\pi\delta x\, dxdy,$

where δ is the density. For each particle of the ring the centrifugal acceleration is the same, being equal to $\omega^2 x$, where ω is the constant angular velocity in radians per second.

The element of centrifugal force, dF, exerted by the ring is then

(4) $dF = \omega^2 x\, dM = 2\pi\delta\omega^2 x^2 dxdy.$

Integrating equation (4) with respect to x and y, there results

(5) $F = 2\pi\delta\omega^2 \displaystyle\int_{-b}^{b} \int_{\frac{a}{b}\sqrt{b^2-y^2}}^{\frac{a}{b}\sqrt{b^2-y^2}} x^2 dxdy = \dfrac{\pi^2\delta\omega^2}{4} b(a^3-b^3).$

In equation (5) F is expressed in dynes when δ is given in grams per cubic centimeter, and a and b in centimeters. The quantity ω may be replaced by $2\pi n$ where n is revolutions per second. The earth makes one complete revolution in 86,164.09 mean solar seconds.

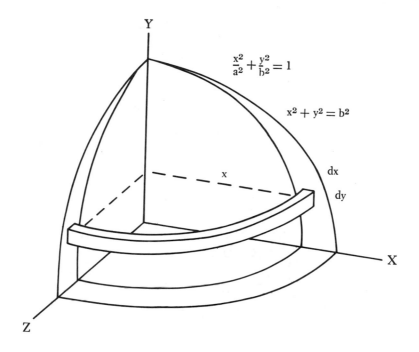

$$\frac{x^2}{a^2} + \frac{y^2}{b^2} = 1$$

$$x^2 + y^2 = b^2$$

Mrs. Deininger's computation based on Gallen's calculus

I. Computation of centrifugal force produced by rotation of the bulge.
   A. Essential Data:
      1. The attached formula should apply to the bulge taken as 13.3443 miles at the equator, not to the bulge as it would be if there were no flattening at the poles.
      2. In making the calculation, I asked Mrs. Harriet Deininger, of the Smith College faculty, to subtract three miles from the depth of the bulge, because we are concerned with a purely mechanical action of stabilization, in which water could have no effect. (We know that we subtracted about three times too much, because we disregarded isostasy, which in this case makes it probable that the rock under the oceans has a density higher than the density of the rock of the continents; so we should have subtracted the weight rather than the volume of the water. This however, is a minor correction.)
      3. Mrs. Deininger actually took the depth of the bulge as nine miles, without the water.

B. *Calculation*

(1) $F = \frac{\pi^2 s w^2}{4} b(a^3 - b^3)$ where s = density in gm/cc
   a = radius of earth at bulge in cm.
   b = radius of earth at poles in cm.
   w = 2 − n   r = rps

(2) $F = \pi^4 sn^2 b(a^3-b^3)$ where $\pi = 3.1416$

$\qquad\qquad\qquad\qquad\qquad s = 2.7$ gm/cc

$$n = \frac{1}{86,164}$$

$\qquad\qquad\qquad\qquad\quad b = 640,200,000$ cm

$\qquad\qquad\qquad\qquad\quad a = 641,650,000$ cm   using nine miles or
$\qquad\qquad\qquad\qquad\qquad\qquad\qquad\qquad\qquad\quad$ 1,450,000 cm as
$\qquad\qquad\qquad\qquad\qquad\qquad\qquad\qquad\qquad\quad$ depth of bulge

(3) $F = (3.1416)^4 \; (2.7) \dfrac{(640,200,000) \, [(641,650,000)^3-(640,200,000)^3]}{(86,164)^2}$

(4) $F = 4.0368$ times 10^{25} dynes, or 40,368,000,000,000,000,000,000,000 dynes

$\qquad$ or $\dfrac{40,368,000,000,000,000,000,000,000 \text{ Kg}}{980}$

$\qquad$ or 4.1192 times 10^{19} Kg

$\qquad$ or 4.1192 times 10^{16} metric tons

$\qquad$ or 41,192,000,000,000,000 metric tons

Note 7.

SOME EVIDENCES OF TECTONIC ADJUSTMENTS AT THE END OF THE ICE AGE

DARWIN'S RISING BEACHLINE ON THE WEST COAST OF SOUTH AMERICA

A singularly impressive piece of evidence for a recent displacement of the crust may be found in the journal of Charles Darwin. Sir Archibald Geikie summarized Darwin's findings thus:

On the west coast of South America, lines of raised terraces containing recent shells have been traced by Darwin as proofs of a great upheaval of that part of the globe in modern geological time. The terraces are not quite horizontal but rise to the south. On the frontier of Bolivia they occur from 60 to 80 feet above the existing sea-level, but nearer the higher mass of the Chilean Andes they are found at one thousand, and near Valparaiso at 1300 feet. That some of these ancient sea margins belong to the human period was shown by Mr. Darwin's discovery of shells with bones of birds, ears of maize, plaited reeds and cotton thread, in some of the terraces opposite Callao at a height of 85 feet. Raised beaches occur in New Zealand and indicate a greater change of level in the southern than in the northern end of the country. . . . (170:288).

If we attempt, by analyzing this evidence in accordance with the assumptions of the displacement theory, to reconstruct the course of events, we reach the following conclusions: Since the evidence of human occupation is found at an elevation of 85 feet, it seems reasonable to suppose that a fall of the sea level of that extent may have occurred within historical times. On the other hand, the continuously rising strandline down the coast to Valparaiso, continued in New Zealand, indicates a tilting of the earth's crust, involving South America and New Zealand, but not involving a general change in the sea level. The magnitude of the upheaval suggests that it may have occurred earlier than the 85-foot general fall in sea level, and may have required much more time. The 85-foot fall in the general sea level we may explain as the result of the withdrawal of water to Antarctica. The uptilting of the continent may be seen as the result of its poleward displacement.

The effect postulated by Gutenberg (p. 228) to account for uplift of areas displaced poleward cannot account for the tilting, but another effect may. This is the increasing compression of the poleward-moving sector as the result of the progressive shortening of the radius and circumference of the earth in the higher latitudes. The compressions resulting from this have been discussed. They result inevitably from the increasing arc of the surface and the increasing convergence of the meridians.

GLOSSARY

ANOMALY, Positive: An excess of mass at a point on the crust, as compared with the average distribution of mass.

    Negative: A deficiency of mass, similarly.

ANTICLINE: An archlike folding of rocks or rock strata so that the lower beds or strata are enclosed in the upper.

ASTHENOSPHERE: A layer of materials in the earth's interior extending from the bottom of the lithosphere to a depth of possibly several hundred miles; it is thought to be weak because its rocks are too hot to crystallize.

BASEMENT ROCKS: Rocks of great obscurity and complexity lying beneath the upper rock layers; thus, the lowermost rocks of the known series.

CENTRIFUGAL FORCE (or EFFECT): The force or effect tending to throw a body away from the center of a rotating body, in a straight-line direction of flight.

CENTRIPETAL FORCE (or EFFECT): The force or effect tending to throw a body toward the center of the same.

CLIMATIC OPTIMUM: See "Hipsithermal."

CONTINENTAL BLOCK: The sum of the rocks comprising a continent.

CONTINENTAL SHELF: The margin of the continental block that is submerged a few hundred feet, extending out varying distances from the coast.

CONVECTION CURRENT: The transmission of heat by the mass movement of heated particles of water or of liquid (or viscous) rock.

CRUSHING STRENGTH: The limit of the ability of any solid to resist crushing under the pressure of equal and opposite forces.

CRUST (OF THE EARTH): Originally applied to the entire thickness of the lithosphere (ca. 30–40 miles); now usually applied to the rock strata lying above the Moho (q.v.) discontinuity.

CRYSTALLINE ROCK: Rock in the solid state, in which it is composed of crystals of various minerals.

DIASTROPHISM: The process or processes by which major features of the earth's crust are formed through rock movements and displacements.

ECOLOGICAL: Pertaining to the mutual relationship between organisms and their environment.

EPEIROGENESIS: A grander form of diastrophism, forming the broader features of crustal relief, such as continents and ocean beds.

EPICONTINENTAL: Pertaining to regions along the continental shelf.

EQUIPOTENTIAL SURFACE: An approximate or imaginary surface of the earth, sometimes coincident with the real surface, at which the poleward and equatorward components of the centripetal and centrifugal effects of the earth's rotation are in balance with the force of gravitational attraction; e.g., the surface of the ocean.

ERRATIC BLOCKS: Large rocks detached from their original formations and carried varying distances by glacial ice.

EUSTATIC: Pertaining to a land mass that has not undergone elevation or depression.

GEANTICLINE: Where the rock strata have been arched by compressive forces in the lithosphere, a part of the process of mountain building.

GEOID: The figure of the earth (an oblate sphere) with the average sea level conceived of as extending through the continents.

GEOMAGNETISM: The force of the magnetic field of the earth. Its lines of force may be imprinted on particles of ferrous minerals, which may show the direction of the earth's field at the time the rocks were formed.

GEOSYNCLINE: A great downward flexing of the crust, a part of the process of mountain building.

GRAVITATIONAL BALANCE: A condition in which any feature of the earth's surface stands at the elevation relative to the geoid that is appropriate to its mass.

GRAVITY, Center of: An imaginary point at which, for reasons of computation, the entire weight or mass of a body is imagined to be cencentrated.

HEAT GRADIENT: The rate at which the temperature of the earth increases with depth.

HIPSITHERMAL: Pertaining to a short period after the last glaciation when the average temperature of the atmosphere was slightly higher than at present.

HORSE LATITUDES: A belt in the neighborhood of 30° N. or S. Lat., characterized by high pressure, calms, and baffling winds.

HYDROSTATIC BALANCE: A condition in which any feature of the earth's surface stands at the elevation appropriate to its density, as if floating in a liquid medium.

IGNEOUS ROCKS: Rocks that have been cooled and solidified from a molten state.

INSOLATION CURVE: A graphic representation of changes in worldwide temperatures assumed to result from the combination of the effects of various astronomical factors, such as precession and variation of the shape of the orbit of the earth about the sun.

INTERFACE: The boundaries of two strata, or of two different states or movements of materials within the earth, such as the boundary between the lithosphere and the asthenosphere.

INTERSTADIAL: A short period of relative warmth and ice recession during a glacial period.

IONIUM DATING: A method of "absolute" dating of geological deposits of the deep sea. It depends on the differential rates of decay of three radioactive elements that are found in seawater: uranium, radium, and ionium. When enclosed in bottom sediments the proportions of the elements gradually change with time so that their final ratios can be used to determine the age of the sediments.

ISOSTASY: The condition of gravitational or hydrostatic balance (q.v.) of the solid surface of the earth, whether continental or ocean bottom.

ISOTOPE: One of two or more forms of an element having the same or closely related properties and the same atomic number but different atomic weights.

LIFE NICHE: An ecological living space for a particular organism; the area where it finds the conditions necessary for its existence.

LITHOSPHERE: The outer shell of the earth, composed of hard crystalline rock, extending down to the point where the effects of heat and pressure destroy the crystalline structure, thought to occur at a depth of 30 to 40 miles, or about 60 kilometers.

MAGMA: Molten rock material within the earth.

MANTLE: The rock materials extending from the bottom of the lithosphere or from the bottom of the Moho discontinuity (as variously supposed) to the earth's core.

MESOSPHERE: An older term equivalent to the "mantle."

METAMORPHIC ROCKS: Rocks that have been physically changed by heat or other means since their formation.

MID-ATLANTIC RIDGE: A submarine system of mountains in the Atlantic Ocean recently found to extend to all the oceans.

MILLIGAL: A mass equal to a thickness of ten meters of granite; the gravitational effects produced by such a mass.

MOHO: A discontinuity in the earth strata at a shallow depth under the oceans and at a slightly greater depth under the continents, named after its discoverer, Dr. Andrija Mohorovicic.

MORAINE: A mass of rocks, gravel, sand, etc., carried or deposited by a glacier either along its side (*lateral moraine*) or at its lower end (*terminal moraine*).

MUTATION: A sudden variation in the characteristics of a life form as compared with those of its predecessors.

NEBULAR THEORY: A theory of the origin of the solar system, according to which a gaseous nebula coalesced and cooled to form compacted centers which then further contracted to form the planets.

NUTATION: A slight vibratory movement of the earth's axis.

OOZE: A soft deep-sea deposit composed of shells, debris, meteoric dust, etc. *Argillaceous* ooze is a clayey type.

OROGENESIS: The process of building mountain systems.

PLANETESIMAL: A small, solid planetary body having an individual orbit about the sun.

PLANETESIMAL HYPOTHESIS: A theory of the origin of the solar system supposing that the planets were formed by the collision and coalescence of planetesimals and thus have never been wholly molten.

PLASTICITY: The quality of solid matter that makes it yield to applied pressure and makes it capable of continuous and permanent change of shape in any direction without breaking apart. The *plastic limit* of a solid is that point of increasing pressure at which its strength fails and fracture results.

PLICATION: The folding of rock layers or strata.

POLLEN DIAGRAM: A graph showing changing types of pollen with varying depth in a sedimentary deposit and relating them to climatic changes.

PRECESSION: The change in the direction of the earth's axis as it turns around the axis of the ecliptic so as to describe a complete circle approximately every 20,000 years. It is the result of the action of the sun and the moon upon the earth's equatorial bulge.

RADIATION, Adaptive: The production of a diversified fauna as the result of the availability of new ecological spaces, or life niches.

RADIOCARBON DATING: A technique of dating materials containing organic carbon derived from living things. It is based on the occurrence of an istope of carbon, with an atomic weight of 14, in very small quantities in the atmosphere.

SEA CORES: Sediment from the ocean bottom obtained by lowering coring tubes from ships.

SEAMOUNTS: Individual submerged mountains on the ocean floor.

SEDIMENTATION: The accumulation of debris of all sorts either on land or in the sea, as a consequence of the operation of natural forces such as wind, rain, and ocean currents.

SELECTION PRESSURE: A factor in the evolution of species. High selection pressure would mean conditions forcing species to change rapidly in order to survive.

SHEAR ZONE: The depth in the earth or in a glacier at which a particular stratum may move across another by a shearing action.

SIAL: Silicon-aluminum rocks.

SIMA: Silicon-magnesium rocks.

STRANDLINE: A beach marking a present or past stand of the sea or of a lake.

TANGENTIAL COMPONENT: That portion of the centrifugal effect of a mass deposited on the surface of a rotating sphere which acts not at right angles to the axis of the sphere but tangentially to the surface.

TECTONIC: Pertaining to rock structures resulting from deformation of the crust.

TENSILE STRENGTH: Resistance to lengthwise stress, measured by the greatest load in weight per unit area, pulling in the direction of length, that a given substance can bear without tearing apart.

TRIAXIAL DEVIATION OR ANOMALY: Deviation of the shape of the geoid because of the "third axis" of the earth.

TURBIDITY CURRENTS: Currents in the deep sea often caused by the slumping of materials from the continental shelves, thought to be formative factors of the features of the ocean bottom.

UNIFORMITARIANISM: Accepted geological theory according to which the geological history of the earth can be understood as being the cumulative result of the operation of the geological forces and factors seen to be operative today.

VARVE: A layer in a deposit of sedimentary material, showing seasonal variation caused by differences in summer and winter deposition; characteristic of certain deposits in glaciated regions and sometimes used to estimate the length of the glacial and interglacial periods.

VIRENZPERIOD: A period in the history of a life form in which it experiences an explosive evolution and proliferation.

VISCOSITY: The quality of being able to yield to stress or of being able to flow; the measure of such a property.

WAVE-GUIDE LAYER: A layer within the earth at a depth of about 100 miles, or 150 kilometers, in which a Soviet geophysicist theorizes that lighter materials, produced through chemical reactions, cause a deflection of seismic waves; hence the name given to the layer.

WEGENER THEORY: The theory of drifting continents as first formulated by Alfred Wegener.

BIBLIOGRAPHY

The following alphabetical list serves both as bibliography and as identification for specific sources cited in the text.

1a. Almond, M., Clegg, J. A., and Jaeger, J. C., "Rock Magnetism." *Philosophical Magazine*, Ser. 8, v. 1, pp. 771ff (1956).

1b. Andel, T. H. van and Laborel, J., "Recent High Relative Sea Level Stand Near Recife, Brazil." *Science*, v. 145, No. 3632 (Aug. 7, 1964).

1c. Anderson, Don L., "The Plastic Layer of the Earth's Mantle." *Scientific American*, v. 207, No. 1 (July 1962).

2. Anderson, E. C., Levi, H., and Tauber, H., "Copenhagen Natural Radiocarbon Measurements I." *Science*, 118:9 (1953).

3. Anderson, R. J., "The Anatomy of the Indian Elephant." *Journal of Anatomy and Physiology*, 27:491–94 (1883).

4. Anderson, W., "Quaternary Sea-Levels in the Northern Hemisphere." *Geological Magazine*, 76:489–93 (Nov. 1939).

5. Antevs, Ernest, "The Climatological Significance of Annual Rings in Fossil Woods." *American Journal of Science*, (5) 9:296–302 (1925).

6. Antevs, Ernest, *The Last Glaciation with Special Reference to the Ice Retreat in Northeastern North America*. New York, The American Geographical Society, 1928.

7. Antevs, Ernest, "Modes of Retreat of the Pleistocene Ice Sheets." *Journal of Geology*, 47:503–08 (July 1939).

8. Antevs, Ernest, "Geochronology of the Deglacial and Neothermal Ages." *Journal of Geology*, 61:195–230 (May 1953).

9. Antevs, Ernest, "Geochronology of the Deglacial and Neothermal Ages. A Reply." *Journal of Geology*, v. 62, No. 5 (Sept. 1954).

10. Antevs, Ernest, "Varve and Radiocarbon Chronologies Appraised by Pollen Data." *Journal of Geology*, v. 63, No. 5 (Sept. 1955).

11. Anthony, Harold, "Nature's Deep Freeze." *Science Digest* (Jan. 1950). (Condensed from *Natural History*, Sept. 1949.)

12. Army Observer's Report of Operation Highjump, Task Force 68, United States Navy, 1947. (Unclassified.)

13. Arrhenius, Gustaf. See Pettersson, Hans.

13a. Arrhenius, Gustaf, "Sediment Cores from the East Pacific." *Reports of the Swedish Deep-Sea Expedition,* v. 5, fasc. 1 (1952).

13b. Arrhenius, Gustaf, "Climatic Records on the Ocean Floor." Reprint, Scripps Institution of Oceanography, La Jolla (received Sept. 30, 1958). Contributions of the Scripps Institution of Oceanography, New Series.

13c. Arrhenius, Gustaf, "The Geological Record on the Ocean Floor." Reprint from *Oceanography,* American Association for the Advancement of Science, copyright 1961.

13d. Arrhenius, Gustaf, "Sedimentary Record of Long-Period Phenomena." From *Advances in Physics,* ed. P. M. Hurley, copyright 1966 by MIT. Published by M.I.T. Press.

14. Arrhenius, G., Kjellberg, G., and Libby, W. F., "Age Determination of Pacific Chalk Ooze by Radiocarbon and Titanium Content." *Tellus,* v. 3, No. 4 (Nov. 1951).

15. Ashley, George H., *General Information on Coal.* Mineral Resource Report No. 6, Part I, Geological Survey of Pennsylvania, 1928.

16. Australian Association for the Advancement of Science, *Reports,* v. 10 (1904).

17. Bailey, Thomas L., "Late Pleistocene Coast Range Orogenesis in Southern California." Abstracts, *Bulletin of the Geological Society,* v. 52, Pt. 2(2), pp. 1888–89 (1941).

18. Bain, George W., "Mapping the Climatic Zones of the Geological Past." *Yale Scientific Magazine,* v. xxvii, No. 5 (Feb. 1953) (Reprint.)

19. Bain, George W., personal communication.

19a. "Climatic Zones Throughout the Ages." In *Polar Wandering and Continental Drift,* ed. Arthur C. Munyan (1963).

20. Ball, Sir Robert, *The Cause of an Ice Age.* New York, Appleton, 1892.

21. Bandy, O. L., "Paleotemperatures of Pacific Bottom Waters and Multiple Hypotheses." *Science,* v. 123, No. 3194 (March 16, 1956).

22. Barley, Alfred, *The Drayson Problem.* Exeter, England, Pollard, 1922.

23. Barley, Alfred, *The Ice Ages.* Lewes, Sussex, England, Baxter, 1927.

24. Barnes, Joseph, personal conference.

24a. Bascom, Willard, *A Hole in the Bottom of the Sea.* Garden City, New York, Doubleday, 1961.

25. Bell, Robert, "On the Occurrence of Mammoth and Mastodon Remains around Hudson Bay." *Bulletin of the Geological Society of America,* 9:369–90 (June 22, 1898).

25a. Beloussov, Vladimir V., "Against Continental Drift," *Science Journal,* pp. 56–61 (January 1967).

25b. Beloussov, Vladimir V., "The Upper Mantle Project," *The Unesco Courier* (October 1963).

26. Benedict, Francis G., *The Physiology of the Elephant.* Washington, Carnegie Institution of Washington, 1936.

27. Benedikt, Elliot T., "A Method of Determination of the Direction of the Magnetic Field of the Earth in Geological Epochs." *American Journal of Science*, v. 241, No. 2 (Feb. 1943).
28. Benfield, A. E., "The Earth's Heat." *Scientific American*, v. 183, No. 6 (Dec. 1950).
29. Benioff, H., "Global Strain Accumulation and Release as Revealed by Great Earthquakes." *Bulletin of the Geological Society of America*, v. 62, No. 4 (April 1951).
30. Benioff, H., Interview by Associated Press, Daytona Beach *Evening News*, March 6, 1952.
31. Bergquist, N. O., *The Moon Puzzle. A Revised Classical Theory Correlating the Origin of the Moon with Many Problems in Natural Science.* Copenhagen, Denmark, Grafisk Forlag, 1954.
32. Berry, Edward Wilbur, *The Past Climate of the North Polar Region.* Washington, The Smithsonian Institution, 1930.
33. Birch, F., "Elasticity of Igneous Rocks at High Temperatures and Pressures." *Bulletin of the Geological Society of America*, 54:263–86 (1943).
34. Birch, F., and Bancroft, D., "The Elasticity of Glass at High Temperatures and the Vitreous Basalt Substratum." *American Journal of Science*, 240:457–90 (1942).
35. Birch, F., and Bancroft, D., "New Measurements of the Rigidity of Rocks at High Pressure." *Journal of Geology*, 48:752–66 (1940).
35a. Bisque, Ramon E., and Rouse, George E., "Geoid and Magnetic Field Anomalies; Their Relationship to the Core-Mantle Interface." (Reprint of paper presented in the Sea-Floor Spreading session of the American Geophysical Union annual meeting, Washington, D.C., April 11, 1968 and published in *Mines Magazine*, Vol. 58, No. 5, 1968).
36. Black, Davidson, "Paleogeography and Polar Shift." Reprinted from the *Bulletin of the Geological Society of China*, v. X, Peiping (1931).
37. Blackett, P. M. S., *Lectures on Rock Magnetism.* New York Interscience Publications, 1956.
38. Blanchard, Jacques, *L'Hypothèse du déplacement des pôles et la chronologie du quaternaire.* Paris, Editions Universitaires, 1942.
39. Bliss, W. L., "Radiocarbon Contamination." *American Antiquity*, v. xvii, No. 3 (Jan. 1952).
40. Boas, Franz, ed., *General Anthropology.* Boston, Heath, 1938.
41. Bradley, W. H., et al., *Professional Papers 136A, B.* United States Geological Survey, 1940.
42. Bradley, W. H., *The Varves and Climate of the Green River Epoch.* Professional Paper 158. United States Geological Survey.

43. Bramlette, G., and Bradley, W. H., *Geology and Biology of North Atlantic Deep-Sea Cores.* Professional Paper 196. United States Geological Survey.

Brannon, H. R., *et al.* See "Humble Oil Radiocarbon Dates I."

44. Breen, Walter, personal communication.
45. Brewster, E. T., *This Puzzling Planet.* New York, Bobbs, 1928.
46. Brice, J. C., personal communication.
47. Bridgman, P. W., "Effects of High Shearing Stress Combined with High Hydrostatic Pressure." *Physical Review,* 48:825–47 (1935).
48. Bridgman, P. W., "Some Implications for Geophysics of High Pressure Phenomena." *Bulletin of the Geological Society of America,* 62:533–36 (May 1951).
49. Bridgman, P. W., *Studies in Large Plastic Flow and Fracture.* New York, McGraw-Hill, 1952.
50. Bridgman, P. W., personal conference.
51. Bridgman, P. W., personal communication.
52. Brooks, C. E. P., *Climate Through the Ages.* New York, McGraw-Hill, 1949.
53. Brouwer, Dirk, personal communication.
54. Brown, Hugh Auchincloss, *Popular Awakening Concerning the Impending Flood.* Lithographed manuscript copy of a proposed illustrated book, copyright 1948, with corrections and additions inserted by the author to 1951. Ann Arbor, Michigan, Edwards, 1948.
54a. Brown, Hugh Auchincloss, *Cataclysms of the Earth.* New York, Twayne Publishers, 1967.
55. Brown, J. MacMillan, *The Riddle of the Pacific.* London, Fisher Unwin, 1925.
56. Bruckner, Eduard, "Die Schneegrenze in der Antarctis." *Zeitschrift für Gletscherkunde,* 7:276–79 (1913).
57. Bucher, Walter H., "The Crust of the Earth." *Scientific American* (May 1950).
58. Bucher, Walter H., *The Deformation of the Earth's Crust.* Princeton, University Press, 1933.
59. Bullard, E. C., Review of Jeffreys's *The Earth,* 3d ed. *Science,* v. 119, No. 3081 (Jan. 15, 1954).
59a. Bullen, K. E., "The Interior of the Earth," Reprint, *The Scientific American* (Sept. 1955).
60. Burtt, E. A., *The Metaphysical Foundations of Modern Science,* rev. ed. Garden City, Doubleday, 1954.
61. Byrd, Admiral Richard E., "Exploring the Ice Age in Antarctica." *National Geographic Magazine,* v. LXVIII, No. 4 (Oct. 1935).
62. Byrd, Admiral Richard E., "Our Navy Explores Antarctica." *National Geographic Magazine,* v. XCII, No. 4 (Oct. 1947).

63. Cain, S. A., *Foundations of Plant Geography*. New York, Harper, 1944.
64. Camp, L. Sprague de, *Lost Continents*. New York, Gnome, 1954.
65. Campbell, James H., personal conference.
66. Campbell, James H., personal communication.
67. Campbell, J. H., and Hapgood, C. H., "Effect of Polar Ice on the Crust of the Earth." *Yale Scientific Magazine*, v. XXXI, No. 1 (Oct. 1956).
68. Carsola, Alfred J., "Bathymetry of the Arctic Ocean." *Journal of Geology*, v. 63, No. 3, pp. 274–78 (May 1955).
69. Caster, K. E., and Mendes, J. C., "Geological Comparison of South America with South Africa after Twenty Years." Abstract, *Bulletin of the Geological Society of America*, 58:1173 (1947).
70. Challinor, J., "Remarkable Example of Superficial Folding Due to Glacial Drag." *Geological Magazine*, 84:270–72 (Sept. 1947).
71. Chamberlin, R. T., "Origin and History of the Earth." In Moulton, F. R., ed., *The World and Man as Science Sees Them*. Garden City, Doubleday, 1937.
72. Chaney, Ralph W., "Tertiary Forests and Continental History." *Bulletin of the Geological Society of America*, v. 51, No. 3 (March 1, 1940).
73. Chaney, R. W., "The Ecological Composition of the Eagle Creek Flora." *Bulletin of the Geological Society of America*, v. 31, No. 1 (March 31, 1920).
74. Chapin, Henry, and Smith, Walton, *The Ocean River*. New York, Scribner's, 1952.
75. Charlesworth, J. K., "The Ice Age and the Future of Man." *Science Progress* (London), XLI:161 (Jan. 1953).
75a. Charlesworth, J. K., *The Quaternary Era, With Special Reference to Its Glaciation*. London, Edward Arnold Ltd., 1957. (2 vols.)
76. "Chicago Radiocarbon Dates." *Science*, v. 113, No. 2927 (Feb. 2, 1951).
76a. "Chicago Radiocarbon Dates II." *Science*, v. 114, No. 2960 (Sept. 21, 1951).
77. "Chicago Radiocarbon Dates III." *Science*, v. 116, No. 3025 (Dec. 19, 1952).
78. "Chicago Radiocarbon Dates IV." *Science*, v. 119, No. 3083 (Jan. 29, 1954).
79. "Chicago Radiocarbon Dates V." *Science*, v. 120, N. 3123 (Nov. 5, 1954).
80. Clegg, J. A., "Rock Magnetism." *Nature*, 178:1085 (Nov. 17, 1956).
81. Clegg, J. A., Almond, Mary, and Stubbs, P. H. S., "Remanent Magnetism of Some Sedimentary Rocks in Britain." *Philosophical Magazine*, 45:365 (June 1954).

82. Clegg, J. A., Deutsch, E. R., and Griffiths, D. H., "Magnetism of the Deccan Trap." *Philosophical Magazine,* Ser. 8, v. 1, pp. 419ff (1956).
83. Clisby, Katharyn H., personal communication.
84. Clisby, Katharyn H., and Sears, Paul B., "The San Augustin Plains —Pleistocene Climatic Changes." *Science,* v. 124, No. 3221 (Sept. 21, 1956).
85. Cloos, Hans, *Conversation with the Earth.* New York, Knopf, 1953.
86. Colbert, Edwin H., "The Pleistocene Mammals of North America and Their Relations to Eurasian Forms." In *Early Man,* symposium edited by George G. MacCurdy, published for the Academy of Natural Sciences. Philadelphia, Lippincott, 1937.
87. Coleman, A. P., *Ice Ages Recent and Ancient.* New York, Macmillan, 1929.
87a. *Columbia Research News,* v. VII, No. 2 (Feb. 1957).
88. Conant, James B., *On Understanding Science,* New York, New American Library, 1951.
88a. Cook, Melvin A., "Continental Dynamics." *Institute of Metals and Explosives Research,* University of Utah (Jan. 31, 1961).
89. "Copenhagen Natural Radiocarbon Measurements II." *Science,* v. 124, No. 3227 (Nov. 2, 1956).
89a. Cox, Allan, and Doell, Richard R., "Geomagnetic Polarity Epochs." *Science,* v. 143, No. 3604 (Jan. 24, 1964).
89b. Cox, Allan, et al., "Reversals of the Earth's Magnetic Field." *Scientific American,* v. 216, No. 2 (Feb. 1967).
90. Crane, H. R., "Antiquity of the Sandia Culture. Carbon 14 Measurements." *Science,* v. 122, No. 3172 (Oct. 14, 1955).
90a. Croizat, Leon, *Panbiogeography, or An Introductory Synthesis of Zoogeography, Phytogeography and Geology, with notes on Evolution, Systematics, Anthropology, etc.* Published by Leon Croizat, Caracas, Venezuela; distributed by Wheldon and Wesley, Ltd., Lytton Lodge, Codicote, England, 1958.
91. Croll, James, *Climate and Time,* London, Daldy, Isbister, 1875. *Crust of the Earth, a Symposium on the.* See Poldervaart, ed.
91a. Curtis, Lipson, and Evernden, "Potassium-Argon Dating of Plio-Potassium Intrusive Rocks." *Nature,* 178:4546 (Dec. 15, 1956), p. 1360.
92. Cushman, Joseph A., "Study of the Foraminifera Contained in Cores from the Bartlett Deep." *American Journal of Science,* v. 239, No. 2, pp. 128–47.
93. Daly, R. A., *The Changing World of the Ice Age.* New Haven, Yale, 1934.
94. Daly, R. A., "Pleistocene Glaciation and the Coral Reef Problem." *American Journal of Science,* Ser. 4, v. 30, pp. 297–308.

95. Daly, R. A., "Earth Crust Slides on a Great Sea of Glass." *Science News Letter* (May 9, 1925).

96. Daly, R. A., The Daly Volume of the American Journal of Science. *American Journal of Science*, v. 242A (1945).

97. Daly, R. A., *The Strength and Structure of the Earth*. New York, Prentice Hall, 1940.

98. Daly, R. A., *Our Mobile Earth*. New York and London, Scribner's, 1926.

99. Daly, R. A., "A Recent World Wide Sinking of Ocean Level." *Geological Magazine*, 52:246–61 (1920).

100. Daly, R. A., personal conference.

101. *Dana's Manual of Mineralogy*, 15th ed., rev. by Cornelius S. Hurlbut, Jr. New York, Wiley, 1952.

102. Dana, James D., *Corals and Coral Islands*. New York, Dodd, Mead, 1872.

103. Darwin, Charles, *The Origin of Species*. (Reprinted from the 6th London edition.) New York, Burt, n.d.

104. Darwin, Charles, *Journal of Research*. New York, American Home Library, 1902.

105. Darwin, George H., "On the Influence of Geological Changes on the Earth's Axis of Rotation." *Philosophical Transactions of the Royal Society*, v. 167, Pt. 1.

106. David, T. W. E., "Antarctica and Some of Its Problems." *Geographical Journal*, 43:605–30 (1914).

107. Davies, O., "African Pluvials and European Glaciations." *Nature*, v. 178, No. 4536 (Oct. 6, 1956).

107a. Deevy, E. S., and Flint, R. F., "Postglacial Hypsithermal Interval." *Science*, v. 125, No. 3240 (Feb. 1, 1957).

108. De Geer, Ebba Hult, "Geochronology of the Deglacial and Neothermal Ages: A Discussion." *Journal of Geology*, v. 62, No. 5 (Sept. 1954).

109. De Geer, Gerard Jakob, "On the Solar Curve as Dating the Ice Age, the New York Moraine and Niagara Falls through the Swedish Time Scale." *Geografiska Annaler* (1926).

110. De Geer, Gerard, "Compte rendu." Geological Congress, 1910.

111. Dennis, Clifford E., "Experiments in Planetary Deformation of the Earth." *Pan-American Geologist*, 55:241–58 (1931).

111a. Deutsch, Ernst R., "Polar Wandering and Continental Drift," in volume of the same title, edited by Arthur C. Munyan, Special Publication No. 10, Society of Economic Mineralogists, Tulsa, Oklahoma, July, 1963.

111b. Dicke, R. H., "The Eötvös Experiment." *Scientific American*, v. 205, No. 6 (Dec. 1961).

112. Dietz, Robert S., *Technical Report* ONRL–7–57, Office of Naval Research, London, Jan. 10, 1957. (An informal report of the conference on rock magnetism held in London in December 1957.)

112a. Dietz, Robert S., "Ocean-Basin Evolution by Sea-Floor Spreading," in *Continental Drift*, edited by S. K. Runcorn. Academic Press, 1962.

113. Digby, Bassett, *The Mammoth and Mammoth-Hunting Grounds in Northeast Siberia*. New York, Appleton, 1926.

114. Dillon, Lawrence S., "Wisconsin Climate and Life Zones in North America." *Science*, v. 123, No. 3188 (Feb. 2, 1956).

115. Dodson, Edward O., *A Textbook of Evolution*. Philadelphia and London, Saunders, 1952.

115a. Doell, Richard R., and Cox, Allan, *Paleomagnetism, Advances in Geophysics*, vol. 8. New York and London, Academic Press, 1961.

116. Dorf, E., "Plants and the Geologic Time Scale." See Poldervaart, ed.

116a. Doumani, George A., and Long, William E., "The Ancient Life of the Antarctic." *Scientific American*, v. 207, No. 3 (Sept. 1962).

117. Drayson, A. W., *The Last Glacial Period in Geology*, 1873. *Thirty Thousand Years of the Earth's Past History*, 1888. *Untrodden Ground in Astronomy and Geology*, 1890.

118. Dreimanis, Alexia, "Stratigraphy of the Wisconsin Glacial Stage along the Nothwestern Shore of Lake Erie." *Science*, v. 126, No. 3265 (July 26, 1957).

119. Du Nouy, Pierre Lecomte, *Human Destiny*. New York, Longmans Green, 1947.

120. Du Toit, A. L., *A Geological Comparison of South America with South Africa*. Washington, Carnegie Institution of Washington Publication No. 381, 1927.

121. Du Toit, A. L., "Further Remarks on Continental Drift." *American Journal of Science*, 243:404–08 (1945).

122. Dutton, Clarence, "On Some of the Greater Problems of Physical Geology." (Chapter XII of Bulletin 78, *Physics of the Earth*, of the National Research Council, 1931.)

123. Eardley, Armand J., "The Cause of Mountain Formation: an Enigma." *The American Scientist*, v. 45, No. 3 (June 1957).

124. Eddington, A. E., "The Borderland of Astronomy and Geology." *Smithsonian Reports*, 1923.

125. Edmondson, Charles Howard, "Growth of Hawaiian Corals." *Bernice P. Bishop Museum Bulletin 58*, Honolulu, published by the Museum, 1929.

126. Einarsson, T., and Sigurgeirsson, T., "Rock Magnetism in Iceland." *Nature*, v. 175, No. 892 (1955).

127. Einstein, Albert, "On the Generalized Theory of Gravitation." *Scientific American*, v. 182, No. 4 (April 1950).

128. Einstein, Albert, personal communication.

129. Einstein, Albert, conference.

130. Ekman, Sven, *Zoogeography of the Sea*. London, Sidgwick and Jackson, 1953.

130a. Elvers, Douglas J., "Dyke Swarms in the Pacific Plate Versus the Conveyor Belt Mechanism." (Preprint). U.S. Dept. of Commerce, Environmental Science Services Administration, Coast and Geodetic Survey, Fredericksburg Geomagnetic Center, Corbin, Virginia.

130b. Elvers, Douglas J., Mathewson, Christopher C., Kohler, Robert E. and Moses, Robert L. "Systematic Ocean Surveys by the USC&GSS *Pioneer*, 1961–65." Operational Data Report C&GSDR–1, Environmental Science Services Administration, U.S. Department of Commerce, Washington, D.C.

131. Emiliani, Cesare, "Temperatures of Pacific Bottom Waters and Polar Superficial Waters during the Tertiary." *Science*, v. 119, No. 3103 (June 19, 1954).

132. Emiliani, Cesare, "Pleistocene Temperatures." *Journal of Geology*, v. 63, No. 6, pp. 538–78 (Nov. 1955).

133. Emiliani, Cesare, "On Paleotemperatures of Pacific Bottom Waters." *Science*, v. 123, No. 3194 (March 16, 1956).

134. Emiliani, Cesare, "Note on the Absolute Chronology of Human Evolution." *Science*, v. 123, No. 3204 (May 25, 1956).

135. Emiliani, Cesare, "Temperature and Age Analysis of Deep-Sea Cores." *Science*, v. 125, No. 3244 (March 1, 1957).

136. Emiliani, Cesare, personal communication.

136a. Engel, A. E. J., "Geologic Evolution of North America." *Science*, v. 140, No. 3563 (April 12, 1963).

137. Ericson, David B., "North Atlantic Deep-Sea.Sediments and Submarine Canyons." *Transactions of the New York Academy of Sciences*, Ser. II, v. 15, No. 2 (Dec. 1952).

138. Ericson, David B., "Sediments of the North Atlantic." *Lamont Geological Observatory Technical Report on Submarine Geology*, No. 1 (Nov. 1953).

139. Ericson, D. B., Ewing, Maurice, and Heezen, B. C., "Deep-Sea Sands and Submarine Canyons." *Bulletin of the Geological Society of America*, 62:961–65 (1951).

140. Ericson, D. B., Ewing, Maurice, and Heezen, B. C., "Turbidity Currents and Sediments in the North Atlantic." *Bulletin of the American Association of Petroleum Geologists*, v. 36, No. 3 (March 1952). (Reprint.)

141. Ericson, D. B., Ewing, Maurice, Heezen, B. C., and Wollin, Goesta, "Sediment Deposition in the Deep Atlantic." *Lamont Geological Observatory Contribution No. 130*. Geological Society of America Special Paper 62 (1955).

141a. Ericson, David B., Ewing, M., and Wollin, G., "Extinctions and evolutionary changes in microfossils clearly define the abrupt onset of the Pleistocene." *Science*, v. 139, No. 3556 (Feb. 22, 1963).

142. Ericson, D. B., and Wollin, Goesta, "Correlations of Six Cores from the Equatorial Atlantic and the Caribbean." *Deep-Sea Research*, 3:104–25 (1956).

143. Ericson, D. B., and Wollin, Goesta, "Micropaleontological and Isotopic Determinations of Pleistocene Climates." *Micropaleontology,* v. 2, No. 3 (July 1956).

143a. Ericson, David B., and Wollin, Goesta, "Micropaleontology and Lithology of Arctic Sediment Cores." Reprint from *Geophysical Research Papers No. 63.* "Scientific Studies at Fletcher's Ice Island, T–3, 1952–1955," v. 1 (Sept. 1959). Terrestrial Sciences Laboratory, United States Air Force, Bedford, Mass.

143b. Ericson, David B., and Wollin, Goesta, "Micropaleontology." *Scientific American,* v. 207, No. 1 (July 1962).

143c. Ericson, David B., and Wollin, Goesta, *The Deep and the Past.* New York, Knopf, 1964.

143d. Ericson, David B., and Wollin, Goesta, *The Ever-Changing Sea.* New York, Knopf, 1967.

144. Ericson, D. B., et al., "Late Pleistocene Climates and Deep-Sea Sediment." *Science,* v. 124, No. 3218 (Aug. 31, 1956).

145. Ericson, D. B., personal communication.

146. Ewing, Maurice, and Donn, W. L., "A Theory of Ice Ages." *Science,* v. 123, No. 3207 (June 15, 1956).

146a. Ewing, Maurice, and Donn, William L., "Polar Wandering and Climate." In *Polar Wandering and Continental Drift,* ed. Arthur C. Munyan (see Munyan).

146b. Ewing, M., and Donn, W., "A Theory of Ice Ages." *Science,* v. 123, pp. 1061–1066 (1956).

146c. Ewing, M., and Donn, W., "A Theory of Ice Ages II." *Science,* v. 127, pp. 1059–1062 (1958).

147. Ewing, Heezen, Ericson, Northrup, and Dorman, "Exploration of the Northwest Atlantic Mid-Ocean Canyon." *Bulletin of the Geological Society of America,* 64:865–68 (July 1953).

148. Farrington, William, personal communication.

149. Faul, Henry, ed., *Nuclear Geology, a Symposium.* New York, Wiley, 1954.

150. Fenner, C. N., "Pleistocene Climate and Topography of the Arequipa Region, Peru." *Bulletin of the Geological Society of America,* v. 59, No. 9 (Sept. 1948).

151. Fenton, C. L., "Paleontology and Mathematical Evolution." *Pan-American Geologist,* v. LV, No. 3, pp. 162–74 (April 1931).

152. Finnegan, H. E., personal communication.

153. Fisk, H. N., and McFarlan, E., "Late Quaternary Deltaic Deposits of the Mississippi River." See Poldervaart.

154. Fleming, J. A., ed., *Terrestrial Magnetism and Electricity.* (Vol. VIII of "Physics of the Earth," published under the auspices of the National Research Council.) New York, Dover, 1949.

155. Flint, R. F., Knopf, A., and Longwell, R., *Outline of Physical Geology,* 2d ed. New York, Wiley, 1941.

156. Flint, Richard Foster, *Glacial Geology and the Pleistocene Epoch.* New York, Wiley, 1947.

157. Flint, Richard Foster, "Rates of Advance and Retreat of the Margin of the Late Wisconsin Ice Sheet." *American Journal of Science*, 253:649–58 (1955).

158. Flint, Richard Foster, personal communication.

159. Flint, R. F., and Deevey, E. S., "Radiocarbon Dating of Late Pleistocene Events." *American Journal of Science*, v. 249, No. 4 (April 1951).

160. Flint, R. F., and Dorsey, H. G., "Iowan and Tazewell Drifts and the North American Ice Sheet." *American Journal of Science*, v. 243, No. 11, p. 627 (1945).

161. Flint, R. F., and Dorsey, H. G., "Glaciation of Siberia." *Bulletin of the Geological Society of America*, 56:89–106 (1945).

162. Flint, R. F., and Dorsey, H. G., "Radiocarbon Dates of Pre-Mankato Events in Eastern and Central North America." *Science*, v. 121, No. 3149 (May 6, 1955).

163. Forbes, W. H., "On the Anatomy of the Indian Elephant." *Proceedings of the Royal Society of London, 1879*, pp. 420–35.

164. Forrest, H. Edward, *The Atlantean Continent*. London, Witherby, 1933.

165. Foster, J. W., *Pre-Historic Races of the United States*. Chicago, Griggs, 1887.

165a. Foster, Theodore, "Convection in a Variable Viscosity Fluid Heated from Within." *Journal of Geophysical Research*, Vol. 74, No. 2 (Jan. 15, 1969).

166. Frank, Phillip, review of *Essay in Physics*, by Lord Samuel. New York *Times* Book Review (Feb. 17, 1952).

167. Frankland, John M., personal communication.

168. Frankland, John M., conference.

168a. Frisch, Bruce H., "Our Moving Continents." *Science Digest*, v. 63, No. 1 (Jan. 1968).

169. Gamow, George, *Biography of the Earth*. New York, Viking, 1948.

169a. Garland, G. D., ed., *Continental Drift*. The Royal Society of Canada, Special Publications No. 9, University of Toronto Press, 1967.

169b. Gaskell, T. F., "Comparison of Pacific and Atlantic Ocean Floors in Relation to Ideas of Continental Displacement." (See Runcorn, ed.)

170. Geikie, Sir Archibald, *Text Book of Geology*, 3d ed. New York and London, Macmillan, 1893.

171. George, T. Neville, "Geology." *Science Progress*, v. XLIII, No. 169 (Jan. 1955).

172. Gerard, R. W., "Experiments in Micro-Evolution." *Science*, v. 120, No. 3123 (Nov. 5, 1954).

173. Gidley, James Williams, *Notice of the Occurrence of a Pleistocene Camel North of the Arctic Circle*. Smithsonian Miscellaneous

Collections, v. 60, No. 26 (March 21, 1913). Publication number 2173.

174. Gilligan, Albert, "A Contribution to the Geologic History of the North Atlantic Region." Presidential address before the Yorkshire Geological Society. Reprinted in the Annual Report of the Smithsonian Institution, 1932, pp. 207–22. Washington, Government Printing Office, 1933.

175. Gilluly, James, "Geologic Contrasts Between Continents and Ocean Basins." See Poldervaart.

175a. Glass, Billy, and Heezen, Bruce, "Tektites and Geomagnetic Reversals." *Scientific American,* v. 217, No. 1 (July 1967).

176. Gold, T., "Instability of the Earth's Axis of Rotation." Nature, v. 175, No. 4456, pp. 526–29 (March 26, 1955).

177. Goldring, Winifred, "Algal Barrier Reefs in the Lower Ozarkian of New York." *New York State Museum Bulletin, No. 315.* Albany, University of New York, 1938.

178. Goldring, Winifred, "Handbook of Paleontology for Beginners and Amateurs, Part I. The Fossils." *New York State Museum Handbook 9.* Albany, University of New York, 1950.

179. Goldschmidt, Richard, "Different Philosophies of Genetics." *Science,* v. 119, No. 3099 (May 21, 1954).

180. Goldschmidt, Richard, *The Material Basis of Evolution,* New Haven, Yale, 1940.

181. Good, Ronald, "The Present Position of the Theory of Continental Drift." *Nature,* v. 166, No. 4223 (Oct. 7, 1950).

182. Gordianko, in the *Red Star,* May 18, 1954.

183. Grabau, Amadeus W., *The Rhythm of the Ages; Earth History in the Light of the Pulsation and Polar Control Theories.* Peking, Vetch, 1940.

184. Grabau, Amadeus W., *A Textbook of Geology. Part II: Historical Geology.* New York, Heath, 1920.

185. Graham, John W., "Evidence of Polar Shift Since Triassic Time." *Journal of Geophysical Research,* v. 60, No. 3 (Sept. 1955).

186. Graham, John W., "Rock Magnetism." In Tuve, *Annual Report for 1955–1956* (which see).

187. Graham, John W., personal communication.

188. Graham, J. W., Torreson, O. W., and Bowles, E., "Magnetic Polarization of Silurian Sediments of the Eastern United States." Abstract, *Transactions of the American Geophysical Union,* v. 31, No. 328 (1950).

189. Gray, George W., "The Lamont Geological Observatory." *Scientific American,* v. 195, No. 6 (Dec. 1956).

189a. Greene, Walter, personal communication.

190. Gregory, Joseph T., "Vertebrates in the Geologic Time Scale." See Poldervaart.

191. Gregory, J. W., "Geological History of the Pacific Ocean." Nature, v. 125, No. 3159 (May 17, 1930).

192. Griffiths, D. H., and King, R. F., "Natural Magnetization of Igneous and Sedimentary Rocks." Nature, v. 173, No. 4415, pp. 1114–16.

193. Griggs, David, "A Theory of Mountain Building." American Journal of Science, 237:611–50.

194. Gutenberg, Beno, ed., Internal Constitution of the Earth. New York, Dover, 1951.

195. Gutenberg, B., and Richter, C. F., Seismicity of the Earth and Associated Phenomena. Princeton, University Press, 1950.

195a. Hales, A. L., in The Earth Today, ed. A. H. Cook et al., pp. 312–19. Royal Astronomical Society, 1961.

195b. Hall, H. Tracy, "Ultrahigh Pressures." Scientific American, v. 201, No. 5 (Nov. 1959).

196. Halle, T. B., "On the Geological Structure of the Falk Islands." Bulletin of the Geological Institute of the University of Upsala, 11:115–229.

197. Hamilton, Edwin L., "Upper Cretaceous, Tertiary, and Recent Planktonic Foraminifera from the Mid-Pacific Flat-Topped Sea Mounts. Journal of Paleontology, 27:207–37.

198. Hansen, Henry P., "Postglacial Forest Succession and Climate in the Oregon Cascades." American Journal of Science, v. 244, No. 10 (Oct. 1946).

199. Hansen, L. Taylor, Some Considerations of and Additions to the Taylor-Wegener Hypothesis of Continental Displacement. Pamphlet copyrighted July 1946 by L. Taylor Hansen, 1158 W. 35th St., Los Angeles 7, Cal. (In the New York Public Library.)

199a. Hapgood, Charles H., Maps of the Ancient Sea Kings. Philadelphia, Chilton Books, 1966.

200. Hapgood, Mrs. Norman (Elizabeth Reynolds), translation of the report on the stomach contents of the Beresovka mammoth (unpublished). See Sukachev.

201. Hardy, M. E., The Geography of Plants. Oxford, University Press, 1920.

201a. Harland, W. B., Smith, A. Gilbert and Wilcock, B. "The Phanerozoic Time Scale." London, The Geological Society of London, 1962.

202. Harris, Herbert, "The Amazing Frozen Foods Industry." Science Digest, v. 29, No. 2 (Feb. 1951). (Condensed from The Reporter, October 24, 1950.)

203. Hartnagel, C. A., and Bishop, Sherman C., "The Mastodons, Mammoths and Other Pleistocene Mammals of New York State." New York State Museum Bulletins 241–242. Albany, 1922.

204. Haskell, N. A., "The Motions of a Viscous Fluid under a Surface Load." Physics, 6:265–69 (1935); 7:56–31 (1936).

205. Heezen, Bruce C., and Ewing, Maurice, "Turbidity Currents and Submarine Slumps, and the 1929 Grand Banks Earthquake." Lamont Geological Observatory Contribution No. 65. *American Journal of Science*, v. 250 (Dec. 1952).

205a. Heezen, Bruce C., "The Rift in the Ocean Floor." *Scientific American*, v. 203, No. 4 (Oct. 1960).

205b. Heezen, Bruce, "The Deep Sea Floor." In *Continental Drift*, edited S. K. Runcorn, Academic Press, 1962.

206. Henry, Thomas R., *The White Continent*. New York, William Sloane Associates, 1950.

207. Henry, Thomas R., "Poles Wander; Earth Crust Falls Off." North American Newspaper Alliance (Jan. 27, 1952).

208. Herz, O. F., "Frozen Mammoth in Siberia." *Smithsonian Reports*, 1903.

209. Hess, H. H., "Comments on Mountain Building." *Transactions of the American Geophysical Union*, V:528–31.

210. Hess, H. H., "Drowned Ancient Islands of the Pacific Basin." *Smithsonian Reports*, 1947.

210a. Heusser, Calvin J., "Late Pleistocene Pollen Diagrams from the Province of Llanquihue, Southern Chile." Reprint from the *Proceedings of the American Philosophical Society*, v. 110, No. 4 (Aug. 1966).

210b. Heusser, Calvin J., "Polar Hemispheric Correlation: Palynological Evidence from Chile and the Pacific North-west of America." Reprint from the *Royal Meteorological Society Proceedings of the International Symposium on World Climate from 8000 to 0 B.C.* (no date).

211. Hibben, Frank C., "Evidence of Early Man in Alaska." *American Antiquity*, VIII:256 (1943).

212. Hibben, Frank C., *The Lost Americans*. New York, Crowell, 1946.

213. Hibben, Frank C., *Treasure in the Dust*. Philadelphia and New York, Lippincott, 1951.

214. Hillaby, John, "Earth as Magnet Said to Weaken." New York *Times* (Dec. 26, 1956).

215. Hobbs, William H., *Earth Evolution and Its Facial Expression*. New York, Macmillan, 1922.

216. Hobbs, William H., *Earth Features and Their Meaning*, New York, Macmillan, 1935.

217. Hobbs, William H., "Repeating Patterns in the Relief and in the Structure of the Land." *Bulletin of the Geological Society of America*, v. 22 (1911).

218. Hobbs, William H., "The Correlations of Fracture Systems and the Evidence for Planetary Dislocations within the Earth's Crust." Reprinted from *Transactions of the Wisconsin Academy of Sciences, Arts and Letters*, XV:15–29 (Aug. 1905).

219. Hobbs, William H., "Eurasian Continental Glacier of the Late Pleistocene." *Science*, v. 104, No. 2692 (Aug. 2, 1946).

220. Hodgman and Lange, *Handbook of Chemistry and Physics*. Cleveland, Chemical Rubber Publishing Co., 1927.
221. Hoffman, Bernhard C., "Implications of Radiocarbon Datings for the Origins of Dorset Culture." *American Antiquity*, v. XVIII, No. 1 (July 1952).
221a. Hollin, J. T., "On the Glacial History of Antarctica." *Journal of Glaciology*, 4(32):173–193, 1962.
221b. Hooker, Dolph Earl, *Those Astounding Ice Ages*. New York, Exposition Press, 1958.
222. Horberg, Leland, "Radiocarbon Dates and Pleistocene Geological Problems of the Mississippi Valley Region." *Journal of Geology*, v. 63, No. 3 (May 1955).
223. Hospers, J., "Rock Magnetism and Pole Wandering." *Journal of Geology*, v. 63, No. 1 (Jan. 1955).
224. Hough, Jack, "Pacific Climatic Record in a Pacific Ocean Core Sample." *Journal of Geology*, v. 61, No. 3 (May 1953).
225. Hough, Jack, "Pleistocene Lithology of Antarctic Ocean Bottom Sediments." *Journal of Geology*, 58:257–59.
225a. Howorth, H. H., "The Mammoth and the Flood," *Geological Magazine*, New Series, 8:309–15 (1881).
226. Hubbert, M. King, and Melton, F. A., "Isostasy, A Critical Review." *Journal of Geology*, v. 38, No. 8 (Nov.–Dec. 1930).
227. Hubbert, M. King, personal communication.
228. "Humble Oil Radiocarbon Dates I." H. R. Brannon, Jr., et al. *Science*, v. 125, N. 3239 (Jan. 25, 1957).
229. "Humble Oil Radiocarbon Dates II." Brannon, Simons, Perry, Daughtry, and McFarlan. *Science*, v. 125, No. 3254 (May 10, 1957).
230. Humphreys, A. A., and Abbot, H. L., *Report on the Physics and Hydraulics of the Mississippi River*. Washington, Government Printing Office, 1867.
231. Humphreys, W. J., *Physics of the Air*. Philadelphia, Lippincott, 1920.
232. Huntington, Ellsworth, *The Pulse of Asia*. Boston and New York, Houghton Mifflin, 1907.
232a. Inglis, D. R., "The Shifting of the Earth's Axis of Rotation." *Review of Modern Physics*, Vol. 29, pp. 9–19.
233. International Bathymetric Chart of the Oceans. Monte Carlo, Monaco, International Hydrographic Bureau.
233a. Irving, Edward, *Paleomagnetism and Its Application to Geological and Geophysical Problems*. New York, London, Sidney, Wiley and Sons, 1964.
233b. Irving, Edward, and Robertson, W. A., "Test for Polar Wandering." *Journal of Geophysical Research*, Vol. 74 (Feb. 15, 1969).
234. Jaggar, Thomas A., "Living on a Volcano." *National Geographic Magazine*, v. LXVIII, No. 1 (July 1935).

235. Jaggar, Thomas A., *Volcanoes Declare War*. Honolulu, Paradise of the Pacific, 1945.
236. Jannsen, Raymond E., "The History of a River." *Scientific American* (June 1952).
237. Jardetsky, W. S., "The Principal Characteristics of the Formation of the Earth's Crust." *Science*, 119:361–65 (March 19, 1954).
238. Jeffreys, Harold, *The Earth*, 2d ed. New York, Macmillan, 1929.
239. Jeffreys, Harold, *The Earth*, 3d ed. New York, Cambridge University Press, 1953.
240. Jeffreys, Harold, *The Theory of Probability*. Oxford, Clarendon, 1939.
241. Jeffreys, Harold, personal communication.
242. Johnson, Frederick, ed., "Radiocarbon Dating." *Memoirs of the Society for American Archeology*, No. 8 (1951). Salt Lake City, by the Society.
243. Johnson, L. H., "Men and Elephants in America." *Scientific Monthly* (Oct. 1952).
244. Joly, J., "The Theory of Thermal Cycles." *Gerlands Beiträge zur Geophysik*, XIX:415–41 (1928).
245. Kahn, Fritz, *Design of the Universe*. New York, Crown, 1954.
246. Kalb, Bernhard, "Adventure on the Mysterious Continent." *New York Times* Magazine (Feb. 19, 1954).
247. Karlstrom, Thor, "Tentative Correlation of Alaskan Glacial Sequences." *Science*, v. 125, No. 3237 (Jan. 11, 1957).
248. Kelly, Allan O., and Dachille, Frank, *Target Earth*. Pensacola, Florida, Pensacola Engraving, 1953. (In the library of the American Museum of Natural History, New York.)
249. Kent, William, *The Mechanical Engineer's Pocket-Book*, 8th ed. New York, Wiley, 1913.
249a. Khramov, A. N., "Polar Wandering During the Neogene and Quaternary Periods Based on Paleomagnetic Investigations in Western Turkmenia and Some Other Regions," in *Chronicle, Third All-Union Conference on Paleomagnetism*, at A. A. Zhdanov Leningrad State University, October, 1959. Izv. Geophys. Ser. 1960, pp. 1091–1096, translated by I. A. Mamantov.
250. Kimball, Arthur L., *College Physics*, 2d rev. ed. New York, Holt, 1917.
251. King, Lester C., "The Necessity for Continental Drift." *Bulletin of the American Association of Petroleum Geologists*, 37:2163–77 (July–Sept. 1953).
252. Kingdon-Ward, F., "Caught in the Assam-Tibet Earthquake." *National Geographic Magazine*, v. CI, No. 3 (March 1952).
253. Knowlton, F. H., "The Evolution of Geological Climates." *Bulletin of the Geological Society of America*, 30:499–566 (1919).
254. Knowlton, F. H., *The Fossil Forests of Yellowstone National Park*. Washington, Department of the Interior, 1914.

254a. Kobayashi, Kunio, "Late Quaternary Chronology of Japan." Reprinted from the "Earth Science (Chikyu Kagaku), the Journal of the Association for Geological Collaboration in Japan, No. 79 (1965), pp. 1–17.

254b. Kobayashi, Kunio, "Significance of the Ikenotairo Interstadial Indicated by Moraines on Mt. Kumazawa of the Kiso Mountain Range, Central Japan." Reprinted from the *Journal of the Faculty of Science*, Shinshu University, v. 1, No. 2 (Dec. 1966).

254c. Kobayashi, Kunio, "Problems of Late Pleistocene History of Central Japan." *The Geological Society of America, Special Paper 84* (1965). (Reprint.)

254d. Kobayashi, Kunio, "Problems of the Quaternary Chronology of Japan and the Change of Sea Level." Reprinted from "Papers presented at the 1964 Peking Symposium." *Natural Science*, II, pp. 1187–1205 (1965).

255. Köppen, Vladimir, "Das Klima Patagoniens im Tertiär und Quartiär." *Gerlands Beiträge zur Geophysik*, 17:391–94 (1927).

256. Kreichgauer, Damian, *Die Aequatorfrage in der Geologie*. 2d ed. Steyl, Haldenkirche, 1926.

257. Kroeber, Alfred Louis, *Anthropology*, rev. ed. New York, Harcourt, Brace, 1948.

258. Krumbein, W. C., and Sloss, L. I., *Stratigraphy and Sedimentation*. San Francisco, Freeman, 1951.

259. Kulp, J. Laurence, and Carr, Donald R., "Surface Area of Deep-Sea Currents." *Journal of Geology*, v. 60, No. 2 (March 1952).

259a. Kulp, J. Laurence, "Geologic Time Scale." Reprinted from *Science*, v. 133, No. 3459 (April 14, 1961).

260. Lack, D. L., "Darwin's Finches." *Scientific American*, v. 188, No. 16 (April 1953).

261. Ladd, Harold S., and Brown, Roland W., "Fossils Lift the Veil of Time." *National Geographic Magazine*, v. CIX, No. 3 (March 1956).

262. "Lamont Natural Radiocarbon Measurements I," J. Laurence Kulp. *Science*, v. 114, No. 2970 (Nov. 30, 1951).

263. "Lamont Natural Radiocarbon Measurements II," J. Laurence Kulp *et al*. *Science*, v. 116, No. 3016 (Oct. 17, 1952).

264. "Lamont Natural Radiocarbon Measurements III," Broecker, Kulp, and Tucek. *Science*, v. 124, No. 3213 (July 27, 1956).

265. Landes, Kenneth K., *Petroleum Geology*. New York, Wiley, 1951.

266. Lane, F. C., *The Story of Mountains*. New York, Doubleday, 1950.

267. Lane, F. C., *The Mysterious Sea*. Garden City, N. Y., Doubleday, 1947.

268. Laplace, Marquis P. S. de, *Mécanique céleste*. Translated with a commentary by Nathaniel Bowditch. Boston, Little Brown, 1839. 4 vols.

268a. Lear, John, "Canada's Unappreciated Role as Science Innovator."
[J. Tuzo Wilson] *Saturday Review of Literature*, (Sept. 2, 1967).

269. Leighton, M. M., and Wright, H. E., "Radiocarbon Dates of Mankato Drift in Minnesota." *Science*, v. 125, No. 3256 (May 24, 1957).

270. Levitt, J., "The Mechanics of Freezing." *Science*, v. 125, No. 3240 (Feb. 1, 1957).

271. Levy, Hyman, *Modern Science*. New York, Knopf, 1938.

272. Libby, Willard F., *Radiocarbon Dating*. Chicago, University of Chicago, 1952. See also "Chicago Radiocarbon Dates."

272a. Libby, Willard F., "The Accuracy of Radiocarbon Dates." *Science*, v. 140, No. 3564 April 19, 1963), pp. 278–80.

272b. Lisitzyn, A. P., "Bottom Sediments of the Antarctic." In *Antarctic Research, Geophys. Mon.* No. 7, NAS-NRD, Wesley *et al*. Pub. 1036 (1962).

273. Löffelhoz von Colberg, Carl Freiherr, *Die Drehungen der Erdkruste in geologischen Zeiträumen*, 2d ed. Munich, 1895.

274. The London *Times* (Sept. 9, 1954).

275. Longfellow, Dwight W., "Continental Drift and Earth's Magnetic Poles and Foci." *Pan-American Geologist*, v. LV, No. 3 (April 1931).

276. Louisiana Department of Conservation, Bulletin No. 8.

277. Lucas, Frederick A., "The Truth about the Mammoth." *Annual Report of the Smithsonian Institution*, 1899, pp. 353–59.

278. Lull, Richard Swann, *Organic Evolution*. New York, Macmillan, 1927.

279. Luyet, B. J., and Gehenio, P. M., *Life and Death at Low Temperatures*. Normandy, Mo., Biodynamica, 1940.

280. Lydekker, Richard, "Mammoth Ivory." *Smithsonian Reports*, 1899, pp. 361–66.

281. Lyell, Sir Charles, *Principles of Geology*, rev. (9th) ed. New York, Appleton, 1854.

282. Lyell, Sir Charles, *The Antiquity of Man*, 4th ed., rev. Philadelphia, Lippincott, 1873.

283. Lyell, Sir Charles, *Travels in North America*. New York, Wiley, 1852.

284. Lyman, Charles P., personal communication.

285. Ma, Ting Ying H., *Research on the Past Climate and Continental Drift, Vol. I. The Climate and Relative Positions of Eurasia and North America during the Ordovician Period as Determined by the Growth Rate of Corals*. Fukien, China, published by the author, July 1943. (Professor Ma was at this time head of the Department of Oceanography of the China Institute of Geography.)

286. Ma, Ting Ying H., Research on the Past Climate, etc. Vol. II. The Climate and Relative Positions of the Continents during the Silurian Period as Determined by the Growth Rate of Corals. Fukien, published by the author, August 1943.
287. Ma, Ting Ying H., Research on the Past Climate, etc. Vol. III. The Climate and Relative Positions of the Continents during the Devonian Period. Fukien, published by the author, November 1943.
288. Ma, Ting Ying H., Research on the Past Climate, etc. Vol. IV. The Equator and the Relative Positions of the Continents during the Cretaceous Period as Deduced from the Distribution and Growth Values of Reef Corals. Taipei, Taiwan (Formosa), published by the author, August 1951. (Professor Ma has been connected since this time with Taiwan National University.)
289. Ma, Ting Ying H., Research on the Past Climate, etc. Vol. V. The Shifting in Pole-Positions with Diastrophisms since the End of the Cretaceous, and the Accompanying Drift of Continents. (22 plates, 1 chart, 7 figures.) Taipei, Taiwan (Formosa), published by the author, May 1952.
290. Ma, Ting Ying H., Research on the Past Climate, etc. Vol. VI. The Sudden Total Displacement of the Outer Solid Earth Shell by Slidings Relative to the Fixed Rotating Core of the Earth. Taipei, Taiwan (Formosa).
    NOTE: These volumes are available at the Schermerhorn Library, Columbia University. They may also be obtained from the World Book Co., Ltd., 99 Chung King Road, 1st section, Taipei, Taiwan (Formosa), China.
291. Ma, Ting Ying H., "Alteration of Sedimentary Facies on the Ocean Bottom and Shortness of the Period of Diastrophism after a Sudden Total Displacement of the Solid Earth Shell." Oceanographica Sinica, v. II, Fasc. I (Sept. 1955).
291a. MacDonald, Gordon J. F., "The Deep Structure of Continents," Science, Vol. 143, No. 3609 (Feb. 28, 1964).
291b. Malaise, René, "Oceanic Bottom Investigations and Their Bearings on Geology." Reprinted from Geologiska Foreningens I. Stockholm, Förhandlingar (March–April 1957).

291c. Mallery, Arlington H., Lost America. Washington, Overlook, 1956.
291d. Maney, Charles A., "Experimental Study of Sliding Friction," American Journal of Physics (April 1952).
291e. Mansinha, L. and Smylie, D. E., "Effect of Earthquakes on the Chandler Wobble and the Secular Pole Shift." Journal of Geophysical Research, Vol. 72, No. 18 (Sept. 15, 1967).

291f. Mansinha, L. and Smylie, D. E., "Earthquakes and the Earth's Wobble." Science, Vol. 161, No. 3846 (Sept. 13, 1968).
291g. Mansinha, L. and Smylie, D. E., "Earthquakes and the Observed Motion of the Rotation Pole." Journal of Geophysical Research, Vol. 74, No. 24 (Dec. 15, 1968).

292. Manson, Marsden, "The Evolution of Climates." Revised and reprinted from *The American Geologist*. Minneapolis, Franklin,
1903.

292a. Markowitz, William, and Guinot, B., *Continental Drift, Secular
Motion of the Pole, and Rotation of the Earth*. Symposium No.
2. International Astronomical Union, Dordrecht, Holland, D.
Reidel; New York, Springer-Verlag, 1968.

293. Marriott, R. A., *Warmer Winters and the Earth's Tilt*. Exeter,
England, Pollard, 1921.

294. Mather, Kirtley F., conference.

294a. Matthes, François E., "Rebirth of the Glaciers in the Sierra
Nevada during Late Post-Pleistocene Time." *Bulletin of the
Geological Society of America*, v. 52, No. 12, Pt. 2, p. 2030.

295. Matthew, W. D., *Climate and Evolution*, 2d ed. New York, New
York Academy of Sciences, 1939.

296. Maxwell, James Clerk, "On a Dynamical Top." *The Scientific
Papers of James Clerk Maxwell*, Vol. I, edited by W. D. Niven,
M.A., F.R.S., Cambridge, University Press, 1890. (2 vols.)

297. Mayo, Captain Charles A., conference.

298. Mayor, A. G., "Growth Rate of Samoan Corals." Papers from the
Department of Marine Biology of the Carnegie Institution of
Washington, Vol. XIX. *Some Posthumous Papers of A. G. Mayor*,
pp. 51–72.

299. Meinesz, F. A. Vening, "Indonesian Archipelago: A Geophysical
Study." *Bulletin of the Geological Society of America*, v. 65, No.
2, pp. 143–64 (Feb. 1954).

300. Meinesz, F. A. Vening, "Major Tectonic Phenomena and the
Hypothesis of Convection Currents in the Earth." *Quarterly
Journal of the Geological Society of London*, 103:191–207 (Jan.
31, 1948).

301. Meinesz, F. A. Vening, "Shear Patterns of the Earth's Crust."
Transactions of the American Geophysical Union, 28:1–61
(1947).

301a. Mellor, M., "Ice Flow in Antarctica." *Journal of Glaciology*,
3(25):377–384.

302. Menard, Henry W., "Fractures in the Pacific Floor." *Scientific
American*, v. 193, No. 1 (July 1955).

302a. Menard, Henry W., "The East Pacific Rise: Convection currents
in the mantle may account for this bulge on the ocean floor."
Scientific American, v. 205, No. 6 (Dec. 1961).

303. Merrill, Elmer D., *Plant Life of the Pacific World*. New York,
Macmillan, 1945.

304. Meryman, Harold T., "Mechanics of Freezing in Living Cells and
Tissues." *Science*, v. 124, No. 3221 (Sept. 21, 1956).

305. Meryman, Harold T., [Reply to Levitt]. *Science*, v. 125, No. 3240
(Feb. 1, 1957).
Michigan Radiocarbon Dates. See "University of Michigan Radio-
carbon Dates."

306. Miller, Arthur Austin, *The Skin of the Earth*. London, Methuen, 1953.
307. Miller, Hugh, *The Old Red Sandstone*, 14th ed. Edinburgh, Nimmo, 1871.
308. Millis, John, "The Glacial Period and Drayson's Hypothesis." *Popular Astronomy*, v. 30, No. 10. Reprinted for the Fortean Society by K. M. McMahon, 1945.
309. Moodie, Roy L., *A Popular Guide to the Nature and the Environment of the Fossil Vertebrates of New York*. Albany, University of the State of New York, 1933.
310. Moody, Paul A., *Introduction to Evolution*. New York, Harper, 1953.
311. Moore, Raymond C., "Late Paleozoic Cyclic Sedimentation in the Central United States." 1950. Reprinted from the *International Geological Congress Report of the 18th Session, Great Britain, 1948*, Pt. IV.
312. Muench, O. B., "Determining Geological Age from Radioactivity." *Scientific Monthly*, v. LXXI, No. 5 (Nov. 1950).
313. Munk, W. H., "Polar Wandering: A Marathon of Errors." *Nature*, v. 177, No. 4508, p. 551.
313a. Munk, W. H., and MacDonald, G. J. F., *The Rotation of the Earth*. Cambridge, The University Press, 1960.
313b. Munk, W. H., and MacDonald, G. J. F., "Continentality and the Gravitational Field of the Earth." *Journal of Geophysical Research*, v. 65, pp. 2169–72 (1960).
314. Munk, Walter, and Revelle, Roger, "Sea Level and the Rotation of the Earth." *American Journal of Science*, 250:829–33 (Nov. 1952).
315. Munk, Walter, and Revelle, Roger, "Remarks at the Geophysical Congress in Rome, 1954." *New York Herald Tribune* (Nov. 9, 1954).
315a. Munyan, Arthur C., editor, *Polar Wandering and Continental Drift*. Society of Economic Paleontologists and Mineralogists, Special Publication, No. 10. Tulsa, Oklahoma, July 1963.
316. Murray, Raymond C., "Directions of Ice Motion in South-Central Newfoundland." *Journal of Geology*, v. 63, No. 3 (May 1955).
316a. Nadai, Arpad, *Plasticity; a Mechanics of the Plastic State of Matter*. New York, McGraw-Hill, 1931.
317. Nagamiya, T., Yosida, K., and Kubo, R., "Antiferromagnetism." *Advances in Physics*, v. 4, No. 13 (Jan. 1955).
318. Nagata, T., *Rock Magnetism*. Tokyo, Maruzen, 1953.
318a. Nagata, T., Akimoto, S., Shimizu, Y., Kobayashi, K., and Kuno, H., "Palaeomagnetic Studies in Tertiary and Cretaceous Rocks in Japan." *Proceedings of the Japanese Academy*, v. 35, pp. 378–83 (1959).

318b. Nagata, T., Akimoto, S., Uyeda, S., Shimizu, Y., Ozima, M., and Kobayashi, K., "Paleomagnetic Study on a Quaternary Volcanic Region in Japan." *Advances in Physics*, vol. 6, No. 23, pp. 255–263 (July 1957).

319. Nares, Capt. Sir G. S., *Narrative of a Voyage to the Polar Sea during 1875–6*, 2d ed. London, Samson Low, Marston, Searle, and Rivington, 1878, 2 vols.

320. Nathorst, A. G., "On the Value of the Fossil Floras of the Arctic Regions as Evidence of Geological Climates." *Smithsonian Reports*, 1911.

321. *National Geographic Magazine*, v. CI, No. 1 (Jan. 1952). Meen and Stewart, "Solving the Riddle of the Chubb Crater."

322. *National Geographic Magazine*, v. XCVIII, No. 4 (Oct. 1950). Rees and Bell, "Sky-High Bolivia."

323. Neel, L., "Inversion de l'aimantation permanente des roches." *Annales géophysiques*, 7:90–102 (1951).

324. Neel, L., "Some Theoretical Aspects of Rock Magnetism." *Advances in Physics*, v. 4, No. 14, pp. 191–243 (April 1955).

324a. Negris, P., "Atlantide." *Revue scientifique*, v. 60, pp. 614–17 (Sept. 23, 1922).

325. Neuville, H., "On the Extinction of the Mammoth." *Smithsonian Reports*, 1919.

326. *New and Old Discoveries in Antarctica* (Discussion of the Piri Reis Map). The Georgetown University Forum, verbatim text (mimeographed) of the radio broadcast of August 26, 1956.

326a. Newell, Norman D., "Crises in the History of Life. Why do whole groups of animals disappear simultaneously from the fossil record? *Scientific American*, v. 208, No. 2 (Feb. 1963).

327. The New York Times (Dec. 9, 1949).

328. The New York Times (Sept. 8, 1954).

329. The New York Times (Oct. 16, 1955).

330. The New York Times (May 6, 1956).

331. The New York Times (Sept. 23, 1956).

332. The New York Times (Dec. 26, 1956).

333. Nicholls, G. D., "The Mineralogy of Rock Magnetism." *Advances in Physics*, v. 4, No. 14 (April 1955).

333a. Nichols, R. L., "Geomorphology of Marguerite Bay area, Palmer Peninsula, Antarctica." *Bulletin of the Geological Society of America* 71(10), 1960, pp. 1421–50.

333b. Nordeng, Stephan C., "Precambrian Stromatolites as Indications of Polar Shift." In *Polar Wandering and Continental Drift*, ed. Munyan, 1963.

334. Nordenskjöld, N. A. E., *The Voyage of the Vega round Asia and Europe*. Translated by Alexander Leslie. London, Macmillan, 1881. 2 vols.

335. Nordenskjöld, Dr. N. O. G., and Anderson, Dr. J. G., *Antarctica, or Two Years Amidst the Ice of the South Pole*. London and New York, Macmillan, 1905.
336. North American Newspaper Alliance. "Shifts in Earth's Crust Millions of Years Ago a Puzzle to Geologists." Dec. 15, 1949.
336a. *Oceanographic Atlas of the Polar Seas*. Part I—Antarctica. H. O. Pub. 705, Washington, D.C.
337. Olivier, Charles P., *Meteors*. Baltimore, Williams and Wilkins, 1925.
338. O'Neill, John J., "Clues to Life's Secret." New York *Herald Tribune*, Sect. 2, p. 12 (April 13, 1952).
339. Opdyke, N. D., and Runcorn, S. K., "New Evidence for the Reversal of the Geomagnetic Field Near the Plio-Pleistocene Boundary." *Science*, v. 123, No. 3208 (June 22, 1956).
340. Osborn, Henry Fairfield, *The Proboscidea*. New York, American Museum of Natural History, 1936–42.
340a. Oxburgh, E. R. and Turcotte, D. L., "Mid-Ocean Ridges and Geotherm Distribution During Mantle Convection." *Journal of Geophysical Research*, Vol. 73, No. 8 (April 15, 1968).
341. Parks, J. M., "Corals from the Brazer Formation (Mississippian) of Northern Utah." *Journal of Paleontology*, 25:171–86 (March 1951).
342. Pauly, K. A., "The Cause of Great Ice Ages." *Scientific Monthly* (Aug. 1952).
343. Peattie, Roderick, ed., *The Inverted Mountains: Canyons of the West*. New York, Vanguard, 1948.
343a. Pettersson, Hans, "The Swedish Deep-Sea Expedition [1947–1948]." *Geological Journal*, v. 114, Nos. 4–6 (1948); v. 114, No. 406 (1949).
344. Piggott, C. S., and Urry, W. D., "Time Relations in Ocean Sediments." *Bulletin of the Geological Society of America*, v. 53, No. 8, pp. 1187–1210 (Aug. 1942).
345. Pirsson, Louis V., and Schuchert, Charles, *A Textbook of Geology*. New York, Wiley, 1929.
346. Plass, Gilbert N., "Smoky World Gets Hotter." *Science Digest*, back cover (August 1953).
347. Plumb, Robert K., New York *Times*, Science Review (Jan. 4, 1953).
348. Poddar, M. C., "Preliminary Report of the Assam Earthquake of August 15, 1950." Bulletin, Series B, Engineering Geology and Ground Water, No. 3. The Geological Survey of India.
349. Poldervaart, Arie, ed., *Symposium on the Crust of the Earth*. Special Paper 62. Geological Society of America, 1955.
349a. Pollock, James B., "Fringing and Fossil Reefs of Oahu." *Bernice P. Bishop Museum Bulletin No. 55*. Honolulu, 1928.
349b. Press, Frank, personal communication.
349c. Press, Frank, "Studies Support Lava-Sea Theory." New York *Times* (September 17, 1961) and personal correspondence.

349d. Priestly, Raymond E., *Antarctic Adventure*. New York, Dutton, 1915.

350. Pringle, R. W., *et al.*, "Radiocarbon Age Estimates Obtained by an Improved Liquid Scintillation Technique." *Science*, v. 125, No. 3237 (Jan. 11, 1957).

351. *Proceedings of the British Association for the Advancement of Science*, p. 277 (Dec. 1950).

351a. "Radiocarbon." *American Journal of Science*, Sterling Tower, Yale University, New Haven, Conn., vols. I-X (Nos. 1, 2.) 1959–1968.

351b. Raeside, J. D., "Some Postglacial Climatic Changes in Canterbury (New Zealand)." *Royal Society of New Zealand Transactions and Proceedings*, v. 77, Part 1 (1948), pp. 153–71.

352. Ramsay, William, "Probable Solution of the Climatic Problem in Geology." *Smithsonian Reports*, 1924.

352a. Rapport, Samuel, and Wright, Helen, "The Crust of the Earth." In *An Introduction to Geology*. A Mentor Book (MD 264) 2d printing, 1959.

352b. Rasch, G. O., ed., *Geology of the Arctic*. University of Toronto Press, 1961.

353. Reibisch, P., "Ein Gestaltungsprinzip der Erde." *Jahresber. d. Ver. Erdkunde zu Dresden*, 27:105–24 (1904). (Second part appeared in 1905, and third part, dealing with the ice age, in 1907.)

354. Reid, Harry Fielding, *The Influence of Isostasy on Geological Thought*. Bulletin 78, The National Research Council, Washington, 1931. (Ch. VIII.)

355. *Report of the Committee on the Measurement of Geologic Time*. National Research Council, Publication 212, Washington, 1952.

356. Rich, J. L., "The Origin of Compressional Mountains." *Bulletin of the Geological Society of America*, 62:1179–1222 (1951).

357. Robertson, Eugene, personal communication.

357a. Robin, Gordon de Q., "The Ice of the Antarctic." *Scientific American*, v. 207, No. 3 (Sept. 1962).

358. Rode, K. P., "A Theory of Sheet Movement and Continental Expansion." *Memoirs of Rajputana University Department of Geology*, No. 1. Udaipur, 1953.

359. Rotta, H., "Elementary Particles, Atomic Power, and Drift of Continents." (Report of the 1955 meeting of winners of the Nobel Prize.) *Science*, v. 125, No. 3258 (June 7, 1957).

359a. Rouse, George E., and Bisque, Ramon E., "Global Tectonics and the Earth's Core." *The Mines Magazine* (March 1968).

360. Runcorn, S. K., "Rock Magnetism—Geophysical Aspects." *Advances in Physics*, v. 4, No. 14 (April 1955).

361. Runcorn, S. K., "Theory of Polar Change through Subcrustal Currents." *Advances in Physics*, v. 4, No. 4 (1955).

362. Runcorn, S. K., "Rock Magnetism" (reply to Clegg). *Nature*, v. 179, No. 4565 (April 27, 1957).

363. Runcorn, S. K., address at Columbia University.

363a. Runcorn, S. K., ed., *Continental Drift*. New York and London, Academic Press, 1962.

364. Saks, N. V., Belov, N. A., and Lapina, N. N., "Our Present Concepts of the Geology of the Central Arctic." Translated by E. R. Hope from *Priroda*, Defense Scientific Information Service, Defense Research Board, Ottawa, Canada. Oct. 4, 1955. (T 196 R.)

365. Sanderson, Ivan T., personal communication.

366. Sarton, George, personal communication (see Appendix).

366a. Sayles, R. W., "Bermuda during the Ice Age." *Proceedings of the American Academy of Arts and Sciences*, v. 66, No. 11 (1931).

366b. Schaeffer, O. A., ed., *Potassium Argon Dating*. New York, Springer-Verlag.

367. Scholander, P. F., "The Wonderful Net." *Scientific American*, v. 196, No. 4 (April 1957).

368. Scholander, P. F., and Kanwisher, John W., "Gases in Icebergs." *Science*, v. 123, No. 3186 (Jan. 20, 1956).

369. Schuchert, Charles, "Paleography of North America." *Bulletin of the Geological Society of America*, 20:427–606 (Feb. 5, 1910).

369a. Schuchert, Charles, *Textbook of Geology. Part II. Historical Geology*. New York, Wiley, 1924.

369aa. Schulthess, Emil, *Antarctica*. New York, Simon & Schuster, 1960.

369b. Schwinner, R., *Lehrbuch der physikalischen Geologie*, I:242. Borntraeger, 1936.

370. *Science News Letter* (May 19, 1951).

371. "History in Bricks." *Scientific American*, v. 196, No. 2 (Feb. 1957).

372. Scott, W. B., *A History of Land Mammals in the Western Hemisphere*. New York, Macmillan, 1937.

373. Seward, A. C., *Plant Life through the Ages*. New York, Macmillan, 1931.

374. Shackleton, E. H., *The Heart of the Antarctic*. Philadelphia, Lippincott, 1909. 2 vols.

375. Shapley, Harlow, ed., *Climatic Change*. Cambridge, Harvard University Press, 1953.

376. Shapley, Harlow, personal communication.

377. Shepard, F. P., and Suess, H. E., "Rate of Postglacial Rise of Sea-Level." *Science*, v. 123, No. 3207 (June 15, 1956).

378. Shull, A. Franklin, *Principles of Animal Biology*. New York, McGraw-Hill, 1929.

379. Simpson, George Gaylord, "Periodicity in Vertebrate Evolution." *Journal of Paleontology*, 26:359–70 (1952).

380. Simpson, George Gaylord, *The Major Features of Evolution*. New York, Columbia University Press, 1953.

381. Simpson, George Gaylord, *Tempo and Mode in Evolution*. New York, Columbia University Press, 1944.

382. Simpson, George Gaylord, *The Meaning of Evolution*. New Haven, Yale, 1952. New York, New American Library, 1952.

383. Simroth, H., *Die Pendulationstheorie*. Leipzig, 1907.
384. de Sitter, L. A., "Pliocene Uplift of Tertiary Mountain Chains." *American Journal of Science*, 250:297–307 (April 1952).
384a. Sleep, Norman H., "The Sensitivity of Heat Flow and Gravity to the Mechanism of Sea-Floor Spreading." *Journal of Geophysical Research*, Vol. 74, No. 2 (Jan. 15, 1969).
385. Slichter, L. B., Rancho Santa Fe conference concerning the evolution of the earth: report of topics discussed. *Proceedings of the National Academy of Sciences*, 36:511–14 (Sept. 1950).
386. Smart, W. M., *The Origin of the Earth*. Cambridge, University Press, 1951.
387. Smith, Albert C., in the New York *Times* (Dec. 4, 1955).
388. Smith, J. L. B., "Investigation of the Coelacanth." *Science*, v. 117, No. 3040 (April 3, 1953).
389. Smith, Stanley, *Upper Devonian Corals of the Mackenzie River Region, Canada*. Geological Society of America, Special Paper 59, 1945.
389a. Smith, Warren, *The Geologic Structure of the Philippine Archipelago*. 'S-Gravenhage, Mouton, 1925.
390. Southwell, R. V., *An Introduction to the Theory of Elasticity for Engineers and Physicists*. Oxford, Clarendon, 1936.
391. Spitaler, Rudolf, *Die Eiszeiten und Polschwankungen der Erde*. Vienna, Aus der Kaiserlich-Königlichen Hof- und Staatsdruckerei, 1912.
391a. *Sputnik*. Monthly Digest of Current Soviet Writing (November, 1968).
392. Stearns, Harold T., "An Integration of Coral Reef Hypothesis." *American Journal of Science*, v. 244, No. 4 (April 1946).
393. Steffens, Lincoln, *Autobiography*. New York, Harcourt, Brace, 1931.
393a. Stehli, F. G., and Helsley, C. E., "Paleontologic Technique for Defining Ancient Pole Positions." *Science*, v. 142, No. 3595 (Nov. 22, 1963).
394. Stetson, H. C., "Geology and Paleontology of the Georges Bank Canyon." *Bulletin of the Geological Society of America*, XLVII: 339–66 (March 31, 1936).
394a. Steuerwald, B. A., Clark, D. L. and Andrew, J. A., "Magnetic Stratigraphy and Faunal Patterns in Arctic Ocean Sediments." *Earth and Planetary Science Letters* 5, pp. 79–85 (1968).
395. Stokes, William Lee, "Another Look at the Ice Age." *Science*, v. 122, No. 3174 (Oct. 28, 1955).
396. Stoner, E. C., "Rocks and the Earth's Magnetic Field." (Review of Blackett, P. M. S., *Lectures on Rock Magnetism*.) *Nature*, v. 179, No. 4557 (March 2, 1957).
397. Stutzer, Otto, *The Geology of Coal*. Translated and revised by Adolph C. Noe. Chicago, University of Chicago Press, 1940.
398. Suess, Eduard, *The Face of the Earth*. Oxford, Clarendon, 1904. 5 vols.

399. Suess, H. E., "Absolute Chronology of the Last Glaciation." *Science*, v. 123, No. 3192 (March 2, 1956).

400. Sukachev, V. N., *Scientific Findings of an Expedition Outfitted by the Imperial Academy of Sciences for the Purpose of Excavating the Mammoth Found by the Beresovka River in 1901*, v. III. Petrograd, 1914. (The article by Sukachev was translated for this book by Mrs. Norman Hapgood.)

401. Sullivan, Walter, "U.S. Team Studies of Ice in Antarctica." New York *Times* (Feb. 1, 1957).

401a. Sullivan, Walter, "Studies Support Lava-Sea Theory." New York *Times* (Sept. 17, 1961).

402. Sverdrup, H. U., Johnson, M. W., and Fleming, R. H., *The Oceans, Their Physics, Chemistry, and General Biology*. New York, Prentice-Hall, 1942.

403. Taber, Stephen, "Perennially Frozen Ground in Alaska: Its Origin and History." *Bulletin of the Geological Society of America*, 54:1433–1548 (1943).

404. Tanner, William F., "North-South Asymmetry of the Pleistocene Ice Sheet." *Science*, v. 122, No. 3171 (Oct. 7, 1955).

405. Tanni, L., *The Continental Undulations of the Geoid as Determined by Present Gravity Material*. Publication 18 of the Isostatic Institute of the International Association of Geodesy. Helsinki, Finland, 1948.

406. Tatel, H. E., and Tuve, M. A., *Seismic Studies (The Earth's Crust)*. Carnegie Institution of Washington. Year Book No. 52, for the year 1952–1953. Annual Report of the Director. December 11, 1953. (Reprint.)

407. Tazieff, Haroun, *Craters of Fire*. New York, Harper, 1952.

408. *Technical Report of the International Conference on Rock Magnetism*. See Dietz, Robert S.

409. Termier, Pierre, "The Drifting of Continents." *Smithsonian Reports*, 1924.

409a. Tilton, G. R., and Hart, S. R., "Geochronology." *Science*, v. 140, No. 3565 (April 26, 1963).

410. *Time* Magazine (Dec. 26, 1949).

411. *Time* Magazine (Sept. 8, 1954).

412. Tolmachev, I. P., "The Carcasses of the Mammoths and Rhinoceroses Found in the Frozen Ground of Siberia." *Transactions of the American Philosophical Society*. New Series, v. XXIII, Pt. 1. Philadelphia, 1929.

413. Torreson, O. W., Murphy, T., and Graham, J. W., "Magnetic Polarization of Sedimentary Rocks and the Earth's Magnetic History." *Journal of Geophysical Research*, v. 54, No. 2 (June 1949).

414. Turekian, Karl K., "The Stratigraphic and Paleoecological Significance of the SR/CA Ratio in Sediments and Fossils." *Petroleum Geologist* (March 28, 1955).

415. Turner, F. J., and Verhoogen, Jean, *Igneous and Metamorphic Petrology.* New York, McGraw-Hill, 1951.
416. Tuve, M. A., *Annual Report, Director of the Department of Terrestrial Magnetism,* Carnegie Institution of Washington, 1948–1949. Carnegie Institution Yearbook, No. 49.
417. Tuve, M. A., *Annual Report, Director of the Department of Terrestrial Magnetism,* Carnegie Institution of Washington, 1949–1950. Carnegie Institution Yearbook, No. 50.
418. Tuve, M. A., *Annual Report, Director of the Department of Terrestrial Magnetism,* Carnegie Institution of Washington, 1955–1956. Carnegie Institution Yearbook, No. 56.
418a. Twenhofel, W. H., ed., "Evironmental Significance of Dwarfed Faunas, A Symposium." *Journal of Sedimentary Petrology,* v. 18, No. 2 (Aug. 1948).
418b. Twenhofel, W. H., *Principles of Sedimentation.* New York and London, McGraw-Hill, 1939.
419. Umbgrove, J. H. F., "Recent Theories of Polar Displacement." *American Journal of Science,* v. 224, No. 2 (Feb. 1946).
420. Umbgrove, J. H. F., *The Pulse of the Earth,* 2d ed. The Hague, Martinus Nijhoff, 1947.
421. "U. S. Geological Survey Radiocarbon Dates I." H. E. Suess, *Science,* v. 120, No. 3117 (Sept. 24, 1954).
422. "U. S. Geological Survey Radiocarbon Dates II." Rubin, Meyer, and Suess. *Science,* v. 121, No. 3145 (April 8, 1955).
423. "U. S. Geological Survey Radiocarbon Dates III." *Science,* v. 123, No. 3194 (March 16, 1956).
424. "University of Michigan Radiocarbon Dates I." H. R. Crane. *Science,* v. 124, No. 3224 (Oct. 12, 1956).
425. "University of Pennsylvania Radiocarbon Dates I." *Science,* v. 121, No. 3136 (Feb. 4, 1955).
426. Urey, Harold C., "Measurement of Paleotemperatures and Temperatures, etc." *Bulletin of the Geological Society of America,* v. 62, No. 4 (April 1951).
427. Urey, Harold C., *The Planets, Their Origin and Development.* New Haven, Yale, 1952.
428. Urry, W. D., "Radioactivity of Ocean Sediments VI." *American Journal of Science,* 247:257–75.
429. Urry, W. D., "The Radioelements in Non-Equilibrium Systems." *American Journal of Science,* 240:426–36.
429a. Ushakov, S. A., and Lazarev, G. Y., "The Question of Isostatic Equilibrium in Antarctica." *Information Bulletin, Soviet Antarctic Expedition,* 1959, p. 138–143.
430. Victor, Paul Emile, *Rapport préliminaire.* Expéditions polaires françaises, Campagne au Groenland, 1950. (Mimeographed.)
431. Visher, S. S., "Tropical Cyclones and the Dispersal of Life from Island to Island in the Pacific. *Smithsonian Report,* 1925.

431a. Volchok, H. L., and Kulp, J. L., "The Ionium Method of Age Determination." *Geochimica et Cosmochimica Acta*, 11:219–24 (1957). (Pergamon Press, Ltd., London.)

432. Wade, A., "The Geology of the Antarctic Continent." *Proceedings of the Royal Society of Queensland*, v. 52 (1941).

433. Wallace, Alfred Russel, *The Malay Archipelago*. New York, Harper, 1869.

434. Wallace, Alfred Russel, *Darwinism*. London and New York, Macmillan, 1891.

435. Wallace, Alfred Russel, *Island Life*, 3d rev. ed. London, Macmillan, 1902.

436. Wanless, Harold R., and Weller, J. Marvin, "Correlation and Extent of Pennsylvania Cyclotherms." *Bulletin of the Geological Society of America*, 43:1003–16 (1932).

437. Washington, H. S., "Deccan Traps and Other Plateau Basalts." *Bulletin of the Geological Society of America*, 33:765 (1922).

437a. Watkins, N. D., and Richardson, A., "Comments on the Relationship Between Magnetic Anomalies, Crustal Spreading and Continental Drift." *Earth and Planetary Science Letters* 4 (1968), pp. 257–64.

438. Webster, A. G., *The Dynamics of Particles and of Rigid, Elastic and Fluid Bodies*, 3d ed. Leipzig, Teubner, 1925.

438a. Weertmen, J., "Equilibrium Profile of Ice Caps." *Journal of Glaciology*, 3(30):953–964, 1959.

439. Wegener, Alfred, *The Origins of Continents and Oceans*. Translated from the third German edition by J. G. A. Skerl. London, Methuen, 1924.

440. Weller, J. Marvin, "Cyclical Sedimentation of the Pennsylvanian Period and Its Significance." *Journal of Geology*, v. XXXVIII, No. 2 (Feb.–March 1930).

441. Westergaard, H. M., *Theory of Elasticity and Plasticity*. Cambridge, Harvard, 1952. New York, Wiley, 1952.

441a. White, Harvey E., *Classical and Modern Physics*. New York, Van Nostrand, 1946.

442. Whitley, D. Gath, "The Ivory Islands of the Arctic Ocean." *Journal of the Philosophical Society of Great Britain*, v. 12 (1910).

442a. Whitney, P. R., *The rubidium-strontium geochronology of argillaceous sediments*. Ph.D. thesis, M.I.T., 1962.

443. Wilson, J. H., *The Glacial History of Nantucket and Cape Cod*. New York, Columbia University Press, 1906.

444. Wilson, J. Tuzo, "On the Growth of Continents." *Papers and Proceedings of the Royal Society of Tasmania*. Hobart, Australia, 1951.

445. Wilson, J. Tuzo, "The Origin of the Earth's Crust." *Nature*, v. 179, No. 4553 (Feb. 2, 1957).

445a. Wilson, J. Tuzo, "Advice for the Establishment." In *The Saturday Review of Literature* (Sept. 2, 1967).

445b. Wilson, J. Tuzo, "Continental Drift." (Reprint.) *Scientific American* (April 1963).

445c. Wilson, J. Tuzo, "Pattern of Uplifted Islands in the Main Ocean Basins." *Science*, v. 129, No. 3555 (Feb. 15, 1963).

446. Woerkom, A. J. van, in Shapley, *Climatic Change*.

447. Wool Industries Research Association, Leeds, England, personal communication.

448. Wright, G. F., *The Ice Age in North America, and Its Bearing upon the Antiquity of Man*. New York, Appleton, 1889.

449. Wright, G. F., *Greenland Ice Fields and Life in the North Atlantic*. New York, Appleton, 1896.

450. Wright, W. B., *The Quaternary Ice Age*. London, Macmillan, 1937.

451. Wright, C. S., and Priestly, R. E., *Glaciology*. London, Harrison, 1922.

452. Wulff, E. V., *An Introduction to Historical Plant Geography*. Waltham, Mass., Chronica Botanica, 1943.

453. "Yale Natural Radiocarbon Dates I." Blau, Deevey, and Gross. *Science*, v. 118, No. 3052 (July 3, 1953).

454. "Yale Natural Radiocarbon Dates II." Preston, Person, and Deevey. *Science*, v. 122, No. 3177 (Nov. 18, 1955).

455. Zeuner, Frederick E., *Dating the Past*. 4th ed. revised and enlarged. London, Methuen, 1958. Reprinted, 1964.

456. Zhivago, A. V., and Lisitzyn, A. P., "New Data on the Bottom Relief and Sediments of the Antarctic Seas." *Izv. Acad. Sci., USSR, Geogr. Ser. No. 1, 1957*.

INDEX OF NAMES

INDEX OF SUBJECTS

NEW BOOKS

THE TIME TRAVEL HANDBOOK
A Manual of Practical Teleportation & Time Travel
edited by David Hatcher Childress
In the tradition of *The Anti-Gravity Handbook* and *The Free-Energy Device Handbook*, science and UFO author David Hatcher Childress takes us into the weird world of time travel and teleportation. Not just a whacked-out look at science fiction, this book is an authoritative chronicling of real-life time travel experiments, teleportation devices and more. *The Time Travel Handbook* takes the reader beyond the government experiments and deep into the uncharted territory of early time travellers such as Nikola Tesla and Guglielmo Marconi and their alleged time travel experiments, as well as the Wilson Brothers of EMI and their connection to the Philadelphia Experiment—the U.S. Navy's forays into invisibility, time travel, and teleportation. Childress looks into the claims of time travelling individuals, and investigates the unusual claim that the pyramids on Mars were built in the future and sent back in time. A highly visual, large format book, with patents, photos and schematics. Be the first on your block to build your own time travel device!
316 PAGES. 7X10 PAPERBACK. ILLUSTRATED. $16.95. CODE: TTH.

PATH OF THE POLE
Cataclysmic Pole Shift Geology
by Charles Hapgood
Maps of the Ancient Sea Kings author Hapgood's classic book *Path of the Pole* is back in print! Hapgood researched Antarctica, ancient maps and the geological record to conclude that the Earth's crust has slipped in the inner core many times in the past, changing the position of the pole. *Path of the Pole* discusses the various "pole shifts" in Earth's past, giving evidence for each one, and moves on to possible future pole shifts. Packed with illustrations, this is the sourcebook for many other books on cataclysms and pole shifts such as *5-5-2000: Ice the Ultimate Disaster* by Richard Noone. A planetary alignment on May 5, 2000 is predicted to cause the next pole shift—a date that is less than a year away! With Millennium Madness in full swing, this is sure to be a popular book.
356 PAGES. 6X9 PAPERBACK. ILLUSTRATED. $16.95. CODE: POP.

IN SEARCH OF ADVENTURE
A Wild Travel Anthology
compiled by Bruce Northam & Brad Olsen

An epic collection of 100 travelers' tales—a compendium that celebrates the wild side of contemporary travel writing—relating humorous, revealing, sometimes naughty stories by acclaimed authors. Indeed, a book to heat up the gypsy blood in all of us. Stories by Tim Cahill, Simon Winchester, Marybeth Bond, Robert Young Pelton, David Hatcher Childress, Richard Bangs, Linda Watanabe McFerrin, Jorma Kaukonen, and many more.
459 PAGES. 6X9 PAPERBACK. ILLUSTRATED. $17.95. CODE: ISOA

ECCENTRIC LIVES AND PECULIAR NOTIONS
by John Michell
The first paperback edition of Michell's fascinating study of the lives and beliefs of over 20 eccentric people. Published in hardback by Thames & Hudson in London, *Eccentric Lives and Peculiar Notions* takes us into the bizarre and often humorous lives of such people as Lady Blount, who was sure that the earth is flat; Cyrus Teed, who believed that the earth is a hollow shell with us on the inside; Edward Hine, who believed that the British are the lost Tribes of Israel; and Baron de Guldenstubbe, who was sure that statues wrote him letters. British writer and housewife Nesta Webster devoted her life to exposing international conspiracies, and Father O'Callaghan devoted his to opposing interest on loans. The extraordinary characters in this book were—and in some cases still are—wholehearted enthusiasts for the various causes and outrageous notions they adopted, and John Michell describes their adventures with spirit and compassion. Some of them prospered and lived happily with their obsessions, while others failed dismally. We read of the hapless inventor of a giant battleship made of ice who died alone and neglected, and of the London couple who achieved peace and prosperity by drilling holes in their heads. Other chapters on the Last of the Welsh Druids; Congressman Ignacius Donnelly, the Great Heretic and Atlantis; Shakespearean Decoders and the Baconian Treasure Hunt; Early Ufologists; Jerusalem in Scotland; Bibliomaniacs; more.
248 PAGES. 6X9 PAPERBACK. ILLUSTRATED. $14.95. CODE: ELPN.

THE CHRIST CONSPIRACY
The Greatest Story Ever Sold
by Acharya S.
In this highly controversial and explosive book, archaeologist, historian, mythologist and linguist Acharya S. marshals an enormous amount of startling evidence to demonstrate that Christianity and the story of Jesus Christ were created by members of various secret societies, mystery schools and religions in order to unify the Roman Empire under one state religion. In developing such a fabrication, this multinational cabal drew upon a multitude of myths and rituals that existed long before the Christian era, and reworked them for centuries into the religion passed down to us today. Contrary to popular belief, there was no single man who was at the genesis of Christianity; Jesus was many characters rolled into one. These characters personified the ubiquitous solar myth, and their exploits were well known, as reflected by such popular deities as Mithras, Heracles/Hercules, Dionysos and many others throughout the Roman Empire and beyond. The story of Jesus as portrayed in the Gospels is revealed to be nearly identical in detail to that of the earlier savior-gods Krishna and Horus, who for millennia preceding Christianity held great favor with the people. *The Christ Conspiracy* shows the Jesus character as neither unique nor original, not "divine revelation." Christianity re-interprets the same extremely ancient body of knowledge that revolved around the celestial bodies and natural forces. The result of this myth making has been "The Greatest Story Ever Sold."
256 PAGES. 6X9 PAPERBACK. ILLUSTRATED. $14.95. CODE: CHRC.

24 HOUR CREDIT CARD ORDERS—CALL: 815-253-6390 FAX: 815-253-6300
EMAIL: AUPHQ@FRONTIERNET.NET HTTP://WWW.ADVENTURESUNLIMITED.CO.NZ

NEWLY RELEASED

LOST CONTINENTS & THE HOLLOW EARTH

I Remember Lemuria and the Shaver Mystery
by David Hatcher Childress & Richard Shaver

Lost Continents & the Hollow Earth is Childress' thorough examination of the early hollow earth stories of Richard Shaver and the fascination that lost continents and the hollow earth have had for the American public. Shaver's rare 1948 book *I Remember Lemuria* is reprinted in its entirety, and the book is packed with illustrations from Ray Palmer's *Amazing Stories* magazine of the 1940s. Palmer and Shaver told of tunnels running through the earth—tunnels inhabited by the Deros and Teros, humanoids from an ancient spacefaring race that had inhabited the earth, eventually going underground, hundreds of thousands of years ago. Childress discusses the famous hollow earth books and delves deep into whatever reality may be behind the stories of tunnels in the earth. Operation High Jump to Antarctica in 1947 and Admiral Byrd's bizarre statements, tunnel systems in South America and Tibet, the underground world of Agartha, UFOs coming from the South Pole, more.
344 PAGES. 6x9 PAPERBACK. ILLUSTRATED. $16.95. CODE: LCHE

INSIDE THE GEMSTONE FILE

Howard Hughes, Onassis & JFK
by Kenn Thomas & David Hatcher Childress

Steamshovel Press editor Thomas takes on the Gemstone File in this run-up and run-down of the most famous underground document ever circulated. Photocopied and distributed for over 20 years, the Gemstone File is the story of Bruce Roberts, the inventor of the synthetic ruby widely used in laser technology today, and his relationship with the Howard Hughes Company and ultimately with Aristotle Onassis, the Mafia, and the CIA. Hughes kidnapped and held a drugged-up prisoner for 10 years; Onassis and his role in the Kennedy Assassination; how the Mafia ran corporate America in the 1960s; more.
320 PAGES. 6x9 PAPERBACK. ILLUSTRATED. $16.00. CODE: IGF

KUNDALINI TALES

by Richard Sauder, Ph.D.

Underground Bases and Tunnels author Richard Sauder on his personal experiences and provocative research into spontaneous spiritual awakening, out-of-body journeys, encounters with secretive governmental powers, daylight sightings of UFOs, and more. Sauder continues his studies of underground bases with new information on the occult underpinnings of the U.S. space program. The book also contains a breakthrough section that examines actual U.S. patents for devices that manipulate minds and thoughts from a remote distance. Included are chapters on the secret space program and a 130-page appendix of patents and schematic diagrams of secret technology and mind control devices.
296 PAGES. 7x10 PAPERBACK. ILLUSTRATED. BIBLIOGRAPHY. $14.95. CODE: KTAL

LIQUID CONSPIRACY

JFK, LSD, the CIA, Area 51 & UFOs
by George Piccard

Underground author George Piccard on the politics of LSD, mind control, and Kennedy's involvement with Area 51 and UFOs. Reveals JFK's LSD experiences with Mary Pinchot-Meyer. The plot thickens with an ever expanding web of CIA involvement, from underground bases with UFOs seen by JFK and Marilyn Monroe (among others) to a vaster conspiracy that affects every government agency from NASA to the Justice Department. This may have been the reason that Marilyn Monroe and actress-columnist Dorothy Killgallen were both murdered. Focusing on the bizarre side of history, *Liquid Conspiracy* takes the reader on a psychedelic tour de force.
264 PAGES. 6x9 PAPERBACK. ILLUSTRATED. $14.95. CODE: LIQC

ATLANTIS: MOTHER OF EMPIRES

Atlantis Reprint Series
by Robert Stacy-Judd

Robert Stacy-Judd's classic 1939 book on Atlantis. Stacy-Judd was a California architect and an expert on the Mayas and their relationship to Atlantis. Stacy-Judd was an excellent artist and his book is lavishly illustrated. The eighteen comprehensive chapters in the book are: The Mayas and the Lost Atlantis; Conjectures and Opinions; Cro-Magnon Man; East Is West; And West Is East; The Mormons and the Mayas; Astrology in Two Hemispheres; The Language of Architecture; The American Indian; Pre-Panamanians and Pre-Incas; Columns and City Planning; Comparisons and Mayan Art; The Iberian Link; The Maya Tongue; Quetzalcoatl; Summing Up the Evidence; The Mayas in Yucatan.
340 PAGES. 8x11 PAPERBACK. ILLUSTRATED. INDEX. $19.95. CODE: AMOE

COSMIC MATRIX

Piece for a Jig-Saw, Part Two
by Leonard G. Cramp

Leonard G. Cramp, a British aerospace engineer, wrote his first book *Space Gravity and the Flying Saucer* in 1954. *Cosmic Matrix* is the long-awaited sequel to his 1966 book *UFOs & Anti-Gravity: Piece for a Jig-Saw.* Cramp has had a long history of examining UFO phenomena and has concluded that UFOs use the highest possible aeronautic science to move in the way they do. Cramp examines anti-gravity effects and theorizes that this super-science used by the craft—described in detail in the book—can lift mankind into a new level of technology, transportation and understanding of the universe. The book takes a close look at gravity control, time travel, and the interlocking web of energy between all planets in our solar system with Leonard's unique technical diagrams. A fantastic voyage into the present and future!
364 PAGES. 6x9 PAPERBACK. ILLUSTRATED. BIBLIOGRAPHY. $16.00. CODE: CMX

24 HOUR CREDIT CARD ORDERS—CALL: 815-253-6390 FAX: 815-253-6300

EMAIL: AUPHQ@FRONTIERNET.NET HTTP://WWW.ADVENTURESUNLIMITED.CO.NZ

THE LOST CITIES SERIES

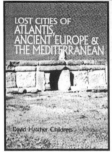

LOST CITIES OF ATLANTIS, ANCIENT EUROPE & THE MEDITERRANEAN
by David Hatcher Childress

Atlantis! The legendary lost continent comes under the close scrutiny of maverick archaeologist David Hatcher Childress in this sixth book in the internationally popular *Lost Cities* series. Childress takes the reader in search of sunken cities in the Mediterranean; across the Atlas Mountains in search of Atlantean ruins; to remote islands in search of megalithic ruins; to meet living legends and secret societies. From Ireland to Turkey, Morocco to Eastern Europe, and around the remote islands of the Mediterranean and Atlantic, Childress takes the reader on an astonishing quest for mankind's past. Ancient technology, cataclysms, megalithic construction, lost civilizations and devastating wars of the past are all explored in this book. Childress challenges the skeptics and proves that great civilizations not only existed in the past, but the modern world and its problems are reflections of the ancient world of Atlantis.

524 PAGES. 6X9 PAPERBACK. ILLUSTRATED WITH 100S OF MAPS, PHOTOS AND DIAGRAMS. BIBLIOGRAPHY & INDEX. $16.95. CODE: MED

LOST CITIES OF CHINA, CENTRAL INDIA & ASIA
by David Hatcher Childress

Like a real life "Indiana Jones," maverick archaeologist David Childress takes the reader on an incredible adventure across some of the world's oldest and most remote countries in search of lost cities and ancient mysteries. Discover ancient cities in the Gobi Desert; hear fantastic tales of lost continents, vanished civilizations and secret societies bent on ruling the world; visit forgotten monasteries in forbidding snow-capped mountains with strange tunnels to mysterious subterranean cities! A unique combination of far-out exploration and practical travel advice, it will astound and delight the experienced traveler or the armchair voyager.

429 PAGES. 6X9 PAPERBACK. ILLUSTRATED. FOOTNOTES & BIBLIOGRAPHY. $14.95. CODE: CHI

LOST CITIES OF ANCIENT LEMURIA & THE PACIFIC
by David Hatcher Childress

Was there once a continent in the Pacific? Called Lemuria or Pacifica by geologists, Mu or Pan by the mystics, there is now ample mythological, geological and archaeological evidence to "prove" that an advanced and ancient civilization once lived in the central Pacific. Maverick archaeologist and explorer David Hatcher Childress combs the Indian Ocean, Australia and the Pacific in search of the surprising truth about mankind's past. Contains photos of the underwater ruins on Pohnpei; explanations on how the statues were levitated around Easter Island in a clockwise vortex movement; tales of disappearing islands; Egyptians in Australia; and more.

379 PAGES. 6X9 PAPERBACK. ILLUSTRATED. FOOTNOTES & BIBLIOGRAPHY. $14.95. CODE: LEM

ANCIENT TONGA
& the Lost City of Mu'a
by David Hatcher Childress

Lost Cities series author Childress takes us to the south sea islands of Tonga, Rarotonga, Samoa and Fiji to investigate the megalithic ruins on these beautiful islands. The great empire of the Polynesians, centered on Tonga and the ancient city of Mu'a, is revealed with old photos, drawings and maps. Chapters in this book are on the Lost City of Mu'a and its many megalithic pyramids, the Ha'amonga Trilithon and ancient Polynesian astronomy, Samoa and the search for the lost land of Havai'iki, Fiji and its wars with Tonga, Rarotonga's megalithic road, and Polynesian cosmology. Material on Egyptians in the Pacific, earth changes, the fortified moat around Mu'a, lost roads, more.

218 PAGES. 6X9 PAPERBACK. ILLUSTRATED. COLOR PHOTOS. BIBLIOGRAPHY. $15.95. CODE: TONG

ANCIENT MICRONESIA
& the Lost City of Nan Madol
by David Hatcher Childress

Micronesia, a vast archipelago of islands west of Hawaii and south of Japan, contains some of the most amazing megalithic ruins in the world. Part of our *Lost Cities* series, this volume explores the incredible conformations on various Micronesian islands, especially the fantastic and little-known ruins of Nan Madol on Pohnpei Island. The huge canal city of Nan Madol contains over 250 million tons of basalt columns over an 11 square-mile area of artificial islands. Much of the huge city is submerged, and underwater structures can be found to an estimated 80 feet. Islanders' legends claim that the basalt rocks, weighing up to 50 tons, were magically levitated into place by the powerful forefathers. Other ruins in Micronesia that are profiled include the Latte Stones of the Marianas, the menhirs of Palau, the megalithic canal city on Kosrae Island, megaliths on Guam, and more.

256 PAGES. 6X9 PAPERBACK. ILLUSTRATED. INCLUDES A COLOR PHOTO SECTION. BIBLIOGRAPHY. $16.95. CODE: AMIC

24 HOUR CREDIT CARD ORDERS—CALL: 815-253-6390 FAX: 815-253-6300
email: auphq@frontiernet.net   http://www.adventuresunlimited.co.nz

THE LOST CITIES SERIES

VIMANA AIRCRAFT OF ANCIENT INDIA & ATLANTIS
by David Hatcher Childress
introduction by Ivan T. Sanderson

Did the ancients have the technology of flight? In this incredible volume on ancient India, authentic Indian texts such as the *Ramayana* and the *Mahabharata* are used to prove that ancient aircraft were in use more than four thousand years ago. Included in this book is the entire Fourth Century BC manuscript *Vimaanika Shastra* by the ancient author Maharishi Bharadwaaja, translated into English by the Mysore Sanskrit professor G.R. Josyer. Also included are chapters on Atlantean technology, the incredible Rama Empire of India and the devastating wars that destroyed it. Also an entire chapter on mercury vortex propulsion and mercury gyros, the power source described in the ancient Indian texts. Not to be missed by those interested in ancient civilizations or the UFO enigma.
334 PAGES. 6X9 PAPERBACK. RARE PHOTOGRAPHS, MAPS AND DRAWINGS. $15.95. CODE: VAA

LOST CONTINENTS & THE HOLLOW EARTH
I Remember Lemuria and the Shaver Mystery
by David Hatcher Childress & Richard Shaver

Lost Continents & the Hollow Earth is Childress' thorough examination of the early hollow earth stories of Richard Shaver and the fascination that fringe fantasy subjects such as lost continents and the hollow earth have had for the American public. Shaver's rare 1948 book *I Remember Lemuria* is reprinted in its entirety, and the book is packed with illustrations from Ray Palmer's *Amazing Stories* magazine of the 1940s. Palmer and Shaver told of tunnels running through the earth—tunnels inhabited by the Deros and Teros, humanoids from an ancient spacefaring race that had inhabited the earth, eventually going underground, hundreds of thousands of years ago. Childress discusses the famous hollow earth books and delves deep into whatever reality may be behind the stories of tunnels in the earth. Operation High Jump to Antarctica in 1947 and Admiral Byrd's bizarre statements, tunnel systems in South America and Tibet, the underground world of Agartha, the belief of UFOs coming from the South Pole, more.
344 PAGES. 6X9 PAPERBACK. ILLUSTRATED. $16.95. CODE: LCHE

LOST CITIES OF NORTH & CENTRAL AMERICA
by David Hatcher Childress
Down the back roads from coast to coast, maverick archaeologist and adventurer David Hatcher Childress goes deep into unknown America. With this incredible book, you will search for lost Mayan cities and books of gold, discover an ancient canal system in Arizona, climb gigantic pyramids in the Midwest, explore megalithic monuments in New England, and join the astonishing quest for lost cities throughout North America. From the war-torn jungles of Guatemala, Nicaragua and Honduras to the deserts, mountains and fields of Mexico, Canada, and the U.S.A., Childress takes the reader in search of sunken ruins, Viking forts, strange tunnel systems, living dinosaurs, early Chinese explorers, and fantastic lost treasure. Packed with both early and current maps, photos and illustrations.
590 PAGES. 6X9 PAPERBACK. PHOTOS, MAPS, AND ILLUSTRATIONS. FOOTNOTES & BIBLIOGRAPHY. $14.95. CODE: NCA

LOST CITIES & ANCIENT MYSTERIES OF SOUTH AMERICA
by David Hatcher Childress
Rogue adventurer and maverick archaeologist David Hatcher Childress takes the reader on unforgettable journeys deep into deadly jungles, high up on windswept mountains and across scorching deserts in search of lost civilizations and ancient mysteries. Travel with David and explore stone cities high in mountain forests and hear fantastic tales of Inca treasure, living dinosaurs, and a mysterious tunnel system. Whether he is hopping freight trains, searching for secret cities, or just dealing with the daily problems of food, money, and romance, the author keeps the reader spellbound. Includes both early and current maps, photos, and illustrations, and plenty of advice for the explorer planning his or her own journey of discovery.
381 PAGES. 6X9 PAPERBACK. PHOTOS, MAPS, AND ILLUSTRATIONS. FOOTNOTES & BIBLIOGRAPHY. $14.95. CODE: SAM

LOST CITIES & ANCIENT MYSTERIES OF AFRICA & ARABIA
by David Hatcher Childress
Across ancient deserts, dusty plains and steaming jungles, maverick archaeologist David Childress continues his world-wide quest for lost cities and ancient mysteries. Join him as he discovers forbidden cities in the Empty Quarter of Arabia; "Atlantean" ruins in Egypt and the Kalahari desert; a mysterious, ancient empire in the Sahara; and more. This is the tale of an extraordinary life on the road: across war-torn countries, Childress searches for King Solomon's Mines, living dinosaurs, the Ark of the Covenant and the solutions to some of the fantastic mysteries of the past.
423 PAGES. 6X9 PAPERBACK. PHOTOS, MAPS, AND ILLUSTRATIONS. FOOTNOTES & BIBLIOGRAPHY. $14.95. CODE: AFA

24 HOUR CREDIT CARD ORDERS—CALL: 815-253-6390 FAX: 815-253-6300
email: auphq@frontiernet.net   http://www.adventuresunlimited.co.nz

ATLANTIS REPRINT SERIES

ATLANTIS: MOTHER OF EMPIRES
Atlantis Reprint Series
by Robert Stacy-Judd

Robert Stacy-Judd's classic 1939 book on Atlantis is back in print in this large-format paperback edition. Stacy-Judd was a California architect and an expert on the Mayas and their relationship to Atlantis. He was an excellent artist and his work is lavishly illustrated. The eighteen comprehensive chapters in the book are: The Mayas and the Lost Atlantis; Conjectures and Opinions; The Atlantean Theory; Cro-Magnon Man; East is West; And West is East; The Mormons and the Mayas; Astrology in Two Hemispheres; The Language of Architecture; The American Indian; Pre-Panamanians and Pre-Incas; Columns and City Planning; Comparisons and Mayan Art; The Iberian Link; The Maya Tongue; Quetzalcoatl; Summing Up the Evidence; The Mayas in Yucatan.
340 PAGES. 8x11 PAPERBACK. ILLUSTRATED. INDEX. $19.95. CODE: AMOE

SECRET CITIES OF OLD SOUTH AMERICA
Atlantis Reprint Series
by Harold T. Wilkins

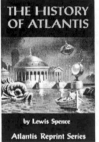

The reprint of Wilkins' classic book, first published in 1952, claiming that South America was Atlantis. Chapters include Mysteries of a Lost World; Atlantis Unveiled; Red Riddles on the Rocks; South America's Amazons Existed!; The Mystery of El Dorado and Gran Payatiti—the Final Refuge of the Incas; Monstrous Beasts of the Unexplored Swamps & Wilds; Weird Denizens of Antediluvian Forests; New Light on Atlantis from the World's Oldest Book; The Mystery of Old Man Noah and the Arks; and more.
438 PAGES. 6x9 PAPERBACK. ILLUSTRATED. BIBLIOGRAPHY & INDEX. $16.95. CODE: SCOS

THE SHADOW OF ATLANTIS
The Echoes of Atlantean Civilization Tracked through Space & Time
by Colonel Alexander Braghine

First published in 1940, *The Shadow of Atlantis* is one of the great classics of Atlantis research. The book amasses a great deal of archaeological, anthropological, historical and scientific evidence in support of a lost continent in the Atlantic Ocean. Braghine covers such diverse topics as Egyptians in Central America, the myth of Quetzalcoatl, the Basque language and its connection with Atlantis, the connections with the ancient pyramids of Mexico, Egypt and Atlantis, the sudden demise of mammoths, legends of giants and much more. Braghine was a linguist and spends part of the book tracing ancient languages to Atlantis and studying little-known inscriptions in Brazil, deluge myths and the connections between ancient languages. Braghine takes us on a fascinating journey through space and time in search of the lost continent.
288 PAGES. 6x9 PAPERBACK. ILLUSTRATED. $16.95. CODE: SOA

RIDDLE OF THE PACIFIC
by John Macmillan Brown

Oxford scholar Brown's classic work on lost civilizations of the Pacific is now back in print! John Macmillan Brown was an historian and New Zealand's premier scientist when he wrote about the origins of the Maoris. After many years of travel thoughout the Pacific studying the people and customs of the south seas islands, he wrote *Riddle of the Pacific* in 1924. The book is packed with rare turn-of-the-century illustrations. Don't miss Brown's classic study of Easter Island, ancient scripts, megalithic roads and cities, more. Brown was an early believer in a lost continent in the Pacific.
460 PAGES. 6x9 PAPERBACK. ILLUSTRATED. $16.95. CODE: ROP

THE HISTORY OF ATLANTIS
by Lewis Spence

Lewis Spence's classic book on Atlantis is now back in print! Spence was a Scottish historian (1874-1955) who is best known for his volumes on world mythology and his five Atlantis books. *The History of Atlantis* (1926) is considered his finest. Spence does his scholarly best in chapters on the Sources of Atlantean History, the Geography of Atlantis, the Races of Atlantis, the Kings of Atlantis, the Religion of Atlantis, the Colonies of Atlantis, more. Sixteen chapters in all.
240 PAGES. 6x9 PAPERBACK. ILLUSTRATED WITH MAPS, PHOTOS & DIAGRAMS. $16.95. CODE: HOA

ATLANTIS IN SPAIN
A Study of the Ancient Sun Kingdoms of Spain
by E.M. Whishaw

First published by Rider & Co. of London in 1928, this classic book is a study of the megaliths of Spain, ancient writing, cyclopean walls, sun worshipping empires, hydraulic engineering, and sunken cities. An extremely rare book, it was out of print for 60 years. Learn about the Biblical Tartessus; an Atlantean city at Niebla; the Temple of Hercules and the Sun Temple of Seville; Libyans and the Copper Age; more. Profusely illustrated with photos, maps and drawings.
284 PAGES. 6x9 PAPERBACK. ILLUSTRATED. TABLES OF ANCIENT SCRIPTS. $15.95. CODE: AIS

ANTI-GRAVITY

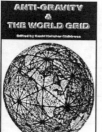

THE ANTI-GRAVITY HANDBOOK

THE ANTI-GRAVITY HANDBOOK
edited by David Hatcher Childress, with Nikola Tesla, T.B. Paulicki, Bruce Cathie, Albert Einstein and others

The new expanded compilation of material on Anti-Gravity, Free Energy, Flying Saucer Propulsion, UFOs, Suppressed Technology, NASA Cover-ups and more. Highly illustrated with patents, technical illustrations and photos. This revised and expanded edition has more material, including photos of Area 51, Nevada, the government's secret testing facility. This classic on weird science is back in a 90s format!
- **How to build a flying saucer.**
- **Crystals and their role in levitation.**
- **Secret government research and development.**
- **Nikola Tesla on how anti-gravity airships could draw power from the atmosphere.**
- **Bruce Cathie's Anti-Gravity Equation.**
- **NASA, the Moon and Anti-Gravity.**

230 PAGES. 7x10 PAPERBACK. BIBLIOGRAPHY. APPENDIX. ILLUSTRATED. $14.95. CODE: AGH

ANTI–GRAVITY & THE WORLD GRID
edited by David Hatcher Childress

Is the earth surrounded by an intricate electromagnetic grid network offering free energy? This compilation of material on ley lines and world power points contains chapters on the geography, mathematics, and light harmonics of the earth grid. Learn the purpose of ley lines and ancient megalithic structures located on the grid. Discover how the grid made the Philadelphia Experiment possible. Explore the Coral Castle and many other mysteries, including acoustic levitation, Tesla Shields and scalar wave weaponry. Browse through the section on anti-gravity patents, and research resources.

274 PAGES. 7x10 PAPERBACK. ILLUSTRATED. $14.95. CODE: AGW

ANTI–GRAVITY
& THE UNIFIED FIELD
edited by David Hatcher Childress

Is Einstein's Unified Field Theory the answer to all of our energy problems? Explored in this compilation of material is how gravity, electricity and magnetism manifest from a unified field around us. Why artificial gravity is possible; secrets of UFO propulsion; free energy; Nikola Tesla and anti-gravity airships of the 20s and 30s; flying saucers as superconducting whirls of plasma; anti-mass generators; vortex propulsion; suppressed technology; government cover-ups; gravitational pulse drive; spacecraft & more.

240 PAGES. 7x10 PAPERBACK. ILLUSTRATED. $14.95. CODE: AGU

ETHER TECHNOLOGY
A Rational Approach to Gravity Control
by Rho Sigma

This classic book on anti-gravity and free energy is back in print. Written by a well-known American scientist under the pseudonym of "Rho Sigma," this book delves into international efforts at gravity control and discoid craft propulsion. Before the Quantum Field, there was "Ether." This small, but informative book has chapters on John Searle and "Searle discs;" T. Townsend Brown and his work on anti-gravity and ether-vortex turbines. Includes a forward by former NASA astronaut Edgar Mitchell.

108 PAGES. 6x9 PAPERBACK. ILLUSTRATED. $12.95. CODE: ETT

Man-Made UFOS 1944-1994
50 Years of Suppression

Renato Vesco &
David Hatcher Childress

MAN-MADE UFOS 1944—1994
Fifty Years of Suppression
by Renato Vesco & David Hatcher Childress

A comprehensive look at the early "flying saucer" technology of Nazi Germany and the genesis of man-made UFOs. This book takes us from the work of captured German scientists to escaped battalions of Germans, secret communities in South America and Antarctica to today's state-of-the-art "Dreamland" flying machines. Heavily illustrated, this astonishing book blows the lid off the "government UFO conspiracy" and explains with technical diagrams the technology involved. Examined in detail are secret underground airfields and factories; German secret weapons; "suction" aircraft; the origin of NASA; gyroscopic stabilizers and engines; the secret Marconi aircraft factory in South America; and more. Not to be missed by students of technology suppression, secret societies, anti-gravity, free energy, conspiracy and World War II! Introduction by W.A. Harbinson, author of the Dell novels *GENESIS* and *REVELATION*.

318 PAGES. 6x9 PAPERBACK. ILLUSTRATED. INDEX & FOOTNOTES. $18.95. CODE: MMU

ETHER-TECHNOLOGY
A Rational Approach to Gravity Control

by Rho Sigma
THE UNDERGROUND CLASSIC IS BACK IN PRINT!

24 HOUR CREDIT CARD ORDERS—CALL: 815-253-6390 FAX: 815-253-6300
EMAIL: AUPHQ@FRONTIERNET.NET HTTP://WWW.ADVENTURESUNLIMITED.CO.NZ

One Adventure Place
P.O. Box 74
Kempton, Illinois 60946
United States of America
Tel.: 815-253-6390 • Fax: 815-253-6300
Email: auphq@frontiernet.net
http://www.adventuresunlimited.co.nz

ORDERING INSTRUCTIONS

✓ Remit by USD$ Check, Money Order or Credit Card
✓ Visa, Master Card, Discover & AmEx Accepted
✓ Prices May Change Without Notice
✓ 10% Discount for 3 or more Items

SHIPPING CHARGES

United States

✓ Postal Book Rate { $2.50 First Item
                     50¢ Each Additional Item
✓ Priority Mail { $3.50 First Item
                  $2.00 Each Additional Item
✓ UPS { $5.00 First Item
        $1.50 Each Additional Item
   NOTE: UPS Delivery Available to Mainland USA Only

Canada

✓ Postal Book Rate { $3.00 First Item
                     $1.00 Each Additional Item
✓ Postal Air Mail { $5.00 First Item
                    $2.00 Each Additional Item
✓ Personal Checks or Bank Drafts MUST BE
  USD$ and Drawn on a US Bank
✓ Canadian Postal Money Orders OK
✓ Payment MUST BE USD$

All Other Countries

✓ Surface Delivery { $6.00 First Item
                     $2.00 Each Additional Item
✓ Postal Air Mail { $12.00 First Item
                    $8.00 Each Additional Item
✓ Payment MUST BE USD$
✓ Checks and Money Orders MUST BE USD$
  and Drawn on a US Bank or branch.
✓ Add $5.00 for Air Mail Subscription to
  Future Adventures Unlimited Catalogs

SPECIAL NOTES

✓ RETAILERS: Standard Discounts Available
✓ BACKORDERS: We Backorder all Out-of-
  Stock Items Unless Otherwise Requested
✓ PRO FORMA INVOICES: Available on Request
✓ VIDEOS: NTSC Mode Only
✓ For PAL mode videos contact our other offices:

European Office:
Adventures Unlimited, PO Box 372,
Dronten, 8250 AJ, The Netherlands
South Pacific Office
Adventures Unlimited Pacifica
221 Symonds Street, Box 8199
Auckland, New Zealand

Please check: ☑

☐ This is my first order   ☐ I have ordered before   ☐ This is a new address

Name					
Address					
City					
State/Province		Postal Code			
Country					
Phone day		Evening			
Fax					

Item Code	Item Description	Price	Qty	Total

Please check: ☑

☐ Postal-Surface
☐ Postal-Air Mail
  (Priority in USA)
☐ UPS
  (Mainland USA only)

Subtotal ➡
Less Discount-10% for 3 or more items ➡
Balance ➡
Illinois Residents 6.25% Sales Tax ➡
Previous Credit ➡
Shipping ➡
Total (check/MO in USD$ only) ➡

☐ Visa/MasterCard/Discover/Amex

Card Number

Expiration Date

10% Discount When You Order 3 or More Items!

Comments & Suggestions	Share Our Catalog with a Friend